AF780971

Queerburbia

To subvert the metronormativity of queer urban studies and re-place queer suburbanism, *Queerburbia* examines LGBTQ2S place-making/unmaking/remaking on the peripheries of Canada's three largest city-regions (Vancouver, Toronto, and Montreal), investigating print media and census representations, civic and para-public allyship, individual and collective activism, and everyday practices of living and dreaming as revealed through photo-elicitation interviews and collective counter-mapping that together unmake and remake suburban places as queer.

Queerburbia offers a comparative case study of how large Canadian city-regions become queerer through LGBTQ2S suburban place-making/unmaking/remaking. For urban scholars, it deepens place-making theory with the conceptual introduction of the neologism "queerburbia" as a means to re-envision metropolitan peripheries as sites of queer futures. Practically, it offers civic leaders, urban planners, and policymakers insights into the complex dynamics of municipal LGBTQ2S misrecognition and critical allyship strategies beyond rainbowization. Methodologically innovative, this book combines print media, census, and municipal policy analysis with expert and photo-elicitation interviews, counter-mapping focus groups, and ethnographic fieldwork. It reveals the multiple layers of queerburban place-making/unmaking/remaking, demonstrating how statistical and media representations, municipal services and social inclusion policies, para-public and activist resistance and organizing, and individual living and dreaming emplace sexual and gender minorities in suburbia.

An interdisciplinary book at the interstices of Geography, Urban Studies, Suburban Studies, Urban Planning, and LGBTQ+ Studies, its intended audiences are scholars of cities, queer theory, and sexual and gender minority life extending to Women's and Gender Studies, Cultural Studies, Sociology, Anthropology, and Psychology. It targets upper-level undergraduates, graduate students, researchers, and practitioners of municipal social inclusion, including civic leaders, urban policymakers, and urban planners.

Alison L. Bain is Professor of Geography in the Faculty of Environmental and Urban Change at York University in Toronto, Canada.

Julie A. Podmore is College Professor in Geosciences at John Abbott College and Affiliate Assistant Professor at Concordia University in Montreal, Canada.

Routledge Studies in Urbanism and the City

Uneven Real Estate Development in Romania at the Intersection of Deindustrialization and Financialization
Edited by Enikő Vincze, Ioana Florea and Manuel B. Aalbers

Imagining Urban Complexity
A Humanities Approach in Tropes, Media, and Genres
Frans-Willem Korsten and Anthony Albright

Mapping Legalities
Urbanisation, Law and Informal Work
Edited by Thomas Coggin and Roopa Madhav

Urban Marginality, Racialisation, Interdependence
Learning from Eastern Europe
Edited by Filip Alexandrescu, Ryan Powell and Ana Vilenica

City Re-construction
Urban Policy Innovation Towards Sustainable Cities in MENA Region
Edited by Ali A. Alraouf, Odeh Al-Jayyousi, and Kheir Al-Kodmany

Hybrid Urbanisms in Secondary Cities of the Global South
Insights in urban planning and infrastructure delivery in Ghana and Peru
Edited by Christian Rosen and Nina Gribat

Cities: Inclusive, Liveable, and Sustainable
Edited by Ashish Kumar Srivastava and Iva Ashish Srivastava

Queerburbia
LGBTQ2S Suburban Place-Making
Alison L. Bain and Julie A. Podmore

For more information about this series, please visit https://www.routledge.com/Routledge-Studies-in-Urbanism-and-the-City/book-series/RSUC

Queerburbia

LGBTQ2S Suburban Place-Making

Alison L. Bain and Julie A. Podmore

Routledge
Taylor & Francis Group
LONDON AND NEW YORK

First published 2025
by Routledge
4 Park Square, Milton Park, Abingdon, Oxon OX14 4RN

and by Routledge
605 Third Avenue, New York, NY 10158

Routledge is an imprint of the Taylor & Francis Group, an informa business

© 2025 Alison L. Bain and Julie A. Podmore

The right of Alison L. Bain and Julie A. Podmore to be identified as authors of this work has been asserted in accordance with sections 77 and 78 of the Copyright, Designs and Patents Act 1988.

All rights reserved. No part of this book may be reprinted or reproduced or utilised in any form or by any electronic, mechanical, or other means, now known or hereafter invented, including photocopying and recording, or in any information storage or retrieval system, without permission in writing from the publishers.

Trademark notice: Product or corporate names may be trademarks or registered trademarks, and are used only for identification and explanation without intent to infringe.

British Library Cataloguing-in-Publication Data
A catalogue record for this book is available from the British Library

ISBN: 978-1-032-63250-6 (hbk)
ISBN: 978-1-032-63251-3 (pbk)
ISBN: 978-1-032-63254-4 (ebk)

DOI: 10.4324/9781032632544

Typeset in Times New Roman
by SPi Technologies India Pvt Ltd (Straive)

Dedication

For our much loved and ever-encouraging families: Anne and Tony Bain, Camden Bain, and Cathy Clarke; and the late John Podmore, Gail Podmore, and Laudan Vaezmir.

Contents

Figures

Tables

About the authors

Alison Bain is Professor of Geography in the Faculty of Environmental and Urban Change at York University in Toronto, Canada, and is a North American Managing Editor for the *Urban Studies* Journal. She is an urban social geographer who studies the inequalities of contemporary urban and suburban place-making through the lens of artistic practice and spatialized identity politics. She is the author of *The Creative Margins: Cultural Production in Canadian Suburbs*, and co-editor of the book *The Cultural Infrastructure of Cities*, two editions of the textbook *Urbanization in a Global Context*, and the forthcoming books *Queer Geographies: Key Debates and Contending Perspectives* and *Co-Authoring Feminist and Queer Geographies: Collaborations, Mentorships, Solidarities, Friendships*.

Julie Podmore is College Professor in Geosciences at John Abbott College and Affiliate Assistant Professor at Concordia University in Montreal, Canada. She is an urban socio-cultural geographer who adopts a feminist, queer historical approach to the study of urban/suburban processes. She is co-editor of *Lesbian Feminism: Essays Opposing Global Heteropatriarchies*, *The Cultural Infrastructure of Cities*, and the forthcoming *Queer Geographies: Key Debates and Contending Perspectives*.

Acknowledgements

This book – our escapist labour of friendship and care for nearly a decade – would not have been possible without the pivotal contributions of activists on the peripheries of Vancouver, Toronto, and Montreal, whose passionate political commitments have nurtured and made queerburban life resonant for so many people. In the Vancouver city-region we acknowledge key activists who generously guided us through their community networks: Mz. Adrien, Nancy Kato, Gary Lising, Henry Lorteau, Amanda Luv, Jenn Marchbank, Evelyn McGowan, Shilpa Narayan, Keith Oberding, Don Ralph, Martin Rooney, Alex Sanga, and Alan Woo. In the Toronto and Montreal city-regions, key para-public actors made this research possible: Howie Dayton, Reverend Shaun Fryday, David Hawkins, Vibhuti Mehra, Yoshith Perera, Maude St-Germain, Councillor Alexandre Warnet.

A project of this scope relies heavily on institutional support. It was funded by a Social Science and Humanities Research Council of Canada Insight Grant (435-2016-1142). Special batches of same-sex couple data were made possible by the Canadian Research Data Centre Network. At York University, CITY was the administrative homebase under the direction of Linda Peake and Luisa Sotomayor while the Faculty of Environmental and Urban Change provided sabbatical release and research assistantships. John Abbott College provided research release and the Geography, Planning and Environment Department at Concordia University granted researcher status.

We began this project through collaboration with our colleague Tiffany Muller Myrdahl whose critical work on queering smaller cities and LGBTQ+ municipal social inclusion policies inspired us. Later we found commonalities with Karine Duplan who invited us to share this research at the University of Geneva's Centre Maurice Chalumeau. The mentorship, collaboration, solidarity, and friendship of queer geographers – especially Michael Brown, Kath Browne, Andrew Gorman-Murray, Gilly Hartal, Lynda Johnston, Larry Knopp, and Chen Misgav – have been invaluable.

The data collection and analysis for *Queerburbia* was far from a solitary act. Statistical and cartographic support came from our co-researcher Brian Ray and research assistant Isabella Croft Richmond who extracted Canadian same-sex household data from the census into data tables and maps that informed our initial understanding of queer Canadian suburbanization. Further support came from Taly Drezner and Tarmo Remmel. Over the years, many graduate student research assistants contributed their energy, imaginations, and skills, including Chan Arun-Pina, Bronwyn Clement, Cheryl Costello, Mathilde Laroche, Alexe Lepine-Dubois, Leanne Letourneau, Rae Rosenberg, Wiley Sharpe, and Marie Tabery.

This book was a journey in many senses. We criss-crossed the country and travelled to city-regional peripheries, meeting new people and communities, challenging our own metronormative assumptions, and expanding our geographies. An abundant queerburbia welcomed us in Vancouver, but in Toronto and Montreal finding queerburbia required determination, especially during COVID lockdowns. And, over eight years, our lives changed. In Toronto, one child was born, and in Vancouver one parent passed away. In the later stages, our homes were taken over by the daily rhythms of co-writing Skype calls and we travelled to showcase queerburbia at conferences. And so, for their patience and for accompanying us on this long adventure, we would especially like to thank Camden, Cathy (reference checker extraordinaire), and Laudan.

1 Place-making/unmaking/remaking

Introduction

Just as heterocisnormativity's "sub" is queer, the urban's "Other" is the suburban. Yet this inverted binary is anything but neat. Not only have Anglosphere queer geographies of metropolitan areas long associated LGBTQ+ (Lesbian, Gay, Bisexual, Transgender, and Queer) communities with central city locations, but so too have they neglected queer suburban lives with their persistent gayborhood focus. The ecological association of subcultural formation with urbanity – its aggregation of marginalized populations – has embedded metronormativity in queer geographies (Halberstam, 2005). Along with small towns and rural areas, suburbs are framed as sites of "heteronormative 'impossibility'" where queer subjectivities are awaited until movement towards urban centres brings queerer possibilities (Halberstam, 2005). While "the countryside" has embodied an obscure, dangerous, even rustic erotic (Bell, 2000; Gray et al. 2016; Herring, 2010), suburbs have epitomized the gender binary norms of compulsory nuclear family reproductive heterosexuality (Rich, 1980). Yet within suburbia's hyperbolic heterocisnormativity there have always been private "perversions" (Langford, 2000) and marginal places for the "clandestine encounters" enfolded into its etymological residue (Lepofsky & Fraser, 2003, 129). If reclaiming "perversion" is a form of queer resistance (Bell, 1995), what is hiding in the repressed "closets" of bedroom communities largely awaits interrogation (Brown, 2000). For "queering use" even in suburban places requires "a certain willingness to be perverse, to deviate from the straight path, the right path" (Ahmed 2019, 201). In its commitment to such "relocations" (Tongson, 2011) and "cultural archipelagos" (Ghaziani, 2019), this book treats the suburbs as the ultimate place to queer.

Derived from the Latin meaning "near to the city", the term suburb references a proximate, but discrete, spatial relationship of dependency of a substandard place lying "outside, beyond, or below the 'urb'" (Bourne, 1996, 164). The central city's ambiguous "Other", Anglosphere suburbs have spatialized white, middle-class domestic ideals (as an escape from urban "ills", offering space for children's safe play, family recreation, and supportive neighbouring) but they have also been stigmatized as "shadowlands" between work and home,

DOI: 10.4324/9781032632544-1

as "non-places" where "nothing exciting ever happens and no one interesting, different or creative lives" (Stevenson, 2003, 124; Kunstler, 1993). While late twentieth century suburbs were imagined as perpetually "out of style", stultifying, and normative (Keil, 2020), twenty-first century "planetary suburbanization" has rendered sprawling and polycentric suburbs globally pervasive as they drive metropolitan change from "the outside in" (Keil, 2017). Contemporary suburbs are increasingly appreciated for their infinite variety – their complexity, density, cohesiveness, and uniqueness (Harris & Vorms, 2017; Wyly, 2020) – with the suburban diversity paradigm underscoring shifts in suburban household structure (Hall & Lee, 2010), changing gender roles and norms (Johnson, 2006), the suburbanization of poverty (Burton, 2015), ethnocultural discrimination (Ray & Preston, 2015), immigrant and refugee settlement, homeownership, and business practices (Teixeira, 2007), and the formation of "ethnoburbs" (Li, 2009). Nevertheless, many place-contingent facets of suburbs and "suburbanism" as a way of life (Walks, 2013; Filion, 2015) remain misunderstood and under-researched, no more so than with respect to sexual and gender diversity (Podmore & Bain, 2020).

Although often invisible and usually underserved, queer lives "take place" in suburbs (Seamon, 2018, 1). Queer and trans suburbanites (whether as questioning or transitioning youth, closeted commuters, same-sex or transgender parents, or single seniors) inhabit and encounter suburbs like anyone else, through their senses, place attachments, and memories. In quotidian ways, however, they are also place-makers/unmakers/remakers who "do" all kinds of small and large things to transform where they live into affirming places of belonging through relationships with others (Schneekloth & Sibley, 1995). By seeking out particular suburban places to foster "belonging, meaning, attachment, inclusiveness, and community" (Fincher et al., 2016, 518), sexual and gender minorities knit the queer and suburban together and bring other possibilities into being. Through the making, unmaking, and remaking of suburbs, sexual and gender minorities publicly lay claim to places to sustain themselves and build connections, while also unravelling suburbia's heterocisnormative belongings and remaking them as queer.

Queer suburbanites thus contribute to the creation of what this book terms *queerburbia* and the *queerburban*. Queerburbia refers to the reconceptualization of what has previously been understood as an impossibility, the synergies between the queer and the suburban. Offering a counterargument to metronormativity, this neologism directs attention to the many queer lives lived in suburbs, transforming them into places inhabited by queer and trans bodies and animated by their lives and dreams. In Lefebvrian terms (Lefebvre, 1991), queerburbia is also a place where sexual and gender minority citizenship is negotiated; where sexual and gender minorities strive to secure "the right to the city" (or in this case suburb) with claims to municipal redistribution, recognition, and encounter (Fincher & Iveson, 2008). Such "*rights in places*" (Pierce et al., 2016), however, are variable and contingent, produced through socio-cultural tectonics that determine the politics of inclusion and exclusion. They condition how different social groups locally access available discursive and material resources and political opportunity structures to make, unmake, and remake places (Lepofsky & Fraser, 2003). Thus, in every place arise critical questions about who has the "right" to place-making and "whose and what imaginative horizons are shaping place and its making?" (Fincher et al., 2016, 522). For, at its core, place-making involves political contestation of access to resources and power, determining which voices, memories, and stories are embedded in places, and will predominate in visions of their future (Buser et al., 2013; Summers, 2019). Seen in this light, it is not enough to affirm queerness in suburbia since "a world dismantling effort" (Ahmed, 2019, 229) is also required for queerurbia to be a possibility. Linear trajectories of embedding queer meanings and making citizenship claims are also inadequate: queerburbia is formed through an oscillating process of taking small steps forward while encountering resistant backtracking and sideways detours on the road to alternate suburban futures (Marrades et al., 2021).

This book critically leverages the conceptual tools of place-making to investigate how practices of LGBTQ2S spatial agency rupture the heterocisnormativity of the suburban imaginary and remake suburban places as queerburban – significant sites of sexual citizenship on the peripheries of city-regions. The addition of the geographically specific 2S to the LGBTQ+ acronym signals the long-standing presence of Indigenous Two-Spirit communities (a term referring to the intersection of a masculine and feminine spirit in First Nation sexual, gender, and spiritual identities) on Turtle Island (North America).[1] The book unpacks suburbia's representations of sexual and gender minorities in the census, print media, and municipal public-facing communication to establish the landscape of heterocisnormativity that queerburban place-unmaking remakes in its layered local contingency, negotiated complexity, and interweaving with social justice ambitions for collective action. It treats queerburban place-making as a contested, social, and unfinished aspirational process of connecting to, rooting in, and transforming places on the peripheries of city-regions where LGBTQ2S residents live and build communities outside of gayborhoods and in the absence of queer infrastructural concentration.

Interrogating the relationship between place-making and social equity for marginalized communities, this book asks what social inclusion innovations are required to sustain queer suburbanisms. It questions the limits of civic social inclusion agendas to serve suburban sexual and gender minorities through place-making, honing in on the misalignment of the structural ordering of place with what is envisioned by queer resisting, organizing, living, and dreaming. In terms of representing, including, and serving LGBTQ2S citizens, this book questions municipal failures to rework heterocisnormative power relations in ways that substantially redistribute resources beyond symbolic representations of inclusion. It demonstrates that while greater inclusion for LGBTQ2S citizens is seldom a civic service priority, it is a crucial aspect of livability for queerburbanites. It asserts that remaking suburban places to serve sexual and gender minorities should seek to destigmatize LGBTQ2S populations, combat homotransphobic discrimination and violence, recreate programmes to serve their needs, and reconfigure municipal infrastructure in support of more liveable queerburban lives. Ultimately, the book is an extended invitation to understand the remaking of suburbs through the LGBTQ2S allyship, activism, living, and dreaming that occurs there – the remaking of the insides of suburban places so as to welcome suburbia's ultimate "outside[rs] in" (Fincher et al., 2016, 535).

Queering place within and beyond the discipline of geography

Place is central to the geographical imagination and lexicon (Daniels, 1992), yet it is rarely queerly theorized. For, "[t]o live in society is to live in heterosexuality" wherein "heterosexuality is always already there within all mental categories" (Wittig, 1992 as cited in Warner, 1993, xxi), including theorizations of place. To queer is to question conceptual and methodological orthodoxies (Brim & Ghaziani, 2016), to destabilize and undo the normative binaries that uphold them, and to invent new modes of writing, representing, and performing to capture the productive possibilities of the "mess" created by queer bodies in normative spaces (Manalansan IV, 2015). As political imagination, queer is non-linear and aspirational, representing an opening rather than an endpoint. It stems from a utopian longing for a world that "is not yet here but it approaches like a crashing wave of potentiality" and serves as "a relational and collective modality of endurance and support" (Muñoz, 2009, 121 & 185). Queering place, therefore, involves theorizing for and with "difference" by prioritizing the generative embodied possibilities of people who have been excluded or dispossessed within heterocisnormative landscapes (Muñoz, 2009) and reworking established disciplinary understandings of place to foreground queer's scrappy illegibilities.

Place was a foundational concept for humanistic geographers of the 1970s who strove to develop a less "objective" understanding of spatiality. These primarily cis-masculine scholars reinterpreted the theoretical "betweenness" of place across the universal registers of the subjective/objective, imagined/material, individual/collective, and specific/general (Entrikin, 1991), but differences wrought by the binaries of masculine/feminine, straight/queer, and cis/trans were not considered. In Tuan's (1974) book, *Topophilia*, place is produced through lived experiences and conceptualizations (e.g., values, perceptions, memories, and traditions) within physical settings but is never disaggregated by socially produced differences such as sex, gender, or sexuality. Furthermore, as Rose (1993) argued, humanistic geographers objectified place "as woman", inscribing an aestheticized heterocismasculinity on the object of their conceptual desire and erasing others. Temporalities of reproductive heteronormativity are also inscribed in humanistic geography's understanding of place in ways that sideline queer imaginaries. In *Place and placelessness*, Relph (1976, 43) wrote:

> …there is for virtually everyone a deep association with and consciousness of the places where we were born and grew up, where we live now or where we have had particularly moving experiences. This association

seems to constitute a vital source of both individual and cultural identity and security.

For queers and many others, such forms of place affinity and security, however, are tempered by hostility, trauma, a denial of the queer and transgender self, and a profound sense of invisibility and disappearance. As Ahmed (2019, 201) asserts, places inscribed with heterocisnormativity offer little ontological security for sexual and gender minorities: "No one has willed or intended your disappearance … [but] just slowly, as talk of family, of heterosexuality as the future, of lives that you do not live, just slowly, just slowly, you disappear". In phenomenological terms, places offer "fields of care", both securing and dissolving individual and collective identities on a spectrum of insideness and outsideness (Relph, 1976). This inside-outside spectrum is enacted in place through "lived bodies" that, for Casey (2009, 327), belong to and co-constitute place. If the main "arena of action" for theorizing place is "the immediate environment of my lived body" (Casey, 2001, 683), then whose body matters in such universalized accounts of place? For, as Ahmed (2006) queerly underscores, there are always other bodies present that expand to fill places up, "saturating … space with bodily matter" (Ahmed, 2006, 11) and making queer attachments to place possible.

Seamon's (2018, 32) *Life takes place* provides a more multidimensional account, viewing place "in contrasting and complementary ways via different vantage points". Its conceptual frame targets the variance of place through the interplay between three systemic dynamics: monadic (wholeness), dyadic (difference), and triadic (relatedness). While the monad describes "place" as a totality and considers the features that make it unique, the dyad refers to the complementarity of unresolved binaries (movement/rest, insideness/outsideness, ordinariness/extra-ordinariness, within/without, and homeworld/alienworld). The resolution of their contradictions comes through the relational triad, the "impulses" that generate the actions, processes, or happenings of place change (via interaction, identity, expansion, concentration, order, and freedom). Such place changes occur on an everyday basis through *interaction* in lifeworlds whereby *identity* emerges from repetition as "actions, skills, routines, and situations … become part of who and what we are" (Seamon, 2018, 73). *Expansion* permits individuals and groups to meaningfully re-order places while *concentration* focuses on the physical environment shaping the life of places. *Order* refers to place structures shaped by laws and legal systems and *freedom* engages the "everyday happenstances and surprises" inspiring "unexpected departures from the repetitive ordinariness of place" (Seamon, 2018, 78 & 79). This systematic structure enables a multi-layered analysis of place, but its conceptual hierarchy, definitive classification, and insistence on immutable categories constrain queer possibilities.

Despite Seamon's (2018) more recent intervention, there has been a marked epistemic turn away from humanistic geography since the 1980s resulting in a departure from the phenomenological emphasis on the perceptions of place residing inside the perceiver as the basis of human "placed experiences" (Pierce et al., 2011). With a more socially informed cultural geography, Marxist-inflected postmodern geographies, and the geographical imaginations of the "localities school", came re-evaluations of place (Massey, 1991). Harvey's (1989) argument that globalized geographies of neoliberalism still relied on distinct constellations of "places" stimulated expansive reconceptualizations of place as "locale" (socio-cultural relations) and "location" (physical setting) (Agnew, 1987, 2002). Post-structural geographical scholarship implicated place in numerous "polyvalent processes embedded in broader sets of social relations" (Jessop et al., 2008, 390), including individual and collective identity formation (Keith & Pile, 1993); political activism (Martin, 2003); interactions between individuals, groups, and institutions (Pred, 1984; Paasi, 2002; Pierce et al., 2011); regional economic development (Massey, 1991); global flows of capital, labour, and goods (Harvey, 1989); and modes of representation (Duncan & Ley, 1993; Shields, 1991). Doreen Massey's (2005, 131) "global sense of place" – a critical feminist response to Harvey's privileging of space – further retheorized place, taking it from a closed and static phenomenon to an unbound process, "a particular constellation" that is "woven together out of ongoing stories".

The queerly subversive "juxtapositions of difference" and "throwntogetherness" of Massey's (2005, 94) "global sense of place" invites key questions about who shapes and controls places, why, and how (Mason & Whitehead, 2012; Sibley, 1995). Enfolded in the same discursive turn, queer scholarship on place initially drew upon the language of exclusion and being "out of place" (Sibley, 1995). Cresswell's (1996, 2014) binary of in place/out of place and analysis of the role played by "discourses of normality" demonstrated the dynamic struggle of competing factions over the meanings of place and its reimagination. For non-normative social groups trapped in the binaries of "good or bad, moral or immoral, appropriate or inappropriate, worthy or unworthy", discursive struggles over the meanings of place can be especially intense, yet necessary "in order to effectively articulate alternative visions of the future" (McCann, 2002, 388 & 391). Powerful groups leverage place to generate hegemonic cultures that "systematically define" acceptable "ways of being", generating feelings of (un)belonging within social landscapes (Valentine, 2007, 18). In these ways, places are made, unmade, and remade every day through "reiterative social practices" (Cresswell, 2014, 39) involving imaginative struggles that work dialogically with the places in which they are immersed (Puleo, 2014). They are constituted through rhetorical negotiations about meanings and values (McCann, 2002), shared imaginative "place-frames" (Martin, 2003) informed by political agendas that advance particular material and political interests through discourses about that place (Pierce et al., 2011). Together the in place/out of place dyadic along with Seamon's (2018) phenomenological relationality and Massey's (2005, 149, 9, & 120) process-oriented treatment of place as "coexisting heterogeneities" and "multiple becomings" holds significant conceptual potential for exploring its registers of queerness.

Queering place-making in theory and practice

In its integration of an established geographical concept into the applied professional fields of urban planning and design, place-making potentially offers a dynamic tool for queering places. Yet, in this practitioner literature, place-making is represented as a dualistic and totalizing process of either top-down urban redevelopment governance or bottom-up resistance to its associated erasures, displacements, and replacements (Gonzalves et al., 2021; Özogul & Tasan-Kok, 2018). Place-making's deployment in the design fields (Cilliers & Timmermans, 2015; Coblence, 2021; Ferbrache & Knowles, 2017) is often as a depoliticized "place maintenance" (Rios & Watkins, 2015, 211) that only incorporates "social equity as a rhetorical flourish" (Fincher et al., 2016, 519). Its project-driven, technocratic knowledge straightens places through standards, codes, and formulas that transform the seeming incoherence of the messy "stories so far" (Massey, 2005, 9) into the "engineered, dominant narratives" of places that are more recognizable to speculative investment capital (Arefi, 2014, 20) than to disenfranchised communities (Kelly et al., 2017). Beyond socio-economic inequalities, the top-down, market-driven ethos of place-making can rematerialize heterocisnormativity through urban redevelopment and its discursive promotional strategies (Fincher et al., 2016) and lead to the direct displacement of queer community spaces as waterfronts are renewed and gentrifying districts are rebranded.

As place-making discourses have proliferated in the public arena, so too have scholarly critiques of their limitations when disconnected from local socio-cultural identities and social relations of well-being and empowerment (Palermo & Ponzini, 2015). Fundamentally, place-making raises critical tensions between "inhabitation and exhibition, dwelling and display", tenancy and ownership (Fincher et al., 2016, 518). Within hegemonic capitalist systems, it relies upon property rights, with legal ownership delimiting land use and public investment in supportive infrastructure (Montgomery, 2016). Critical examination of place-making's power dynamics further demonstrates what is "at stake" in place for differently minoritized groups of people (Burns & Berbary, 2021, 647). Homonormativity not only "denigrates 'ordinary' gay lives" (Brown, 2012); its iteration as homonormative entrepreneurialism, for example, is place-making that privileges "corporatized and easily marketable kinds of queer spaces and subjects over others" (Kanai & Kenttamaa-Squires, 2015, 388). By focusing on global tourism and urban redevelopment to the neglect of already displaced queer place histories, it "deems some people as less worthy of forming place-based attachments than others" (Burns & Berbary, 2021, 650). Beyond increasing property values and the cost of living, such globally oriented upscaling ruptures pre-existing social networks, uproots entire communities, and undermines the very "right to live and thrive in place" (Ramírez, 2020, 152).

Place-making, therefore, is intimately and simultaneously connected to processes of "place unmaking", "where every making becomes an unmaking of what has been, will come, and may be returned" (Burns & Berbary, 2021, 651). For Oakley and Johnson (2013, 342), "places are made by taking from others" in a process of "place-taking". Fincher et al. (2016, 534) concur, adding that place-making is perhaps best understood as "masking" the "displacement of those incumbent low-income people seen as transgressive and undesirable, and for their replacement with a better-resourced and presumably better-behaved population in an envisioned, desired future". Yet marginalized social groups do not merely react to power and authority; they too have place-making aspirations (Addie & Fraser, 2019; Andres, 2013). Such place-making is about making places that people feel comfortable using, that satisfy their everyday needs, and where they want to remain (Hewidy & Lilius, 2022).

Through the most ordinary, repetitive, and routine behaviours of living, working, and celebrating, individual residents seek to connect with each other and to belong. Residency, with all its lived experiences and embodiments, strengthened by rights from property ownership, attaches people to places (Lepofsky & Fraser, 2003). For residents who are variously marginalized from property ownership or are invisible in public places and policies, visibility becomes a crucial confirmation of community membership and the legitimation of the right to participate in place-making. "Cultural visibility" in place – the "marks, signs, symbols, languages, and practices of distinctive groups" (Sezer & Maldonado, 2017, 194) – is a "declaration of identity" (Hewidy & Lilius, 2022), an explicit means of making places recognizable and unique in "plural cities" (Harney, 2006). Therefore, the question of who or what should be visible in places – as much for purposes of celebration and protest as for "mundane contacts that shape perceptions and solidarities" (Montgomery, 2016, 781) – engages political questions of recognition and representation. Place-making privileges certain groups, enhancing their visibility in place by actively inserting their voices, stories, symbols, and opportunities into the public realm, but the degree of visibility generated varies by social group, local political cultures, and location within the metropolitan area. Rarely visible in suburban place-making, queerburbanites must creatively aggregate place leftovers into "unstable assemblages" (Laszczkowski, 2015, 138) through place-making/unmaking/remaking possibilities.

This book meshes Seamon's (2018) phenomenological approach to place with scholarly critiques of place-making, arguing that monadic and dyadic interpretations of suburban places can be disrupted through the more relational triadic process of making, unmaking, and remaking. It re-works Seamon's six place-making processes (intensification, realization, release, interaction, identity, and creation) to deepen a critical understanding of how suburban places are made/unmade/remade as queer. For Seamon (2018), place *intensification* references the plans that order place, making its *realization* possible through progressive actualization. Through place *release* generated by "unexpected happenings" and everyday *interactions*, people come to "unself-consciously and self-consciously accept and recognize that place as part of their personal and communal *identity*" (Seamon, 2018, 91). Place *creation* emerges from dedicated involvement to "creatively improve place" through "thoughtful understanding and action" but also imaginative and collective dreaming (Seamon, 2018, 91). Together these phenomenological dimensions of place inform this narration of queerburban place-making interpreted as inseparable from unmaking and remaking. Formal heterocisnormative municipal place-making is associated with place dynamics of intensification and realization that align with civic practices of inclusion and service to sexual and gender minorities. Queer place-unmaking involves the complex dynamics of doubling down to undermine heterocisnormative municipal place-making via place release and identity derived from LGBTQ+ community resistance and organizing. Finally, place-remaking through interaction and creation occurs via living and dreaming.

Canadian suburbs as sites of queer place-making/unmaking/remaking

In Canada, issues of sexuality and gender have been on the public agenda due to debates since the 1980s about sexual and gender equality leading up to the adoption of federal human rights protections in 1996, civil marriage equality in 2005, and anti-discrimination legislation regarding gender identity and expression in 2016. While there is currently a general public condemnation of homotransphobic violence, harassment, and discrimination, in some regions it is countered by heteroactivism that sometimes colludes with conservative municipal political regimes (Nash & Browne, 2020). Regional unevenness in political representation, services regimes, and opportunity structures persists from the centres of large cities out to their peripheries and beyond in mid-sized cities, small towns, and rural areas (Muller Myrdahl, 2013, 2016; Phillips et al., 2000; Stone, 2018). It is in suburbs, however, where most Canadians now reside, that queer reworldings remain underexamined.

As the new suburban studies literature reveals, suburbanism represents not just a distinctive way of life, but also a zeitgeist urban process of our time. Sociologists responded to the "hyperbolically heteronormative" (Lenc, 2020) post-war era of mass suburbanization in the Anglosphere by theorizing suburbanism as rooted in a demographic homogeneity wrought by neighbourhood self-selection (Fava, 1956). The archetypal American sitcom suburb of the mid-twentieth century produced communities of limited liability based on loose-knit secondary relations and fragmented suburban institutional affiliations (Gans, 1967). The "cookie-cutter" moniker described seemingly monotonous rows of nearly identical "starter homes" packed onto small lots in American exurban locations, producing "weak-ties" of "neighbouring" organized around two-parent heteropatriarchal nuclear families and reinforcing whitened middle-class individualism, privacy, and sexual and gender norms (Gans, 1967). By way of contemporary contrast, the political-economy priorities of the new global suburbanism literature emphasize governance frameworks, land development processes, and infrastructural provision (Keil, 2017). Even when reconceptualized as a capitalist regime and an everyday experience, the conceptual space for "new hybrids" (e.g., diversification, innovation, and progressive mobilization) still forecloses queer suburbanism by negating the socio-spatial politics of gender, sexuality, and racialization (Walks, 2013). Such obfuscations of sexuality occur because of how power relations are conceived of with respect to social diversity and domesticity. This blatant scholarly avoidance evidences the "regularity with which even critical urban theorists can dismiss" not just women, but also sexual and gender minorities, "producing disembodied theory" that enacts "epistemic violence" (Peake, 2016, 221 & 222).

Through homologous comparison as part of a large, multi-year Social Sciences and Humanities Research Council of Canada funded project (435-2016-1142), this book traces queer suburban place-making/unmaking/remaking in Canada's three largest metropolitan areas – Toronto (Mississauga, Brampton, Markham, and Ajax), Vancouver (Burnaby, New Westminster, and Surrey), and Montreal (Laval and Longueuil, the West Island – which includes the smaller municipalities of Baie-D'Urfé, Beaconsfield, Dollard-des-Ormeaux, Dorval, Kirkland, Pointe-Claire, Sainte-Anne-de-Bellevue, and Senneville). The case-study peripheral municipalities were chosen because of their relative density of same-sex households and LGBTQ+ print media representations, as well as detectable community organizational and public sphere presence. The secondary data sources include same-sex household census-data (2016), print media (n = 1660) and municipal public-facing communications records (n = 348), informational interviews (n = 182), focus groups (n = 19), and photo-elicitation interviews with LGBTQ2S suburbanites (n = 65). The semi-structured, key informant, informational interviews were undertaken between 2017 and 2022 with civic leaders, service providers, and community organizers to identify key urban actors and types of collaborations, dominant issues, and political priorities. As part of "yourLGBTQ2Scity" case study focus groups (of two to seven queerburbanites), the queerness of suburban environments was collectively mapped and followed up with photo-elicitation interviews to understand how queerness is perceived, experienced, and interpreted in place. Focus groups and photo-elicitation interviews were transcribed and abductively coded using NVivo qualitative data analysis software to permit iterative moves between theory and empirics. Taken together, this data informs the subsequent chapters, each prioritizing one or two data sources. All ten peripheral municipalities are profiled below, narrating their historic, geographic, demographic, political-economic, and place-brand specificity.

Tkaronto: The Toronto city-region

Tkaronto, a Mohawk word for the place where there are trees standing in water, has grown into Canada's largest (at 6.2 million people in 2021) and most ethno-culturally diverse global city, its economic engine, and purportedly its "LGBTQ2S capital" (Bourne et al., 2011). Toronto's suburbanization accelerated in the post-war period, driven by immigration and massive government-led restructuring of urban space (Relph, 2014). In the twenty-first century, inner-suburban disinvestments and outer-suburban hyper-diversification have produced an uneven geography of access to social, cultural, and economic services (Cowen, 2005; Boudreau et al., 2009; Hulchanski, 2010). Neighbourhoods in Mississauga, Brampton, and Markham with the highest proportions of foreign-born residents, for example, have declining household incomes and relatively high low-income household concentrations (Walks, 2011). Of the three case studies, the Toronto CMA (Figure 1.1) has the highest populations of immigrants (47 percent), non-official language speakers (50 percent), and visible minorities (57 percent) (Canada's problematic census term for non-white and non-Indigenous people). It has the highest rates of post-secondary education (38 percent) and average household incomes (CDN$129,000). Across the sprawling city-region, densities are low (1,051 per km^2) with one of the highest property-owning rates (65 percent), predominantly in detached dwellings (39 percent). The 2021 national census shows it is home to the highest numbers of same-gender, transgender, and non-binary couples (n = 21,303). Within Canada, Toronto has been at the centre of LGBTQ2S activism, the foundation of many national LGBTQ2S organizations (e.g., Egale and The ArQuives), and queer urban place-making. From the 1960s onwards, Toronto's LGBTQ2S population gradually received increasing social tolerance and acceptance, consolidating residentially and commercially in a downtown gay village that emerged "despite rather than because of gay movements' efforts" as a "national and international oas[i]s of so-called queer culture" (Nash, 2006, 2; 2012).

Against this demographic backdrop, Toronto's intense pace of growth has produced a polynucleated "pattern of development that is simultaneously dispersed and concentrated, economically integrated but politically fragmented" (Bourne et al., 2011, 237). The 1974 provincial governance restructuring of the Greater Toronto Area (GTA) added regional municipalities to manage common functions (policing, health and social services, major arteries and public transit, waste disposal, sewerage, and water systems, and regional land use planning) fracturing its political structure and limiting inter-municipal cooperation between the City of Toronto and its four two-tier regional government units (Durham, Halton, Peel, and York). Regionalization also fostered inter-municipal competition for new industries, investments, tax revenues, and jobs while unevenly distributing the financial and administrative burden for welfare, social services, housing, infrastructure provision, and public realm enhancement. Within this administrative division, social planning is the responsibility of regional municipalities and most service supports for the LGBTQ2S community are health- and youth-oriented. Limiting LGBTQ2S needs to medicalization and youth services neglects political responsibilities for social inclusion in housing, ageing, sociability, recreation, and community-building that seemingly fall between the cracks of municipalities and regional municipalities.

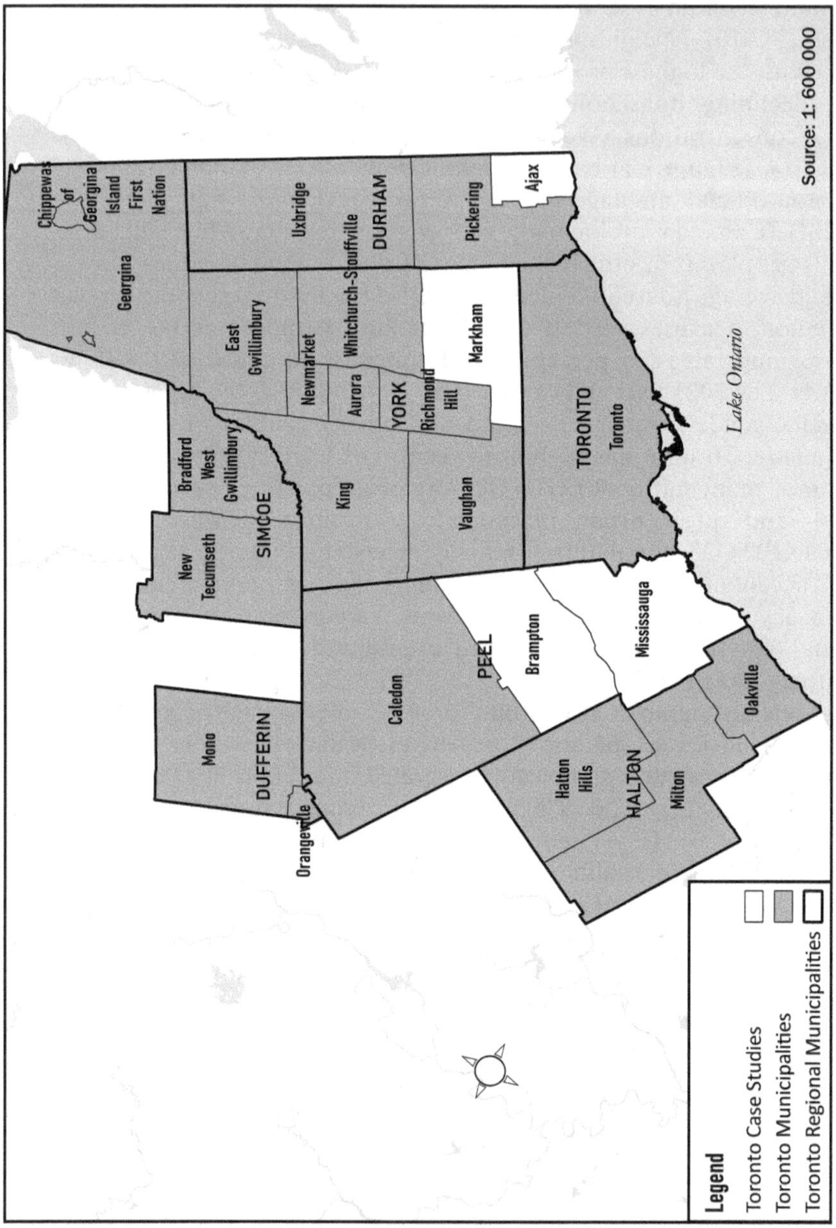

Figure 1.1 The case study municipalities of the Toronto Census Metropolitan Area

(Source: Julie Podmore).

Mississauga

On the western edge of the City of Toronto, Mississauga (an Anishinaabe word meaning people at the "Great River-mouth") is the second largest municipality in the GTA and Canada's sixth largest. Beginning as the "Toronto Township" and developing as the Toronto-Hamilton Highway and the Queen Elizabeth Way opened up land for development between 1917 and 1939, it is now framed by the 410 and 403 Highway corridors. Incorporated as a town in 1968, the amalgamation of Streetsville, Port Credit, and Mississauga brought city status in 1974. Geographically it extends across 290 km^2 and 24 neighbourhoods in the Peel region, with a population of 717,961 in 2021, most of whom are homeowners (70 percent). Over half of Mississaugans are foreign-born and 62 percent are visible minorities speaking over 90 different languages – predominantly South Asian (Hindi, Punjabi, Tamil, and Urdu), Arabic, Chinese (Mandarin and Cantonese), Italian, Polish, and Portuguese. With a moderate density of 2,453 people per km^2, Mississauga's heterogeneous landscapes contain wealthy waterfront neighbourhoods, large condominium developments, winding subdivisions, big box retail and warehouse complexes, office parks, and the Toronto Pearson International Airport complex, all stitched together via multi-lane roads and piecemeal public transit routes. Under the mayoral leadership of Hazel McCallion (1978–2014), Mississauga attracted the headquarters of 57 Fortune 500 companies and 51 Global Fortune 500 companies, developing particular concentrations in advanced manufacturing, financial services, and information communication technologies. With the place moniker "leading today for tomorrow", 40th anniversary civic celebrations in 2014 introduced its iconic "M" logo, representing five pillars (e.g., prosper, connect, move, belong, and green), three brand ingredients (e.g., "welcoming world culture", "inspiring possibilities", and "naturally enriching"), and three amalgamated municipalities (Paulmann, 2023). Civic place-making in Mississauga replicates other globally competitive cities: condominium development around densified centres, the starchitectural rebranding of its "skyline", the revitalization and promotion of its waterfront, and an expansion of its privatized public realm (Bain, 2013).

Brampton

North of Mississauga is Brampton, Canada's ninth largest "city in waiting" (Keil, 2020, 6). A former agricultural fair site, it was incorporated as a village in 1853, a town in 1873, and a city in 1974 when it became the administrative centre of the Regional Municipality of Peel. Housing the greenhouses and nurseries of some of the largest cut flower exporters in North America, today Brampton is also a regional and national construction market leader (Keil & Üçoğlu, 2021) that anchors expanding logistics, warehousing, and airport-related business sectors in what is envisioned by the Toronto Board of Trade (2020) as an "Innovation Corridor". Brampton is growing three times as fast as the rest of the province of Ontario and has a hyper-diverse population of 656,480, 81 percent of whom are racialized. Over half of its residents are immigrants with 48 percent speaking non-official languages and many identifying with non-Judeo-Christian religions (e.g., 25 percent Sikh, 18 percent Hindu, and 9 percent Muslim). Brampton has high rates of married or common-law couples with children (57 percent), large household sizes (3.6 people), and a large population of young people (18 percent under the age of 15). Since 2001, Brampton has doubled in size, and this growth rate is largely driven by migration from South Asia (Statistics Canada, 2022). From this postcolonial ethnoburb (Li, 2009) with a strong discursive identity signified by new nomenclatures (e.g., "Bramladesh", "Browntown", or "Singhdale" (Ahmed-Ullah, 2016)) emerged the political base that elected Canada's first visible-minority leader of a federal political party, Jagmeet Singh of the National Democratic Party (Keil, 2020). While Brampton continues to envision itself beyond large tracts of low-density, predominantly detached housing subdivisions and a "landscape of mortgage indebtedness" (Keil & Üçoğlu, 2021) with few rental housing options (only 23 percent are tenants), its place-brand still trains on its agricultural history as "The Flower City". A white-outlined yellow rose nestling a dove is the city's logo, symbolizing civic success as a grower and exporter of roses, its location within the Credit and West Humber River valleys, and its civic aspirations of care (City of Brampton, 2023). This "flower" strategy informs civic commitment to beautification projects, environmental sustainability, and the maintenance of its cultural and ecological heritage.

Markham

East of Brampton, along the 407 Highway, and north of the City of Toronto is Markham, a peripheral municipality that has grown rapidly since the new millennium. Originally an Iroquois settlement, Markham is an amalgamation of the Township of Markham and surrounding villages (Unionville, Milliken, and Thornhill) incorporated as the Town of Markham in 1971. By the time it joined the eight other towns and cities to form the Regional Municipality of York in 1974, it was growing exponentially because of the provincial government's construction of the York-Durham trunk sewer and water connections (Boudreau et al., 2009). Developers transformed prime agricultural land into subdivisions of detached houses, new urbanist communities, business parks, and shopping centres. Markham is home to a super-majority of visible minorities (82 percent), unofficial language speakers (59 percent), and immigrants (59 percent), the majority of whom are from China, Hong Kong, or Taiwan (51 percent). It is exceptionally irreligious with 41 percent describing themselves as having no religion. Relative to other Toronto suburbs, Markham has high: average dwelling values (CDN$1,208,000); average household incomes (CDN$130,800); and levels of postsecondary education (49 percent). In contrast with other ethnoburbs, it has slightly smaller households (3.0 people), more seniors (18 percent), and fewer youths (15 percent) and is close to the

CMA average for couples raising children. In a 20–30-kilometre arc from the City of Toronto, Markham joins Mississauga and Brampton in being an important GTA employment sub-centre with numerous higher-order and back-office functions. When it became a city in 2012, Markham's place-brand proclaimed itself as "Canada's high-tech capital". Its civic leaders aspire to make it the Silicon Valley of the North with its high-tech image and the moniker of "the mark of excellence".

Ajax

East of Markham on the shores of Lake Ontario is the Town of Ajax in the Regional Municipality of Durham with a population of 126,666. A farming community until the Canadian government expropriated land to build munitions plants, its post-war economy was anchored in aerospace, automobile manufacturing, and post-secondary research laboratories, bringing municipal independence in 1955. By the 1970s, large tracts of farmland were redeveloped as bedroom community subdivisions to house commuters to Toronto, Whitby, and Oshawa. In 1974, Ajax joined with seven other municipalities and one First Nations territory to form the Durham region centred on Oshawa. Historically a working-class community, today Ajax has high average household incomes (CDN$134,000) and homeownership rates (84 percent). While its homes have lower than average dwelling values (CDN$874,000), they have many more rooms per dwelling (6.9 rooms) than the CMA average. Ajax households have lower levels of post-secondary education (30 percent), are younger on average (38.6 years), and have more children (18 percent are under 15 years) and fewer seniors (13 percent). The majority are married and common-law couples raising children (55 percent) making its family sizes significantly larger (3.2 people). The majority of Ajax's population are visible minorities (65 percent), but fewer are immigrants (42 percent) and non-official language speakers (28 percent). The Canadian suburb with the highest concentration of Black residents (16.7 percent) (Census of Canada, 2016), Ajax's Black population is the second largest of its visible minorities (26 percent) (after South Asians at 41 percent). It also has high levels of religiosity with 51 percent identifying as Christians, 14 percent as Muslims, and only 21 percent as non-religious. Inspired by its ongoing demographic changes and upcoming 70th anniversary, civic leaders pragmatically delay a refresh of their corporate "Town by the Lake" place-brand.

Tiohtià:ke: The Montreal city-region

East of Toronto with a population of 4.3 million, Greater Montreal is Canada's second largest metropolitan area and the regional metropolis of the Francophone province of Quebec. The Island of Montreal at its centre is known as Tiohtià:ke ("where the group divided/parted ways") by the Kanien'kehá:ka Nation due to its historic role in diplomacy and exchange (Concordia University, 2017).

From a French missionary colony and fur-trading post, it grew into Canada's most important inland colonial port and railway network hub (Shearmur & Rantisi, 2011), yet remained socially fractured by class, language, and religion. A combination of a North America-wide manufacturing recession and Quebec's neo-nationalist "Quiet Revolution" led Toronto to supersede Montreal in the Canadian urban system in population, capital-intensive industry, and the service and finance sectors (Shearmur & Rantisi, 2011). With the election of the nationalist Parti Québécois in 1976 and Quebec's adoption of The Charter of the French Language in 1977, Montreal officially became a French-speaking city – today the fourth largest in the world – and the "regional metropolis" of a more autonomous province of Quebec (Linteau, 2007). Once dominated by labour-intensive, low-wage traditional industries (e.g., food, tobacco, textiles, and printing), deindustrialization brought job losses that rendered Montreal Canada's "poverty capital" with 41 percent of CMA households deemed low-income in the 1996 census (Lessard & Raynault, 2009). Concomitantly, newer high-technology industries (e.g., aerospace, biopharma, engineering, and transportation) clustered in inner-suburban areas and suburban technopoles (Germain & Rose, 2000). Today, Montreal's economy revolves around the knowledge industries, subsidized "cultural industries" (e.g., media, theatre, film, circus performance, and the arts), and international niche economies (e.g., artificial intelligence, film animation, and video game production) localized in specialized innovation districts (Shearmur & Rantisi, 2011). But given Montreal's "marginal" economy in the 1980s and its social housing and rent controls (Rose, 1996), redevelopment gentrification in inner-city areas and incumbent displacement to suburbs have been slower and later than other large Canadian cities (Rose & Twigge-Molecey, 2013). Its "marginal" economic status is also seen in current social class measures: average household incomes (CDN$98,400), dwelling values (CDN$500,400), and post-secondary education (29 percent) remain lower than those of Toronto (CDN$129,000, CDN$1,112,000, 38 percent, respectively) and Vancouver (CDN$117,300, CDN$1,304,000, 35 percent, respectively).

A "bicultural" city within a "multicultural" federation (Germain & Rose, 2000), Montreal's population remains predominantly Francophone (60 percent) and Christian (58 percent, 76 percent of whom are Catholic). Although it has a small but powerful historic English-speaking minority (11 percent) and a growing population of new immigrant, non-official language speakers (23 percent), the Quebec Francophone majority means proportionally fewer immigrants (24 percent) and visible minorities (27 percent) and higher rates of religiosity (only 29 percent define themselves as secular) than in Toronto or Vancouver. Anglophone suburbia extends westwards from the elite inner-city suburb of Westmount to the suburbs of the West Island, with eastern and off-island suburbs being predominantly Francophone with pockets of second- and third-generation immigrants (Walks, 2011). Home to over 11,000 same-gender, transgender, and non-binary couples (Statistics Canada, 2023b), Montreal's linguistic duality also shapes its LGBTQ2S communities, which have their own "national" French-language institutions to address rights, provide services, and build political constituencies. Moreover, its eastern gayborhood was formed in the early 1980s after gay and lesbian nightlife was displaced from the more Anglophone downtown core following police raids to "clean up" the city in preparation for hosting the 1976 Olympics (Podmore, 2015). The linguistic duality – along with a lack of regional land use planning, provincial government ambivalence about Montreal's economy, and the region's island archipelago – is also partly responsible for its municipal fragmentation (Germain & Rose, 2000). Montreal's CMA is currently composed of 91 separate municipalities and two First Nations territories (Kanesatake and Kahnawake) (Figure 1.2) that remain independent despite attempts at amalgamation that have raised tensions around linguistic power, local democracy, and fiscal equity. The 2001 implementation of "One Island, One City", a provincial plan forcing the merger of the 27 island municipalities with Montreal, became a political "theatre of a struggle for linguistic rights" culminating in the demerger of 15 primarily Anglophone municipalities when referenda were held in 2004 (Boudreau, 2003, 179). A city-regional governance body, the Montreal Metropolitan Community (MMC) was also created in 2001 and two types of supralocal bodies were introduced: urban agglomerations (for Montreal and Longueuil) and 12 Regional and County Municipalities (MRCs) (Beaudet, 2021).

The West Island

This area is composed of eight western on-island municipalities (Baie-D'Urfé, Beaconsfield, Dollard-des-Ormeaux, Dorval, Kirkland, Pointe-Claire, Sainte-Anne-de-Bellevue, and Senneville) and two City of Montreal boroughs (Pierrefonds-Roxboro and L'Île-Bizard-Sainte-Geneviève). Excluding these two Montreal boroughs, the remaining municipalities form a relatively low-density (1,365 per km^2) suburban region with a total population of 149,625 inhabitants. In the nineteenth century, the area was primarily agricultural, but the railway brought the "sumptuous villas" and country clubs of Montreal's English elite and the first commuter suburb in Pointe-Claire (Beaudet, 2021, 237). Planned suburban developments and the opening of Montreal's international airport in Dorval came in the 1940s, later followed by the creation of Autoroutes 20 and 40 that initiated mass suburbanization (Beaudet, 2021). The association of the area with English-speaking communities continues to determine the West Island's demographics and political representation: it retains a relatively high concentration of Anglophones (44 percent) and all but one municipality (Saint-Anne-de-Bellevue) has bilingual status. Its relative affluence is reflected in high average household incomes (CDN$153,638) and above CMA average dwelling values (CND$774,938), high rates of homeownership (77 percent), detached dwellings (69 percent), post-secondary education (44 percent), and average rooms per dwelling (7.3 rooms). Across the ten case studies, it has by far the highest (23 percent) proportion of seniors. But municipal

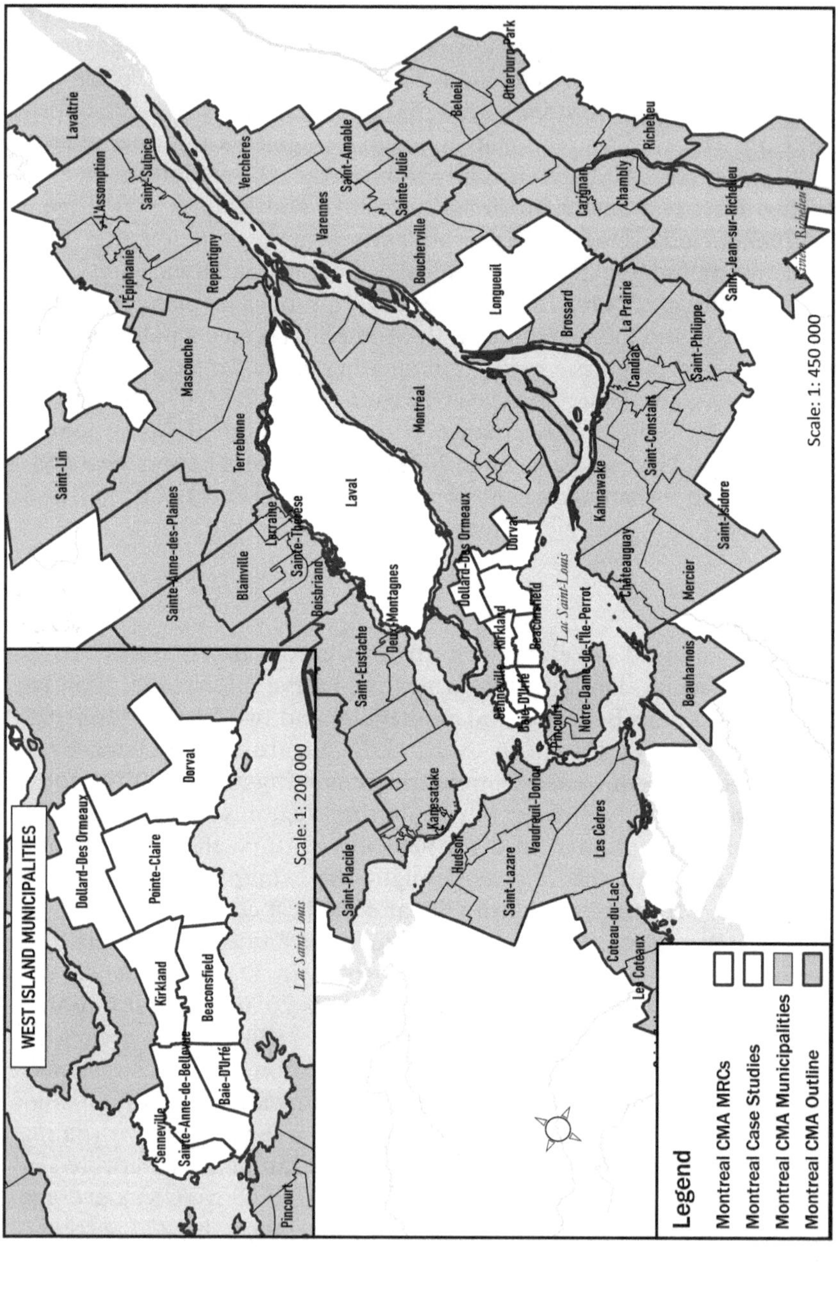

Figure 1.2 The case study municipalities of the Montreal Census Metropolitan Area

(Source: Julie Podmore).

aggregation masks socio-economic differences: Baie d'Urfé, Beaconsfield, and Senneville have some of the city-region's highest dwelling values while Dorval, Kirkland, and Sainte-Anne-de-Bellevue are closer to the CMA average. Although most are predominantly Francophone/Anglophone and Christian, Dollard-des-Ormeaux, Dorval, and Kirkland have higher concentrations of immigrants, visible minorities, minority religions, and mother tongues as well as higher densities, lower incomes, fewer detached dwellings, and more renters. Today, the West Island's educated workforce draws biomedical and pharmaceutical companies while its international airport attracts aerospace, transportation, warehousing, and manufacturing to industrial parks. Challenging the 2001 forced provincial merger, these municipalities officially demerged in 2006 and regained their political autonomy, participating independently in the Agglomeration of Montreal and the MMC (Boudreau, 2003).

Laval

To the north-east, with a population of 438,366 inhabitants, Laval is Montreal's largest suburb and Quebec's third largest municipality. On the North Shore, separated from the l'Île de Montréal by the Rivière des Prairies, Laval was formed in 1965 when 14 smaller municipalities on l'Île Jésus and two smaller islands were amalgamated and later upgraded to an MRC in 1980. Building on its agricultural heritage, in the 1950s the island was branded as "The Garden of Montreal" (Ville de Laval, 2011a, 6) because most of its 240 km^2 of land was dedicated to hay production, dairy farming, and market gardening (Beaudet, 2021). Post-war construction of highways and bridges stimulated bedroom community construction that fueled its retail and manufacturing economy (Lapointe, 2022). Today, Laval is a large, low-density, multi-ethnic, and multilingual (33 percent speak a non-official language as a mother tongue) peripheral municipality with concentrations of second- and third-generation immigrants from Italy, Portugal, and Greece as well as newer immigrants from France, Lebanon, Haiti, Morocco, and Algeria. Its population exhibits high levels of religiosity (only 18 percent reported no religion) and a higher than CMA average Muslim population (13 percent), in a city-region where the vast majority are Catholic or non-religious (83 percent). Laval's households are primarily middle-income earners (CDN$102,900), who are married or common-law couples raising children (48 percent) with moderate levels of post-secondary education (25 percent) and homeownership (67 percent). Its mixed landscape comprises historic villages, industrial parks, office complexes, tracts of single-family detached dwellings, duplex and high-rise neighbourhoods, working-class urban areas, golf courses, agricultural tracts, and large-scale retailing complexes. With the extension of three Montreal Metro subway stops to southern Laval in 2007, urban development is centred on a new transit-oriented downtown with a municipal arts centre, arena, and satellite university campuses. Following a mayoral corruption and fraud scandal, in 2019 Laval changed its moniker from "My city, my future" to "Laval: An island of exceptions" to accentuate its unique "education, investment and employment" opportunities (Ville de Laval, 2019, para. 1).

Longueuil

Located on Montreal's South Shore, this historically working-class, industrial suburb is high-density (2,198 per km^2) suburb with 254,483 people. Across the ten case-studies, it has the lowest average incomes (CDN$87,800), dwelling values (CDN$399,200), and proportion of detached dwellings (32 percent). Ethno-culturally, it has the most homogeneous population: the majority are non-immigrant (79 percent) Quebec Francophones (72 percent), and a few are visible minorities (25 percent). From its rural parish origins, with the opening of Montreal's first airport in Saint-Hubert in 1928, Longueuil became an industrial centre for metal works, chemical and textile production, food processing, and later aeronautics (Collin & Poitras, 2002, 282). The construction of the Jacques-Cartier and Mercier bridges in the 1930s and the opening of the Victoria Bridge to automobiles accelerated Longueuil's suburbanization (Collin & Poitras, 2002). Suburban subdivisions for wealthy English and French families were constructed alongside "shacktowns" of self-built "maisonettes" (Beaudet, 2021, 249). Post-war annexation of surrounding municipalities increased Longueuil's size and diversified its morphology, combining old village landscapes, bedroom communities, and working-class urbanized areas with unplanned suburbs. Following the 2001 merger-demerger process, Longueuil's boundaries were redrawn. Today, it is composed of the historic and urbanized Vieux Longueuil, the bilingual residential suburb of Greenfield Park, and the more industrial Saint-Hubert (Poitras & Linteau, 2020). Vieux Longueuil's central neighbourhoods are experiencing increased immigrant settlement, rising poverty levels, housing shortages, demographic ageing, and, compounding these challenges, gentrification through small-scale condominium redevelopment (Ville de Longueuil, 2019). With the slogan "Longueuil: Where opportunities materialize", "downtown" Longueuil is undergoing a large-scale, mixed-use redevelopment to transform its downtown into "the gateway to the South Shore" (Ville de Longueuil, 2023).

Musqueam territory: The Vancouver city-region

Far west of Toronto and Montreal, situated on the Pacific Ocean where the Fraser River meets the Strait of Georgia in the Lower Mainland of British Columbia on the unceded, ancestral, and traditional territory of the xʷməθkʷəy̓əm (Musqueam), Sḵwx̱wú7mesh (Squamish), and səl̓ílwətaʔɬ (Tsleil-Waututh) peoples, is the Vancouver city-region. It is the smallest of Canada's three largest CMAs with a total population of 2.6 million (Figure 1.3). As Canada's Pacific "gateway city", international capital investment from the Asia Pacific region has synergized with a trans-Pacific property industry and a local

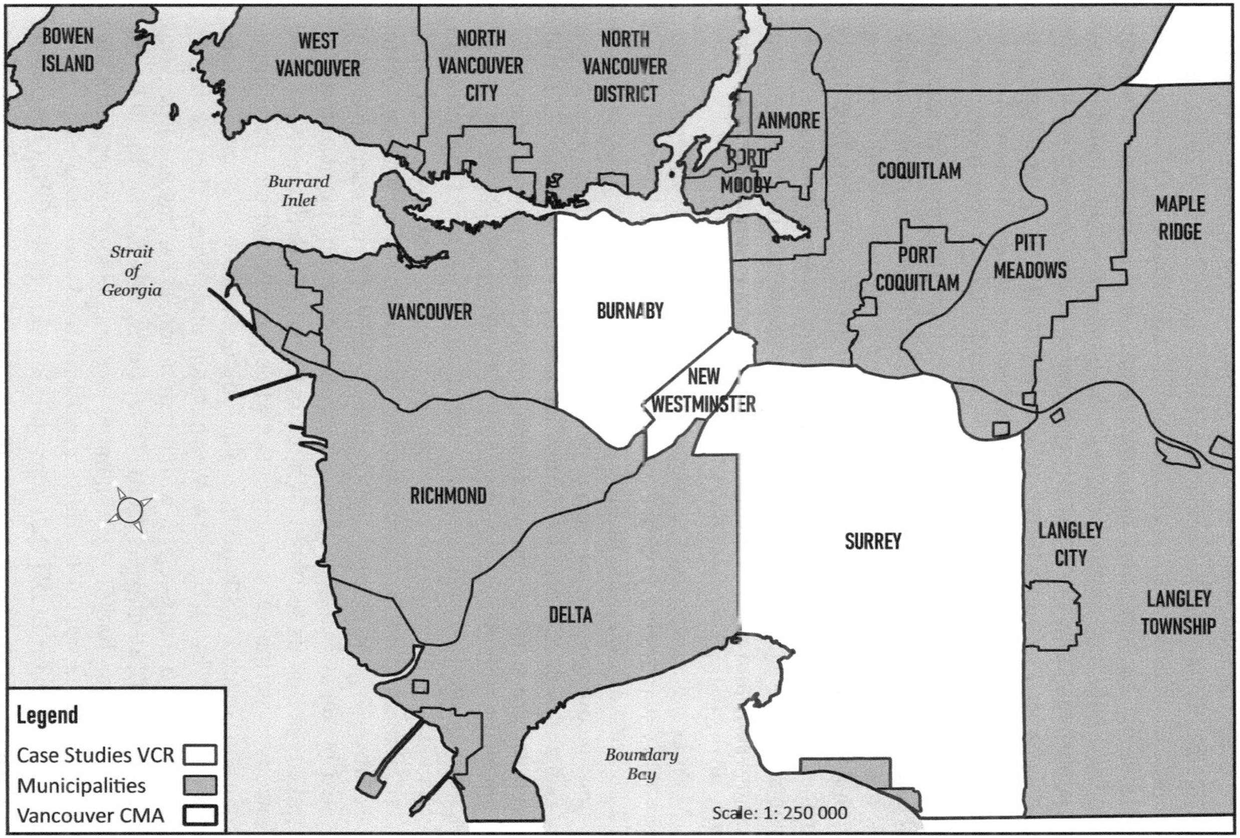

Figure 1.3 The case study municipalities of the Vancouver Census Metropolitan Area (Source: Julie Podmore).

growth coalition to transform the local housing market into one of the least affordable in North America (Ley, 2010). It is characterized by significant population growth (at a five-year rate of 7.3 percent), high levels of immigrant settlement (42 percent), and the country's highest land values (the average dwelling value is CDN$1,304,000) and severe housing unaffordability (Ley, 2021), resulting in internal densification and centrifugal peripheral displacement. Surrounded by mountains on the north and east, the sea on the west, and the American border on the south, these natural and political boundaries have contained and directed suburbanization eastward up the Fraser River Valley forming the Greater Vancouver Regional District (GVRD). Established in 1967 from provincial regional district legislation, the GVRD was renamed Metro Vancouver in 2007 and became the Metro Vancouver Regional District (MVRD) in 2017. A federation of municipal and electoral areas covering nearly 3000 km^2, the MVRD contains 53 percent of British Columbia's population within 21 municipalities, one electoral district, and one treaty First Nation. Within this regional institutional framework, individual municipalities maintain their independence yet have the flexibility to develop ad hoc inter-municipal governance structures. As a regional authority, the MVRD is responsible for water, air, and waste management, housing, and regional planning, as well as maintaining parks and the prosperity of the political federation (MetroVancouver, 2023).

The MVRD has embraced the international renown of "Vancouverism" (Beasley, 2019), a planning approach oriented towards urban liveability and sustainability first associated with the City of Vancouver's expansion of greenspaces, and encouragement of social participation within high-density, yet amenity-rich and walkable, condominium neighbourhoods (Thomas, 2021). In reality, however, any sense of "progressive regionalism" (Bain & Podmore, 2022) is undermined by neoliberal place promotion (Vanwynsberghe et al., 2013), consumer-driven lifestyle initiatives (Hayter et al., 2024), and significant car-dependency within still predominantly low-rise suburbs. Vancouverism seeks to retain a "rural romance of living close to nature" in combination with an "urban romance of diversity and complexity" (Kataoka, 2009, 42), but such diversity is complicated by socio-spatial tectonics. Vancouver's wealth is concentrated in the central city and in its southern (White Rock) and North Shore (North and West Vancouver) peripheral municipalities while poverty and racialized/immigrant exclusions extend eastward in a band along the Kingsway arterial road into the suburbs of Burnaby, New Westminster, and Surrey (Ley & Lynch, 2012). Yet this is also a city-region with a reputation as "an avant-garde paradise for people who identify as LGBTQ" (Murray, 2015, 71). In 2021, the Vancouver city-region had an equal number of same-gender, transgender, and non-binary couples (n = 11,630) to the larger Montreal and it retains a history of nationally significant LGBTQ2S activism. Canada's first LGBTQ organization was formed here in 1974 – the Association for Social Knowledge (ASK) – inspiring the formation of LGBTQ rights and service organizations that have supported decades of citizen activism and shaped public policy across election cycles. Such activism underpinned the formation of the city's Davie Street gay village and lesbian "Eastside", the establishment of Canada's first LGBTQ civic advisory committee, and "openly queer" political representation (Murray, 2015). Outside of the central city, however, there is significant variation in policy enactments of LGBTQ2S-supportive social inclusion. All three case-study peripheral municipalities share a history of working-class residents and politics but differ significantly in size and ethno-cultural composition.

Burnaby

Bordering the City of Vancouver, Burnaby is a mid-sized (249,125), medium-density (2,751per km^2) peripheral municipality. In the 1950s, it was a bedroom community realized through the zoning of large tracts of land for single- and two-family residential development. Over the ensuing decades, Burnaby experienced rapid urban growth via apartment and shopping mall construction across four town centres (Metrotown, Brentwood, Edmonds, and Lougheed). With the extension of the elevated SkyTrain from the City of Vancouver out to its periphery in 1986, South Burnaby's Metrotown became the primary centre for commercial and industrial urban growth. By 1992, Burnaby was granted city status, 100 years after its incorporation. Occupying only 4 percent of the MVRD (City of Burnaby, 2011), its population is predominantly composed of immigrants (50 percent), visible minorities (68 percent), and non-official language speakers (54 percent). While Burnaby promotes itself as a city where the residents speak more than 100 different languages, its non-official mother tongues are predominantly Mandarin (24 percent), Cantonese (21 percent), Korean (6 percent), and Tagalog (5 percent). Its largest immigrant groups come from Asia: China (26 percent), India (13 percent), and the Philippines (8 percent). Reflecting Vancouverism's densification and high land values, only 19 percent of Burnaby's dwellings are detached, and the average number of rooms is low (4.9 rooms). Nevertheless, Burnaby's household size (2.4 people) and percentage of families with children (44 percent) remain close to the CMA average, but with fewer children (13 percent) and more seniors (18 percent). Burnaby's dwelling values are below the CMA average (CDN$1,175,000), but remain high relative to household income (CDN$105,100). Consequently, housing (including the pace and intensity of residential redevelopment and ensuing renovictions) remains politically contentious; so divisive that it contributed to the ousting of the long-term socialist mayor and council in 2018 whose policies diametrically opposed the provincial and federal Liberals. Despite having the largest percentage of a non-religious-affiliated population (48 percent) in the case-study peripheral municipalities, politicians have prioritized the views of the city's powerful Asian Christian lobby to the neglect of other groups and social issues. Ironically, Burnaby is noteworthy for electing Canada's first openly gay members of the federal Parliament, yet this is not part of its place-brand. Instead of showcasing its formative role in shaping national LGBTQ2S rights legislation, civic leaders emphasize its founding natural environment "by river and sea".

New Westminster

On Burnaby's eastern border lies New Westminster, a smaller (78,916), high-density (5,052 per km^2) suburb, and the Vancouver city-region's oldest municipality. It was founded in 1859 as the province's British settler capital and named by Queen Victoria. Located along the Fraser River, it developed as a major freshwater port, producing lumber, canning salmon, brewing beer, operating a penitentiary, and serving as an administrative and commercial centre for the surrounding Fraser Valley. Consequently, its morphology resembles that of a historic industrial city with a riverfront downtown core extending along Columbia Street, an early-twentieth century main street now lined by office buildings, department stores, pubs, and hotels. Columbia Street has a concentration of LGBTQ2S-friendly restaurants, pubs, and theatres, including a gay-owned bathhouse, that has contributed to its characterization as the region's "second gay village". Its newer "uptown" district is composed of high-rise apartments and lower-density heritage homes. The condominiums along its redeveloped waterfront have attracted out-migrants from Vancouver and accelerated residential and commercial gentrification. New Westminster is less ethnically diverse than its municipal neighbours: it has fewer immigrants (38 percent), visible minorities (47 percent), and unofficial language speakers (37 percent). While it has close to the CMA level of seniors (16 percent), their spatial clustering in uptown apartments near the municipal seniors' centre makes them politically important. With few detached dwellings (13 percent), very small household sizes (2.2 people), and few children (13 percent), New Westminster is less family-oriented than either Burnaby or Surrey. Once a working-class city supportive of labour and monarchist movements, its municipal government remains a progressive regional leader in social inclusion policy. As part of its most recent commitment to inclusivity, civic leaders intend to phase out the "Royal City" moniker and develop "a new brand identity that is inclusive and allows for collective pride in our city" (McManus, 2022, para. 2).

Surrey

On New Westminster's eastern border is Surrey, the city-region's fastest growing and second largest municipality (568,322), with a five-year population growth rate of 9.7 percent. Built on agricultural land, its residential subdivisions, developed in the 1980s, are spatially organized around seven town centres and a recently designated downtown core. Over a decade ago, civic leaders commissioned architects to develop a new city hall plan to transform "the architectural landscape of the city center" and reshape "the culture and ethos" of Surrey "from 'suburb' to thriving business and urban center" (Architizer.com, 2021, para. 1). As of 2014, the 210,000 square-foot city hall became the centerpiece of Surrey's new downtown, surrounded by the office buildings, shopping malls, and satellite campuses of post-secondary institutions that displaced its incumbent low-income residents. Within the city-region, Surrey is disproportionately young (31 percent are under 24) and ethno-culturally diverse (47 percent declare a non-official mother tongue). Part of the Lower Mainland's southeastern "Bible Belt", 71 percent are religious (75 percent of whom are Christian, while the rest are Sikh, Muslim, or Hindu) (Statistics Canada, 2019). While the city houses a low proportion of the city-region's suburban same-sex households (2 per 1000) (Statistics Canada, 2017), it has a vibrant history of LGBTQ2S activism combatting religious intolerance that dates back to a 1999 national legal challenge to the Surrey School Board's ban on three books about same-sex families (Bain et al., 2020; Podmore & Bain, 2019).

Together Canada's three largest city-regions and their ten case-study peripheral municipalities offer a vibrant national cross section of suburban life. Each has a different settler history and geography, shaping variegated contemporary socio-economic landscapes, political priorities, and class and cultural cleavages. Canada's ethno-cultural, religious, and racialized tensions complexify its suburban heterocisnormativities, with ethnoburb formation being definitive of the nation's contemporary suburbanization. Linguistic and cultural bifurcation between Montreal – as part of the French-speaking unofficial "nation" of Quebec – and the two Anglo-Canadian city-regions shapes differences in policy priorities, political interpretations of sexual and gender minority identities and communities, para-public LGBTQ2S allyship and their respective tools of social inclusion. Seemingly culturally constructed as sites of monotonous sameness, Canadian suburban comparison ultimately illuminates the diversity of the nation's city-regional peripheries.

Conclusion

What kinds of queer place-making/unmaking/remaking, therefore, are associated with queer suburbanism? Who makes/unmakes/remakes place for LGBTQ2S people in suburbs, where, and how? What forms of activism and institutional allyship exist and with what individual and collective place-making/unmaking/remaking impacts? How do these actions transform the micro-public spaces of suburban everyday interactions that are the lived experiences of places? How do queerburbanites, therefore, produce liveable lives in place and dream queerer suburban futures? These are the questions that this book grapples with through the lens of place-making/unmaking/remaking – the range of social, political, and material processes through which people (re)create the geographies of their everyday lives.

Across nine chapters, the title gerunds of representing, including, serving, resisting, organizing, living, and dreaming, are used to re-place queer suburbanism and theorize the multiple practices of queerburban place-making/unmaking/remaking. Chapter 2 initiates a discussion of discursive representations of queerburbia in the census and print media. Chapter 3 focuses on municipal inclusion discourses and their (mis)recognition of LGBTQ2S constituencies in variable political opportunity structural alignments that are modulated by the commodification of diversity. Civically serving suburban

LGBTQ2S constituencies is the subject of Chapter 4, highlighting municipal first-order allyship practices that exploit the rainbow as a place-making tool. These official place-making practices are held in tension with para-public critical allyship and local LGBTQ2S organizing to remake queerburbia. Chapter 5 focuses on the coalitional moments of second-order para-public organizational allyship that through insider activist place-unmaking resist suburban hetercisnormativity by filling civic programming and service gaps. Place-unmaking is deepened in Chapter 6 by individual and collective practices of LGBTQ2S organizing to sustain interpersonal connections and foster community care. The queerburban remaking of place is introduced in Chapter 7's analysis of hacking heterocisnormative suburban social infrastructure to create queer alternatives that make life liveable. The futurity of queerburbia and its sustained place-remaking potential is recuperated in Chapter 8 through social dreaming and counter-topographies of collective desire. Finally, the conclusion considers what it means to re-place queerburbia, elaborating on the neologism and calling for greater cross-cultural dialogue on queering the metropolitan periphery.

Note

1 The book moves between the specificities of including Two Spirit people in the acronym when referring to Canada and the more generic LGBTQ+ acronym when speaking more generally to an international audience and literature.

2 Representing

Introduction

City centres generally contain higher densities of visible LGBTQ2S places – service resources, institutions, residential and commercial neighbourhoods, and community activities – than suburbs. Thus, LGBTQ2S communality is part of a metronormative "sexual imaginary", equating "coming out" with a "great gay migration" to "the big city" (Weston, 1995) that bypasses queerburbia and "other" impossibly queer (non)metropolitan spaces. Yet countervailing centrifugal forces have drawn sexual and gender minorities to the suburbs, where they navigate the dynamics of spatial dispersal and queer community and infrastructural inadequacies. While the metropolitan landscapes of contemporary queer and trans life can be read for geographical specificity and unevenness, the only secondary sources for tracing them at this scale are removed from their experiences. In biopolitical government data and popular discourses, queerburbanites are anomalies who are neither completely nor systematically counted. They are statistical outliers in census data and underrepresented, mischaracterized, and even pathologized in print media representations.

Geographers examine how places are variously represented through, for example, maps (Jackson, 1989), postcards (Waitt & Head, 2002), monuments (Hall, 1997), paintings (Casey, 2002), photographs (Rose, 2001), and travel writing (Pratt, 1992). Such representations communicate subjective understandings of place by accentuating some dimensions over others, but they also have multiple meanings and are open to interpretation (Holloway & Hubbard, 2014). Necessarily incomplete and contestable, representations can be leveraged by regimes of power to consolidate shared meanings and assert institutional control by legitimizing selected interpretations, uses, and lived experiences of place to the neglect, and even denial, of others (Duncan & Ley, 1993). Representations, then, "shape what can be and what is known or appreciated" about places (Biddulph, 2014, 280). They are "discursive-material formations" that are "fluid, permeable, and affected by the ways in which they are imagined" and interpreted (Davis, 2005, 610).

This chapter examines the intersections between suburban places and LGBTQ2S identities from two of the only macroscale spatial forms of

DOI: 10.4324/9781032632544-2

representation available. First, newspaper databases offer textual and visual representations of "real-world" place-making events, debates, experiences, and narratives (Seamon, 2018, 17). Indicators of "political context and public discussion and debate", newspapers are key information filters that "shape, convey and legitimate a certain view of reality" and influence public opinion on "policy debates, context and decisions" (Raynor et al., 2017, 1525). In this way, they mediate place while leaving historical records that can be traced in digital database searches. Second, statistics derived from national population censuses provide ongoing information about demographic patterns, measurable commonalities, and aberrations, but they are also used to monitor populations and inform policy formation and investment decision-making, giving governmental regimes a calculative agency (Prince, 2020). Such "datafication" – the practices of classification, quantification, and measurement – reductively converts the world into numbers (Prince, 2020). To "know" human collectivities spatially through data is also to "intervene" in them, identify places and groups as problematic or needy, and interpret them within government political priorities (Osborne & Rose, 2004).

Queering big data such as newspaper databases and statistics poses important representational challenges related to scale, classification, and capture. Scavenging for smaller incidents of LGBTQ2S representation only serves to underscore "big data's mythos" as the sole source of objective generalization (Gieseking, 2018b, 151). Representing the LGBTQ2S acronym's subgroups results in unevenness since some are "smaller" and others completely absent. The scale of big data thus reproduces norms of the larger categories based on "objective and scientific claims" to "validity" which is arguably a "measuring stick of masculinist, racist, colonialist, ableist, and heteronormative structural oppressions" (Gieseking, 2018b, 150). Not easily check-boxed, queerness is difficult to capture in censuses that impose an "order onto sexual identities" that does not resemble the "way in which people live their sexual lives" (Fetner, 2024, 22). Big data representations are ultimately challenged by the "complexity, nuance, and fluidity of queer lives and experiences" that potentially destabilize the logic of classification while limiting their "capture" and reducing visibility (Ruberg & Ruelos, 2020, 10).

Drawing on geographical scholarship about place, this chapter analyzes newspaper and census representations of queerburbia and queerburbanites. Two secondary data sources are used: the 2016 Canadian national census; and an assembled print media database (1985 to 2014) of articles from the metropolitan English- and French-language newspapers referencing LGBTQ2S lives and issues. Critical examination of both sources highlights governmental and popular erasures, normalizations, and constructions of the heterocisnormativity of suburban subjectivity and citizenship. In dialogue, these public portraits of queerburbanites raise questions of what lies beyond cartographic dots and how contextual reading affords understanding beyond a frame of "problems" or "threats" to emplaced heterocisnormative socio-spatial orders. Ultimately, by tracing the dispersal of some LGBTQ2S populations and the heterocisnormative discourses emerging about them, an outsider's portrait of queerburbia emerges that nevertheless demonstrates LGBTQ2S socio-spatial ubiquity across city-regions.

Finding queerburbia in the census? Mapping "same-sex" couples

Queer mapping projects render visible historical and contemporary LGBTQ+ experiences of places, whether they collate individual cognitive maps (Gieseking, 2020a, 2020b), crowd-sourced collective memory projects (Ketchum, 2020; Mattson, 2016), (re)present analogue data on digital maps (Knopp & Brown, 2021), or cross-tabulate survey data with urban, suburban, and rural differences in LGBTQ2S well-being (Grant et al., 2024). Such cartographies are facilitated by geographical information systems (GIS) which embed social inequalities in their algorithms and codes (Gieseking, 2018a). The most common method of representing the geographical distribution of LGBTQ+ people across space is through national census data and household survey statistics. However, classifying, counting, and mapping LGBTQ+ populations over time, especially in low-density areas such as suburbs, raises representational challenges (Brown & Knopp, 2006) and spatially fixes queerburbia to the few homes of cohabiting gay and lesbian couples inferring that it is a product of domestic densities (Brennan-Horley, 2010).

Most of the representational issues of LGBTQ+ census data mapping in North American, Australasian, and European countries since the 1990s arise from the classification of these populations under a single, reductionist variable, "the same-sex" household (Podmore & Bain, 2020, 2021; Forrest et al., 2019). Such household constructions underestimate the LGBTQ+ population because they only include self-reported, adult, cisgender, couples living in "married" or common-law relationships and disqualify anyone who is unpartnered or not cohabiting, and/or in a non-monogamous or polyamorous relationship (Anacker & Morrow-Jones, 2005; Brown & Knopp, 2006; Ghaziani, 2014a). Relying on biological "sex" categories, transgender and/or non-binary identities and relationships have until recently been misclassified and underrepresented (Doan, 2016; Frisch, 2021; Ghaziani, 2014a). Two-Spirit, trans, and queer Indigenous people may be underrepresented due to the right to refuse settler government entry to First Nations territories. Reporting reluctance skews data towards younger age categories, people already empowered ethnically, racially, and socio-economically by homonormativities, and residents of "progressive" large central-city areas (Brown & Knopp, 2006). Even the cisgender, same-sex couples included will not constitute a 100 percent sample due to state policies protecting privacy where small numbers do not meet the minimum for non-disclosure (Frisch, 2021; Gates, 2007). Cluster analysis is thus challenged in low-density areas where evidence of same-sex households may be suppressed and unavailable for analysis or cross-tabulation with intersectional variables (Brown & Knopp, 2008). Analysis of relative concentration will exaggerate the absence of same-sex households in less densely populated areas making them appear low (Anacker & Morrow-Jones, 2005). Thus, how such data "represent" LGBTQ+ populations raises many epistemological, methodological, and geographical questions, never more so than when applied to the private, lower-density, ethno-culturally diverse areas that are Canada's suburbs.

Macro-scale analysis in the United States and Australia has nevertheless proven useful for demonstrating how spatial patterns of participating same-sex households change over time and are statistically evidenced across national space (Gates & Ost, 2004; Gorman-Murray et al., 2010). It has disrupted metronormative assumptions about LGBTQ+ inner-city concentration demonstrating that "we are everywhere", including small towns and the suburbs of major metropolitan areas (Gates & Cooke, 2011; Gates & Ost, 2004; Gorman-Murray et al., 2010). American scholars have identified patterns of same-sex household decentralization and congregation, including neighbourhood safety, "amenities", real estate markets, perceived LGBTQ+-friendliness, and creative sector job growth (Anacker & Morrow-Jones, 2005; Smart & Whittemore, 2017; Whittemore & Smart, 2016). There are, however, strong modulations based on sex, class, and ethno-cultural differences. Male same-sex households are more segregated from opposite-sex households (clustering in inner-city areas with high creative sector job growth, urban amenities, and gay cultural infrastructure), while female same-sex households are more dispersed and decentralized in mixed areas (in terms of income, housing stock, and ethno-cultural differences) and living with children in neighbourhoods of predominantly opposite-sex households (Compton & Baumle, 2012; Hayslett & Kane, 2011; Lewis, 2017; Spring, 2013). Intersections of social class, "race" and ethno-cultural identity further shape LGBTQ+ suburbanization, with African American male same-sex households clustering in African American suburbs (Lewis, 2017) while female same-sex households specialize based on age, class, and ethno-cultural identity (Compton & Baumle, 2012). The strongest locational determinant of all, however, is the presence of other same-sex households (Whittemore & Smart, 2016) which signals a perceived "openness" to difference and safety through the co-creation of a "gay/lesbian defined space" even in suburbia (Hayslett & Kane, 2011, 151). In sum, American macro-scale scholarship invaluably locates select same-sex households in suburban areas, but these patterns have yet to be explored in Canada.

Canadian census representations of suburban same-sex couples

Working with the Canadian national census requires attention to categorization and sampling as Statistics Canada adapts the same-sex couple household category to legal shifts and changing understandings of gender and sexual identities (Statistics Canada, 2023c). Cohabiting same-sex couples were detected in the census in the 1990s, but it was not until 2001 that national data on 34,200 same-sex common-law couples was tabulated (Statistics Canada, 2023b). After same-sex marriage was legalized in 2005, the short-form 2006 census recorded a 25 percent increase in same-sex couples (45,300) permitting distinction of households by sex, age, and marital status (Statistics Canada, 2009). The next long-form survey in 2011 was unable to offer a fuller picture of same-sex couples because it was replaced by a voluntary National Household Survey (NHS) resulting in non-response bias and sampling errors due to a decreased participation rate (Smith, 2015, para. 4). Nevertheless, with statistical adjustments, a 42 percent increase in same-sex couples from 2006 was detected (64,575) and sex differences in marriage rates and children were observed (Statistics Canada, 2012a). The 2016 short-form census identified 72,880 same-sex couples and further detailed patterns of marriage, children, and geography (Statistics Canada, 2017). With the 2021 census, however, Canada became the first country to add gender diversity by distinguishing between "sex at birth" and "reported gender" making it possible to identify different-gender couples (cisgender), same-gender couples (cisgender), transgender couples (at least one partner is transgender), or non-binary couples (at least one partner in non-binary) (Rault, 2023). In 2021, the number of same-gender (cisgender) couples totaled 95,435 with the new transgender/non-binary category at 32,205 (Statistics Canada, 2022).

With the tripling of same-sex couples in the 2011 census came a more nuanced portrait of their geographies. Canada's same-sex couples were concentrated in its largest CMAs: 46 percent lived in Toronto, Montreal, or Vancouver (as opposed to 33 percent of opposite-sex couples), a concentration that had dropped slightly from 50 percent in 2006 (Statistics Canada, 2012b, 8). Female same-sex couples (46 percent), however, were less concentrated in these CMAs than their male (55 percent) counterparts (Statistics Canada, 2012b). By 2016, there were nuances in the national geography of same-sex couples. The largest province of Ontario had the most same-sex couples (26,585), but both Quebec and British Columbia had higher numbers relative to population (Statistics Canada, 2017). The metropolitan concentration expanded with half of all same-sex couples in Canada living in four of the five largest CMAs of Toronto, Montreal, Vancouver, and Ottawa-Gatineau (Statistics Canada, 2017) with significant increases in smaller CMAs (Edmonton, Halifax, Hamilton, London, Quebec City, and Victoria) and stabilization in smaller towns and rural areas (Statistics Canada, 2017). While same-gender couples followed similar patterns in the 2021 census (with some decentralization), the new gender categories shifted the portrait: Nova Scotia, Yukon, and British Columbia were the provinces with the most transgender and non-binary individuals who were concentrated in Victoria, Halifax, and Fredericton, now classified as Canada's most "gender diverse" cities (Statistics Canada, 2023a).

Using the 2016 data for same-sex couple households,[1] the three case-study CMAs demonstrate a strong centrality but with significant regional variation. Same-sex couple data by Forward Sortation Area (FSA) was disaggregated by sex and classified by metropolitan location (central city or suburban, with a distinction made between inner and outer suburbs)[2] (Table 2.1). With 65

Table 2.1 Summary of geographical distribution of same-sex couples by sex for Vancouver, Toronto, and Montreal, 2016

	Vancouver			*Toronto*			*Montreal*		
	Males	*Females*	*Total*	*Males*	*Females*	*Total*	*Males*	*Females*	*Total*
Central	**2320 (55.6)**	**1040 (33.9)**	**3360 (46.4)**	**5895 (71.3)**	**2705 (55.5)**	**8600 (65.0)**	**3890 (52.2)**	**1445 (28.9)**	**5335 (42.9)**
Suburbs	**1850 (44.4)**	**2025 (66.1)**	**3875 (53.6)**	**2370 (28.7)**	**2290 (45.5)**	**4660 (35.0)**	**3560 (47.8)**	**3555 (71.1)**	**7115 (57.1)**
Inner	1130 (27.1)	1135 (37.0)	2265 (31.3)	1340 (16.2)	1080 (21.6)	2420 (18.0)	1670 (22.4)	1210 (24.2)	2880 (23.1)
Outer	720 (17.3)	890 (29.0)	1610 (22.3)	1030 (12.5)	1210 (24.2)	2240 (17.0)	1890 (25.4)	2345 (46.8)	4235 (34.0)
Total	**4170**	**3065**	**7235**	**8265**	**4995**	**13260**	**7450**	**5000**	**12450**

Source: Brian Ray for authors.

percent of its same-sex couples residing within the central city, Toronto was the least queerburbanized especially for same-sex male couples, 71 percent of whom resided in the central city. In Vancouver and Montreal, central-city concentration was under 50 percent, indicating some queerburbanization. In Vancouver, this was especially the case for female same-sex couples (66 percent) living in inner suburbs (37 percent). In Montreal, female same-sex couples were also more likely to be suburban (71 percent) but living in outer suburbs (47 percent). Analysis of other dimensions of these households further reveals important differences between the three city-regions (Table 2.2). While the balance between central city and suburban same-sex households with children was relatively even in Toronto (51 percent versus 49 percent), in Vancouver (74 percent) and Montreal (77 percent), most same-sex couples with children lived in suburbs, especially in Montreal's outer suburbs (50 percent). Household income differentials were relatively even, but slightly elevated in Toronto's central area and Vancouver's outer suburbs. In all three CMAs, the queerburban populations were older than that of the central city with people over 50 years of age concentrated at a rate of 65 percent in suburban areas. At rates of over 60 percent, same-sex couples with at least one foreign-born individual were overall more suburban, especially in Toronto's inner suburbs (43 percent), but much lower in Montreal, especially in its outer suburbs (6 percent). Overall, these three CMAs parallel the existing literature with female same-sex couples with children, older couples, lower income, and foreign-born (with the exception of Montreal) being more suburban.

Mapping the location of same-sex couples across these city-regions provides a snapshot of coupled lesbian and gay suburbanization in 2016. With a higher concentration of same-sex couples in its central area, Toronto (Figure 2.1) exhibits high centrality and lower suburbanization compared to Vancouver and Montreal. Same-sex couples aggregate throughout the downtown (including its Church and Wellesley Street gay village) with concentrations also extending into inner-city queer-friendly neighbourhoods (eastwards to Leslieville and westwards to Parkdale and the inner suburb of Etobicoke). Suburbanization occurs northwards along Yonge Street into the inner suburb of North York and continues into northern outer suburbs (Markham, King, Aurora, and East Gwillimbury) with some westward movement along the lakeshore (Mississauga, Oakville, and Milton). Montreal's same-sex couples (Figure 2.2) concentrate in its eastern downtown gay village, adjacent inner-city areas (Plateau Mont-Royal and Hochelaga-Maisonneuve), and western inner suburbs (Verdun, Notre-Dame-de-Grâce, and Westmount). Outer suburban clusters appear in western (Vaudreuil-Dorion, Notre-Dame de l'Île-Perrot, and Oka), northern (Sainte-Anne-des-Plaines), and southern (Chambly) municipalities. Vancouver's (Figure 2.3) same-sex couples gravitate to downtown and its Davie Street gay village with an additional cluster in the East-End Commercial Drive area. Suburban concentrations extend along the SkyTrain corridors south into Richmond and southeast along Kingsway towards Surrey, the location of New Westminster, the suburb with the highest densities of same-sex couples in Canada.

Table 2.2 Cross-tabulation of the geographical distribution of same-sex couples for the Vancouver, Toronto, and Montreal CMAs with selected demographic characteristics, 2016

	Vancouver				*Toronto*				*Montreal*			
	Children[a]	*Income*[b]	*50+*[c]	*Foreign*[d]	*Children*[a]	*Income*[b]	*50+*[c]	*Foreign*	*Children*[a]	*Income*[b]	*50+*[c]	*Foreign*
Central	**26**	**87750**	**36**	**34**	**51**	**99500**	**35**	**30**	**23**	**75750**	**34**	**22**
Suburbs	**74**	**84375**	40	37	**49**	**87000**	**5**	43	**77**	**76000**	42	20
Inner	38	78250	49	25	18	82000	40	32	27	75500	47	6
Outer	36	90500			31	92000	42		50	76000		

Source: Brian Ray for authors.

[a] Couples with at least one dependant child living at home.
[b] Average household income in CDN$.
[c] Couples with one or more adults over 50 years of age.
[d] Couples with one or more adults who is foreign-born.

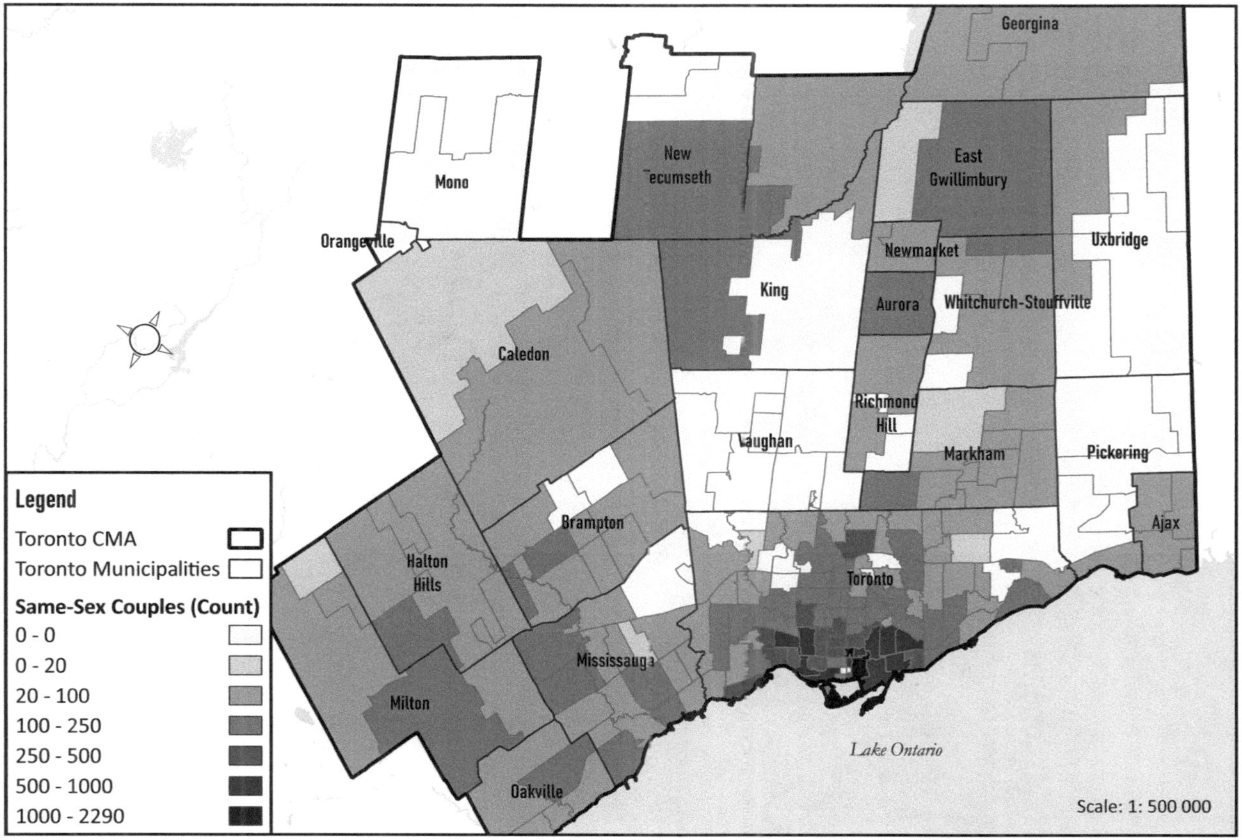

Figure 2.1 The geographical distribution of same-sex couples in the Toronto CMA, 2016
(*Source:* Julie Podmore).

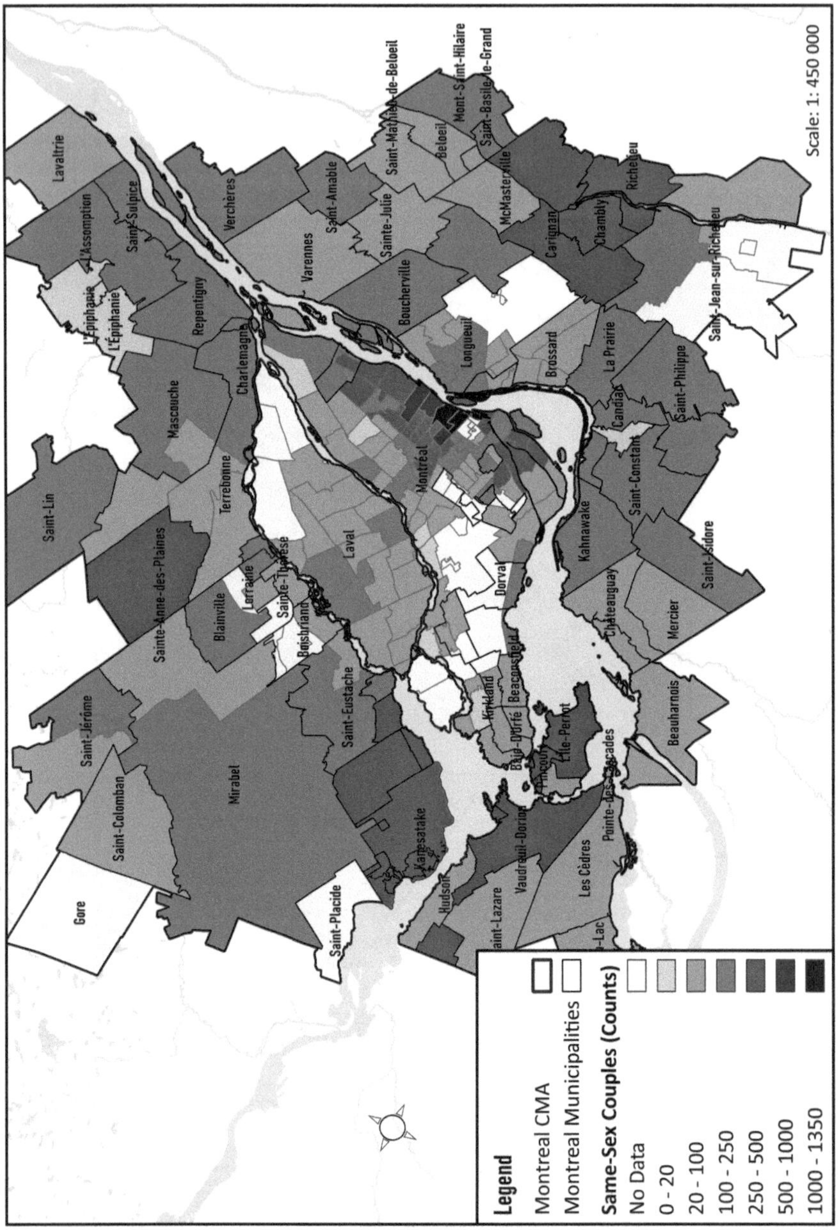

Figure 2.2 The geographical distribution of same-sex couples in the Montreal CMA, 2016
(*Source:* Julie Podmore).

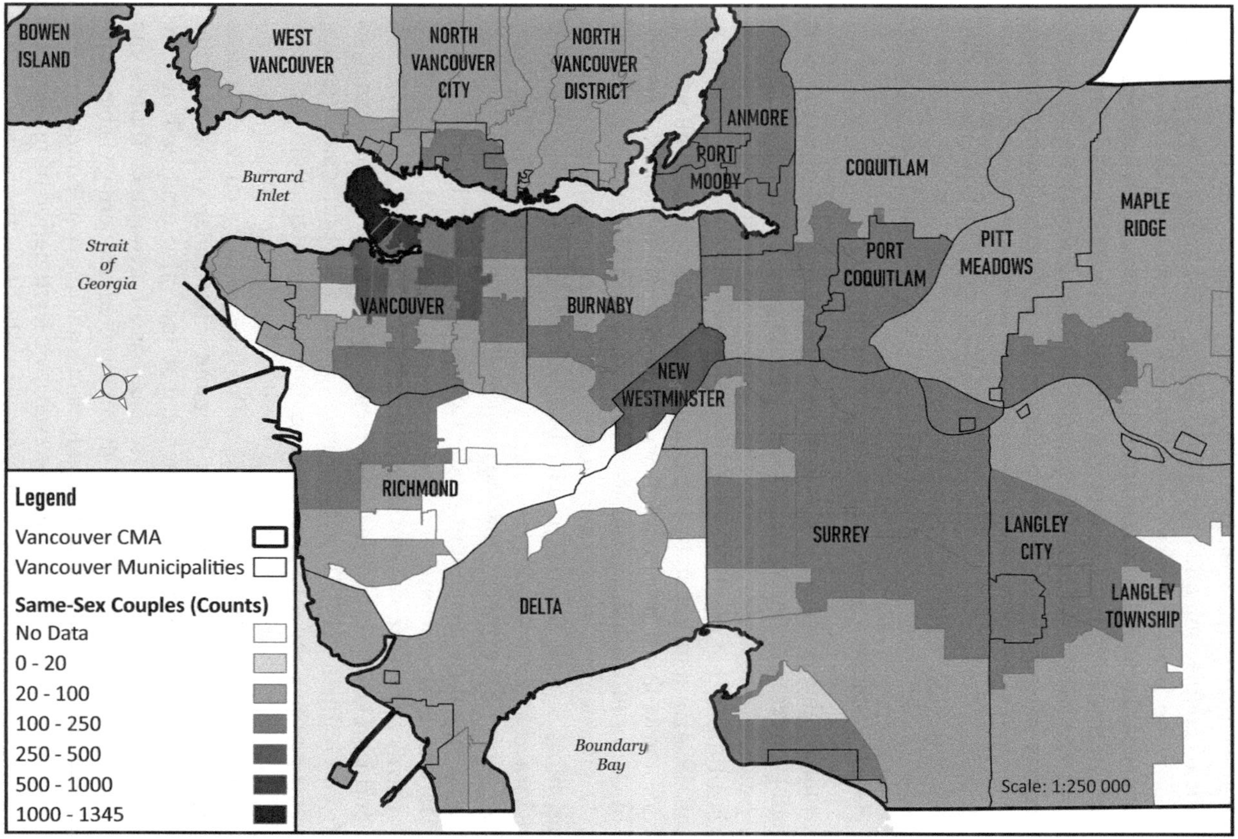

Figure 2.3 The geographical distribution of same-sex couples in the Vancouver CMA, 2016 (*Source:* Julie Podmore).

Acknowledging the data's limitations for representing the full range of LGBTQ2S positionalities, these CMA same-sex couple geographies signal suburban places perceived to be more liveable and tolerant. Setting aside the lack of enumeration of transgender and non-binary households, youth identities outside of coupledom, and alternative household arrangements, same-sex couples form moderate clusters in some suburban areas. However, greater queerburban public visibility, communality, civic awareness, and recognition cannot be inferred from such evidence of clustered private domestic arrangements. Only the Vancouver case study municipalities correspond with higher relative concentrations of same-sex couples, while in Montreal and Toronto, suburban clustering is less distinct. To augment the picture of Canadian queerburban representation, analytical attention now turns to the print media discourses shaping place-based representations.

Out of place? Mediated representations of queerburbia

Print media databases are replete with stories that provide idiosyncratic but meaningful representations of discourses surrounding peoples and places. Privately owned and shaped by commercial interests, newspapers are public service institutions that are beholden to the opinions of their readers (Parisi & Holcomb, 1994) and mediated by journalists, political figures, and informants who shape the political process of news production (D'Angelo & Kuypers, 2010). In the print media, journalists "frame" topics and issues for readers, selectively influencing their message through the organization and presentation of ideas, most visibly, through catchy bylines and sensationalist headlines (Brown-Saracino & Rumpf, 2011).

Geographers have explored how the print media frames urban phenomena, including marginalized neighbourhoods (Kearns et al., 2013; Liu & Blomley, 2013; Martin, 2003; Richardson, 2014), gay villages (Forest, 1995; Miller, 2005), gentrification (Brown-Saracino & Rumpf, 2011; Gin & Taylor, 2010), and suburbia (Harris, 2015; Harris & Hendershott, 2018). Frame analysis stems from Goffman's (1974) "social frameworks", a tool for understanding how people perceive and interpret meanings from social interactions and their settings. The frame distils portrayals of speakers, social movements, political actors, and issues through a specific lens. Products of their cultures, frames highlight meaningful information and promote a particular vision of problems, peoples, and places for an audience, but they also omit and obscure (Entman, 1993). Content and discourse analysis of newspaper frames can reveal how peoples and places are socially constructed, rendering visible seemingly self-evident normativities, value judgments, and plays of power manifest in moral panics, populist opposition, or even civic legitimation of mainstream policies (Bain et al., 2020). With its interpretive authority in public debates, the print media contributes narratives about and intertextual constructions of sexual and gender minority subjects (Miller, 2005). It has reproduced an urban imaginary of sexuality through coverage of LGBTQ2S people and places that spatialized society's othering, policing, and moral panics within cities while also offering LGBTQ+ people covert communication about hidden social worlds (Fejes, 2008; Stein, 2000; Streitmatter, 1995).

This section demonstrates a representational baseline that the print media has established with respect to the relationship between Canadian LGBTQ2S people and suburbs. It focuses on aggregate patterns and trends over time, in relation to LGBTQ2S themes and issues and their respective institutional spheres and scales of influence. It also provides a detailed analysis of the major suburban news stories in each city-region. It argues that mainstream newspaper coverage of LGBTQ2S people and issues in the suburbs of Canada's largest cities frames queerburbanites as spatio-temporal disruptors, unanticipated subjects who challenge suburbia's archetypal institutions of religion, education, recreation, and home to unmake and remake place.

A suburban LGBTQ2S print media database

A print media database (n = 1,589)[3] was assembled from metropolitan and national press coverage between 1985 and 2014 of LGBTQ2S-related themes in Canada's three major CMAs (Montreal, n = 357; Toronto, n = 519; and Vancouver, n = 703). This time span covers important debates regarding social and political LGBTQ2S inclusions during a period of growing sexual rights activism (Grundy & Smith, 2005) that ended with the waning of print media as a quotidian source of information accompanied by digital news and social media platform expansion (Bain et al., 2020). The number of LGBTQ2S-related articles grew annually in the first two decades, peaking in 2002, and then declined over the final period ending in 2014. Across this temporal arc, the database was assembled in two ways. First, articles that reference queerburban subject matter in Canada's three largest metropolitan regions were drawn from a reading of press clipping collections in The ArQuives (formerly the Canadian Lesbian and Gay Archives) and the Archives gaies du Québec.[4] Built through a combination of donations and archivists' clippings, these collections were generated by LGBTQ2S readings of newspapers, but they are neither comprehensive nor complete. Second, to supplement the peoples' archive, two digital newspaper databases, ProQuest for English- and Eureka for French-language sources, were searched for articles referencing LGBTQ2S populations and themes by peripheral municipalities for each city-region.[5] To include local and national coverage, the database derives from both metropolitan (e.g., *Vancouver Sun*, *Vancouver Province*, *Toronto Star*, *Toronto Sun*, *Journal de Montréal*, *La Presse*, and *The Gazette*) and national newspapers (e.g., *National Post*, *Globe & Mail*, and *Le Devoir*), incorporating news reports, editorials, letters to the editor, and opinion pieces.[6]

As Figure 2.4 shows,[7] the number of articles referencing LGBTQ2S themes increased between 1985 and 2014, peaking for all three CMAs in the early 2000s and declining from 2005 to 2014. Between 1985 and 1995, much of the coverage was pathologizing, centring on crime reports that included sexual and

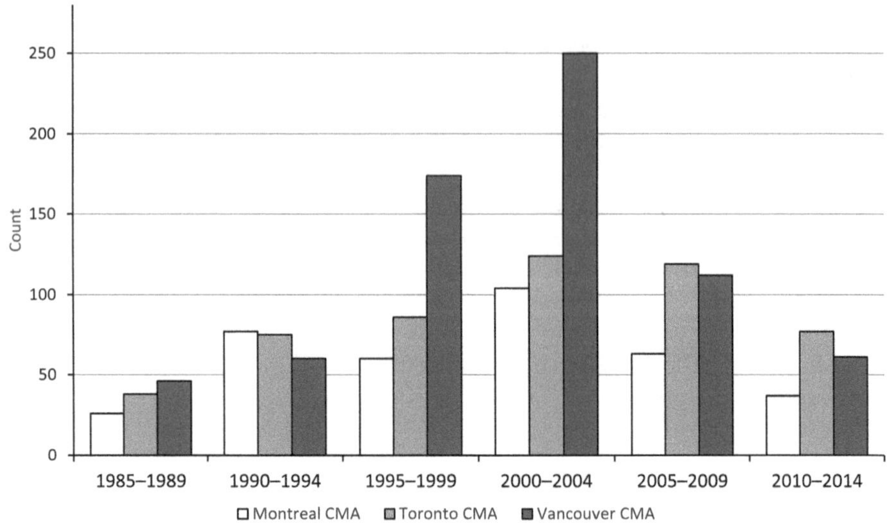

Figure 2.4 Total number of print media articles per five-year period by CMA, 1985–2014 (*Source:* Bronwyn Clement for authors).

gender minorities. With federal changes to the Canadian Human Rights Act in 1996 prohibiting discrimination came a decade of increased print media coverage of LGBTQ2S institutional activism testing this legislation in schools and churches, law and policy, insurance and benefits, and labour relations, all of which extended into suburbia (Podmore & Bain, 2019). From the late 1990s, debates about same-sex civil marriage especially impacted coverage of peripheral municipalities due to the political stances taken by suburban members of parliament until the adoption of the Equal Marriage Act in 2005. Under pressure from provincial court rulings determining that discrimination in civil marriage was unconstitutional, the opposition Conservative Party of Canada made same-sex marriage a "wedge issue" during the 2004 election as they sought to wrest power from the centrist Liberal Party of Canada by winning over suburban ridings.

Iterative thematic coding produced seven categories of "issues": crime; discrimination and equity; relations and kinship; sexual conduct (often regarding public sex or sex work); sexual health and education; social support and allyship; visibility and coming out. Interwoven into the queerburban portraits of the city-regions discussed below are the top four themes of this analysis: discrimination and equality (n = 725), relationships and kinship (n = 365), visibility and representation (n = 233), and crime (n = 231). Discrimination and equality refers to LGBTQ2S discrimination and efforts to protect against it, including human and legal rights, anti-bullying, hate speech, recognition, and social acceptance. Relationships and kinship refers to same-sex partnership recognition (e.g., spousal benefits, same-sex unions, and same-sex marriage) and family formation (e.g., parenting, adoption, and artificial insemination). Visibility and representation refers to LGBTQ2S "publicness" such as personal "coming out" narratives or increasing acceptance through visibility during events such as Pride. Crime refers to criminal acts by (e.g., pedophilia, sexual predation, "inappropriate" sexual conduct, and sexual assault) and against (e.g., gay bashing, police violence, and domestic abuse/assault) LGBTQ2S individuals. While there were no notable inter-metropolitan differences in visibility and representation, Vancouver was highest in discrimination and equality (53 percent), Toronto in relations and kinship (45 percent), and Montreal in crime (39 percent). This thematic variation is reflected in the inter-metropolitan analysis of representation of queerburbanites in relation to peripheral institutional spheres.

Institutional spheres refer to the structures of civic life where citizen encounters and political contentions are organized and connected (Abrutyn, 2016; Turner, 2010). In Figure 2.5, 11 spheres are disaggregated: arts and culture (e.g., arts-based and cultural events like drag shows and Pride parades); commerce and economies (e.g., business and workplace issues such as service refusals, professional misconduct, or spousal benefit demands); education; health and medicine (e.g., physical, sexual, and mental health); kinship and domestic life (e.g., domestic, familial, and interpersonal webs of relations); law and crime; media (e.g., free speech, access to LGBTQ2S publications); politics and legislation; public sphere; religion; and sports and recreation (e.g.,organized sports and sporting and fund-raising events). These spheres overlap as forms

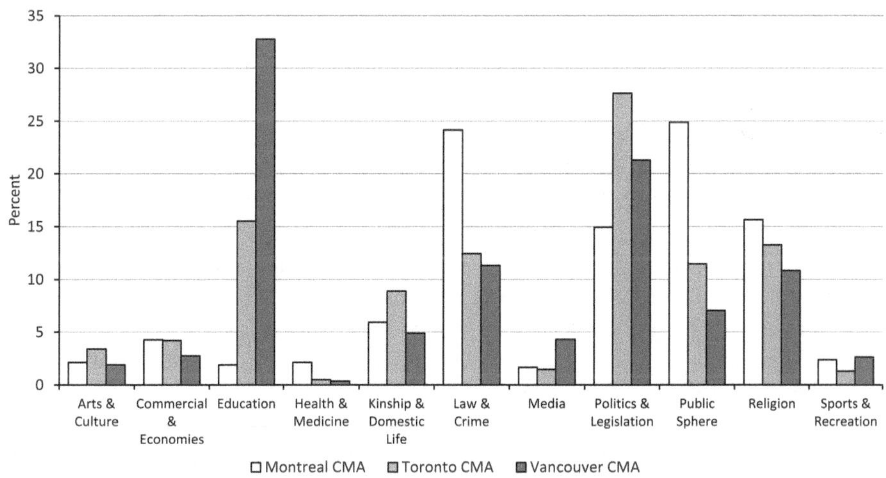

Figure 2.5 Institutional spheres as a percentage of each CMA's print media coverage of LGBTQ2S themes, 1985–2014

(*Source:* Bronwyn Clement for authors).

of recognition (including media coverage) variously connect them to institutional domains through lived reality and action (Abrutyn, 2016). The place-based contours of queerburban spatialities emerge at the intersections of these themes and institutional spheres accentuating LGBTQ2S issues of contention.

The five largest institutional sphere categories in descending order are politics (n = 413), education (n = 379), law (n = 274), religion (n = 239), and the public sphere (n = 235). Politics and legislation refers to party politics (and the actions of individual politicians) and discussions of challenges at various levels of government, including civil law and military service. This category is relatively evenly distributed across Vancouver (43 percent) and Toronto (41 percent), while underrepresented in Montreal (15 percent). Education refers to curriculum, school board decisions, and anti-discrimination policies in schools and is a dominant theme in Vancouver (73 percent). Law and crime speaks to criminal, civil, and personal legal cases, crime reporting, police (mis)conduct, and legal challenges to legislation. Law and crime is slightly higher in Montreal (37 percent) but the category has an even CMA distribution. Religion addresses issues of inclusion within churches in response to legislative and policy changes. Its coverage has even CMA distribution with slightly more articles in Toronto (38 percent). The public sphere refers to events occurring in the public realm such as Pride parades, protests, and debates over public space, public sex, and public safety. Public sphere coverage is highest in the Montreal region (45 percent) and lowest in Vancouver (25 percent) with Toronto at 30 percent.

Aggregate patterns in the timing, themes/issues, and institutional spheres of suburban LGBTQ2S print media vary by CMA. The highest volume occurred between 2000 and 2004 when same-sex unions were nationally debated, seconded by the preceding five-year period when the Charter of Rights provided the opportunity to test LGBTQ2S rights in suburban institutional spheres. The themes and issues correspond with this rights orientation as most articles centre on discrimination and equality as well as relationships and kinship. The institutional spheres of rights formulation dominate the data set, namely politics, law, and the public sphere, but also education, religion, and the family, realms of societal pastoral care where domestic life, child rearing, and morality are reproduced through practices of socialization, teaching, and guidance. In contrast with the major late-twentieth century urban LGBTQ2S activist themes of HIV/AIDS (Brown, 2003), homophobic and transphobic policing (Namaste, 1996) and securing safer spaces free from homophobic violence (Hanhardt, 2013), suburban LGBTQ2S activism prioritizes transforming schools and churches as places of work, membership, and participation, or attaining same-sex marriage rights.

Coverage of queerburban print media centres cisgender, white, middle-class, property-owning gay men with influential positions of community leadership (Podmore & Bain, 2019). Their privilege invites attention to their voices of objection in a place where property, class, and profession, along with monogamy and coupledom, override misalignments of household configuration, procreation, sexual deviance, and potential promiscuity (Warner, 1993). Their dominance must also be read against the backdrop of Canadian suburbia, which by the late-1990s was a place of significant ethno-cultural and social-class diversity with an ever-widening range of household structures. In this coverage, queerburbanites are rarely BIPOC-identified, Two-Spirit, transgender, or lesbiqueer. Canadian queerburban issues furthermore skew towards youth rather than elders, centre gay men rather than lesbians, focus on settlers rather than Indigenous people, and feature visible minorities as suburbia's homophobic "cultural communities" rather than queer subjects (Puar, 2007). In what follows, these generalizations are grounded in CMA portraits, offering overviews of the suburbs with the most coverage. In this way, print media representations of LGBTQ2S people and their issues are revealed as part of queerburban place-making. Nowhere is this more evident than in the Vancouver city-region.

The Vancouver city-region

Within Canada, the contentions generated by queerburban place-unmaking and place-remaking in the Vancouver city-region generated the highest volume of articles ($n = 784$) that, when disaggregated by peripheral municipality, were concentrated in Burnaby ($n = 213$) and Surrey ($n = 299$). This overrepresentation was largely due to the amount of media coverage generated by two highly publicized cases, the 1988 public "coming-out" of Burnaby's Member of Parliament Svend Robinson (1979–2004) and the court case against the Surrey School Board's ban of books representing same-sex parents (1998–2002). Less prolific, singular LGBTQ2S issues arose in the other suburbs such as the 2007 adoption of a mandatory "community covenant" prohibiting staff and students from sexual relationships outside of heterosexual marriage by a local private Christian liberal arts university (Trinity Western) in Langley. Some stories traverse multiple suburbs, notably the debates about politician stances on same-sex marriage (2003–2005) and municipal bans on the LGBTQ2S newspaper *Xtra! West* from Fraser Valley's public libraries. The Vancouver city-region's suburban database is crowded with many small incidents foregrounding queerburban socio-spatial conflicts emerging from debates about renting church basements or legion halls for queer events, queer public visibility in suburban art galleries and public libraries, the creation of rainbow crosswalks, and the civic raising of Pride flags, all of which challenge the heterocisnormativities of suburbia's public realm.

The story of lawyer Svend Robinson's tenure as a Member of Parliament representing Burnaby (the ridings of Burnaby, Burnaby-Kingsway, Burnaby-Douglas) is national, but his repeated re-election and "coming out" raised local debate. Robinson was first elected to Canada's Parliament in 1979 as a member of the left-wing, socially democratic, labour-backed New Democratic Party (NDP). He was raised in Burnaby, which, like many of its surrounding municipalities, was a labour movement stronghold that consistently supported the

NDP party and its predecessor, the Co-operative Commonwealth Federation (CFF). In the print media database, his story begins in 1988, when he publicly identified himself as gay in order to "draw attention to a segment of society" that was "increasingly under attack" and thus "focus Parliament's attention on the need for new legislation to protect homosexual rights" ("MP pleads", 1988, A5). Robinson wanted to "speak 'in the first person' about gay rights" and pleaded "for understanding and acceptance" of gays and lesbians ("MP pleads", 1988, A5). While the national NDP caucus applauded his announcement, the party leader told the press that Mr. Robinson's sexuality was a "private affair" ("Robinson's sex life", 1988, A6). Rumours initially elicited constituent support, but leading up to this public declaration, his office entrance was vandalized, "smashed to smithereens" by bricks ("Sexuality cited", 1988, A5). Robinson told the press: "Quite obviously it's linked to the recent stories with respect to my sexual orientation", brushing off the attack by stating that "Broken windows can be replaced … but … the human spirit is a different matter" ("Sexuality cited", 1988, A5). He later leveraged this vandalism as evidence of homophobic violence, a "graphic illustration" of the need for laws to protect gays and lesbians ("Laws needed to protect", 1988, A1). During the fall 1988 election, Robinson was publicly harassed with letters (one with a bullet enclosed) and complaints, and personally attacked by Conservative party politicians (including the Prime Minister). A Surrey-based evangelical Christian publication, *Life Gazette*, was distributed in his riding with the headline "Sodomite Invasion Planned for 1990" (referencing the forthcoming International Gay Games in Vancouver) and calling homosexuality a "perverse lifestyle" (Pynn, 1988, C12). When Robinson won his seat, he became the first openly gay incumbent to be elected to the Canadian Parliament. He claimed "his convincing election win in Burnaby-Kingsway" as "a victory for homosexuals across Canada" and a "strong statement" in favour of "tolerance" and rejection of "bigotry" on the part of Burnaby's constituents (Pynn, 1988, B1). Robinson remained in office until 2004, the year that his efforts to have sexual orientation added to federal hate crimes legislation came to fruition and the Supreme Court of Canada ruled that the marriage of same-sex couples was constitutional.

For two decades, Burnaby was an important place in LGBTQ2S rights politics, jumping from the municipal to the federal scale due to the significant institutional power and public sphere access of gay white men. A network of openly gay United Church ministers was ordained in Burnaby churches, supported Robinson's constituency office, and allied with educational activists to broker political support to Bill Siksay, Robinson's long-time constituency assistant, who was elected to the seat as the first openly gay non-incumbent elected to Canadian Parliament in 2004. The riding continued to receive significant press coverage as Siksay initiated the first bill to amend the Canadian Human Rights Act to prohibit discrimination on the basis of gender identity or expression (in 2005, 2009, and 2010), until 2011 when Bill C-389 was adopted as a report (with a vote of 143 to 135) coming into law in 2016 as Bill C-16. Initially a private members bill in the lottery system, Siksay advocated in the press for protections in healthcare and the workplace for transexual and transgender people who were "not a large community in Canada" but still "a significant group of people" (O'Neil, 2005, A17). The rights of LGBTQ2S teachers and young people to see themselves represented in school materials and curriculum were demanded alongside such legislation, localized in the Vancouver print media in Burnaby but more persistently in nearby Surrey.

When Gay and Lesbian Educators of British Columbia (GALE-BC) started a campaign against homophobia and heterosexism in schools in 1996, a Surrey kindergarten teacher and GALE-BC member, James Chamberlain, requested classroom resource permission from the local school board to use three children's books depicting same-sex parenting: *Asha's Mums* (Paulse & Elwin, 1990); *Belinda's Bouquet* (Newman, 1991); and *One Dad, Two Dads, Brown Dads, Blue Dads* (Valentine, 1994). The Surrey School Board responded with a district-wide elementary school book ban citing parental religious concerns. Chamberlain then launched a legal challenge through the B.C. Supreme Court with the support of GALE-BC, the B.C. Teachers' Federation (BCTF), and the B.C. Liberties Association. The court overturned the Board's ban in 1998 for violating the principles of secularism and non-discrimination guaranteed in Section 76 of the British Columbia School Act. Two years later, the B.C. Court of Appeal upheld the ban, rejecting board rights to "decertify" learning materials (Smith, 2004). Finally, in 2002, the Supreme Court of Canada upheld the lower court's ruling, arguing that the Board's ban "did not comply with the broad principles of tolerance and secularism underlying the legislation that governed the School Board" (Chamberlain v. Surrey School District No. 36, 2002). Ultimately, the Board rejected the same books in 2003 using arguments about grammar and weak storylines but approved other same-sex family learning resources recommended by the Surrey Teachers' Association.

This story broke in the print media with the headline: "Surrey board vows gay book not approved" (Spencer, 1997b, A8). The print media had already begun to depict Surrey as "backward" a month earlier when its school board rejected a BCTF request to institute anti-homophobia training for teachers (Austin & Spencer, 1997, A4). The book ban produced print media headlines asserting "Board head wants to ban gay books in school" and reprinted the slogans of protest placards: "Welcome to Surrey, the book-banning capital of BC" (Spencer, 1997b, A8). Book-banning also reinforced the suburb's emerging reputation for neoconservative religious views (Collins, 2006). An opinion piece by a mother of a gay teen depicted Surreyites as being at "the same stage of cultural evolution as were the Newfoundland nuns who tied children's left hands behind their backs 40 years ago" (Johnson, 1997, A14). In "Homophobia 101", concerns that books depicting same-sex parents would lead to contagion were discredited with journalistic sarcasm: Murphy (1997, A40) quipped that everyone will be gay in a few years, "[e]xcept for the people of Surrey" who "will be abstinent homophobes". Another described Surrey as a "Hotbed of Homophobia" (Sullivan, 1999, A19).

Press characterizations of Surrey's homophobia were reinforced by quotations from racialized religious figures such as leaders of Surrey's Guru Nanak Sikh Gurdwara Society who asserted:

> In our culture, it is unnatural to have a man and a man together … If some parents want to teach their children what is in their books, that is up to them, but it should not be forced on all children.
>
> ("Depictions divide educators, clerics", 1997, A3)

Underscoring the "unity" of Surrey's various religious groups on this issue, the press reported that "Hindu, Sikh, Catholic and Protestant religious leaders have rallied behind school trustees" ("Gay book ban backed", 1998, A20). But it also gave voice to Surrey high school students who welcomed the opportunity to learn from "gay literature" (Spencer, 1997b, 14) and a lesbian mother whose sons now feared school because they were "afraid of being ostracized, even beaten up" (Clark, 1997, A5). The press published the attestations of anonymous Surrey gay teenagers regarding the homophobic violence they endured. In "Meet Rick", a gay teen described fearing his homophobic father and schoolmates, and defending himself by "acting butch" and bullying others (Spencer, 1997a, A14). His daily survival strategy has contemporary resonance: "In Surrey, you keep your hat on backwards, don't look anybody in the eye and keep going" (Spencer, 1997a, A14).

Reports of Surrey's homophobia endured in the aftermath of the book ban case in reports of a lawsuit against school board trustees by lesbian parents for "allowing a poisonous homophobic and discriminatory atmosphere" in public meetings (O'Connor, 2004, A14) and the board's cancellation of the American play *The Laramie project* (Kaufman, 2001) at Elgin Park Secondary School. The play memorialized the life of gay University of Wyoming student Matthew Shepard who was brutally murdered in Laramie, Wyoming in 1998. The Board argued that it "dealt with a very mature subject matter" and was "not appropriate as family entertainment" (Ramsey, 2005, A4). Suburban Surrey's national reputation for homophobia was confirmed in a column about *The Laramie project*'s staging at a school in Sudbury, Ontario quoting the high school drama teacher's reflections on its implications for homophobia in smaller cities: "At the end of the play we have a question-and-answer session with the audience and I start by asking: 'Could this be our town?' And people inevitably say, 'Of course it could be'" (Mason, 2005, A10). The columnist concluded: "In fact, it could be just about any town in Canada. Including Surrey" (Mason, 2005, A10).

Suburbs with pivotal stories of LGBTQ2S rights struggles, especially when there is significant opposition by local constituencies and institutions, develop a homotransnegative place reputation – collective ideas, impressions, and attitudes towards a place held by external audiences that drive behaviour and shape urban development (Braun et al., 2018). As Braun et al. (2018, 23) contend, "a place's reputation is the composite of its past developments, investments, actions, achievements and place (brand) images". A cultural phenomenon that cannot necessarily be controlled by civic stakeholders, a "negative place reputation" is remarkably enduring and difficult to change (Nelson, 2015, 192). The homophobia of the book ban still haunts Surrey's current reputation as it strives for an "inclusive city" place brand (see Chapter 3). Burnaby's reputation as a progressive suburb aligned with the leftist principles of the labour movement was enhanced by the election of openly gay Members of Parliament, Robinson, and later Siksay. However, its print media record after Siksay stepped down in 2010 was dominated by citizen opposition to the Burnaby School Board's 2011 adoption of its Sexual Orientation and Gender Identity (SOGI) policy (see Chapter 5). During this episode, Burnaby gained notoriety as another suburban "hotbed of homophobia" when Chinese Christian constituencies mobilized in opposition, undermining its "socially sustainable" place brand. Print media imagery and narratives contribute to the composite and conflicting representations of places, embedded in cyclical impulses to make/unmake/remake them as is the case in Toronto.

The Toronto city-region

The municipalities in the Toronto CMA/GTA generated the second highest volume of press coverage with $n = 519$. When disaggregated, the stories concentrate in two adjacent case study sites, Mississauga ($n = 91$) and Brampton ($n = 55$). Also noteworthy is the coverage of the sizable former municipality of Scarborough ($n = 71$) and adjacent Oshawa – due to the 2001 Marc Hall high school prom discrimination case against its Catholic school district ($n = 81$, see Chapter 5). Mississauga and Brampton were overrepresented not because of any single story, but due to their rapid urban development trajectories and significant population sizes. Their largest stories coincide with predominant spheres and issues in the database for this city-region, namely in politics and education. Across the region's many municipalities, there are other patterns. In the early-1990s, the press often reported police entrapment and arrests in suburban parks, under headlines such as "Police fight to reclaim 'perverts' playground'" (Mascoll, 1993), even after "gross indecency" was removed from Canada's Criminal Code in 1985. Homophobic coverage of "criminality" in the 1980s and 1990s shrouds Toronto's suburban churches, schools, and neighbourhoods in the suspicion of perversion by confounding "homosexuality" with "pedophilia". When some gay men serve as teachers, coaches, scout leaders, and priests, they are publicly accused in the press of perpetrating sexual assault and youth predation. The institutional legitimacy of LGBTQ2S people is negotiated through these same spheres during debates about the establishment of Gay-Straight Alliances (GSAs), the ordination of gay ministers, the recognition of same-sex unions, and the extension of spousal benefits. Same-sex households asking for benefits before the Equal Marriage Act was passed in 2005 is one of the few instances when lesbians appear in suburban press coverage, making public their children's parenting testimonies. In a letter to the editor, a conservative professional Mississauga resident congratulates a journalist for reporting on "same-sex couples living the mainstream lifestyle in the suburbs", sharing "boring Saturday nights at home" with their children as they too "just live life" (McNaughton, 2004, A23).

These smaller stories contribute to a broader portrait of homophobia playing out across Toronto's peripheries. A suburban councillor at a Metro Toronto community funding debate refers to LGBTQ2S cultural groups as "unscrupulous S-and-M pigs wallowing in our tax dollars" and contributing to "Sodom and Gomorrah-like destruction" (Royson, 1993, A6). A "redneck" tavern owner in Burlington denies services to two heterosexual women who danced together later testifying at the Ontario Human Rights Commission that they had "never been treated in a minority way before" and "got a quick glimpse of what homosexuals must go through on a daily basis. It was awful" (DiManno, 1994, A7). There are also incidents of transgender persecution in suburban prisons and the census (Raphael, 2001). In the aggregate, these snippets of societal homo-, lesbo-, and transphobia, accusations of pedophilia and predation, write heterocisnormative morality politics into queerburbia.

Such morality politics surface in queerburban public schools across the country, and Mississauga, particularly its Catholic School Board, is no exception. In 2011, following the death by suicide of the son of an Ottawa city councillor, the Government of Ontario passed the Accepting Schools Act to mandate stricter and clearer anti-bullying responses in schools, but also to require publicly-funded schools to permit Gay-Straight Alliances (or Gender and Sexual Alliances) (see Chapter 5). The Mississauga Catholic School Board then violated provincial policy by banning GSAs and perpetuating homophobia (e.g., at dances and proms, in educational materials and posters, classroom marginalization, shunning on class trips, and banning a lesbian comedian from performing at an anti-bullying event). St. Joseph's Secondary School, which followed in the footsteps of the Halton Catholic School Board, was profiled by journalists for its spring 2011 decision to forbid GSAs unless under the generic banner of an anti-bullying student club. One student, Taechun Menns, sought to make St. Joseph's the first Catholic high school in the province with a GSA, skirting the ban with the name the "Open Arms Group". As he told journalists, "school's supposed to be a second home", a place where "if students are in the closet or feel like they can't talk about these things at home, they should be able to talk about them at school" (Mills, 2011, A13). The school also banned the display of rainbows and blocked a student donation to the LGBT Youth Line (Selley, 2011). Profiled in the media, the Open Arms Group founder showcases his Facebook page (St. Joe's GSA) and a letter to the principal explaining that "groups such as these do not in any way impose upon the religious freedom or denominational rights of the

Catholic School Board" (Baluja & Hammer, 2011, A13). Stories of the Mississauga board's discrimination culminated in an Ontario human rights tribunal citing 35 student allegations against the École Secondaire Catholique Sainte Famille (e.g., a refusal to recognize same-sex parents when discussing family, and stalling for 20 months on the approval of a GSA). Here, too, the press documented the school board's denial of discrimination and assertions that the claims were "groundless" even when the Tribunal ended in a settlement (Strapagiel, 2015).

In queerburbia, debates about governmental LGBTQ2S equalities legislation bring homophobic discourses to the fore as part of local power-brokering to win elections. Nowhere is this more apparent than in Brampton, which became central to the federal election and national parliamentary debates on the same-sex marriage Bill C-338 in 2004. A Liberal Party of Canada stronghold, Conservative candidates tried to leverage their opposition to same-sex marriage by "desperately seeking" the votes of politicized religious Sikh, Muslim, and Catholic communities (Ivison, 2005, A6). When the Sri Guru Nanak Sikh Center Brampton barred the Liberal incumbents from speaking because they had voted in favour of the same-sex legislation, leaders of this gurdwara are quoted in the press as being "unambiguous" in their opposition, asserting that "they will not vote for those who support same-sex marriage, and have encouraged their congregants to follow their lead" (Aly, 2005, A15). For some political leaders, however, it was possible to set aside their "personal beliefs and fundamentally stand for the Charter of Rights", but in a climate of fear, individual Bramptonians removed their Liberal lawn campaign signs to avoid being targeted as Liberals and thus LGBTQ2S allies (Aly, 2005, A15). In their vote chase, however, Conservative candidates could not win by countering national equalities legislation nor by mistakenly assuming that ethno-culturally diverse and religious suburbanites opposed equal marriage. For example, a Conservative leadership candidate striving to connect with the Sikh, Portuguese, and Filipino communities "brought up the same-sex issue at every opportunity in his heavily ethnic riding but to no avail" as ultimately this "issue" was not a "vote getter" (Ivison, 2005, A6). Reflecting on the Conservative federal election loss in 2004, Frum (2004, A16) asked readers: "Is it plausible that the Conservatives failed to make inroads into Brampton and Kitchener because voters there are horrified by opposition to same-sex marriage?" The assumption that the suburban ethno-cultural and religious electorate of Brampton would cast aside civil rights in favour of homophobia proved to be a Conservative miscalculation.

With a more variable print media image, this analysis of the nation's largest city, Toronto, surfaces additional representational layers to queerburbia. Stories of police entrapment of MSMs in suburban parks or the conflation of homosexuality with pedophilia when openly gay men are present in suburban institutions are part of queerburbia's criminogenic stigmatization. In suburbia, LGBTQ2S people have faced the same pathologization for being "out-of-place", and largely in the absence of support from downtown activists who initially mobilized against police surveillance and brutality. When mapped across the city-region, queerburban stories make clear that some Toronto suburbs have also acquired homotransphobic place reputations: Mississauga for its Catholic School Board opposition to GSAs; and Brampton for its Conservative politicians' assumptions of ethno-cultural opposition to same-sex marriage.

The Montreal city-region

Relative to Toronto and Vancouver, news coverage of LGBTQ2S issues in Montreal's suburbs generated fewer articles (n = 357) with more than half published in French (where references to sexual and gender minorities were less frequent and linguistically more covert). When disaggregated by peripheral municipality, articles primarily referenced two of the case studies, Laval (n = 44) and Longueuil (n = 32), as well as the West Island municipality of Pointe-Claire (n = 78) and the adjacent former municipality of Lachine (n = 27). The main themes were discrimination and inequality (n = 146), crime (n = 91), and visibility and representation (n = 67) with news coverage most frequently referencing the public sphere (n = 105), law and crime (n = 102), and religion (n = 66).

In the law and crime category, there were many small location-based reports of suburban LGBTQ2S people fighting local discrimination, such as a 1987 story of a transgender woman's lawsuit against a Brossard department store for having been "manhandled" by security and police ("Transsexual sues", 1987, A4) or, in nearby Candiac, a lesbian couple who accused the municipality of discrimination for taxing a partial transfer of home ownership and "thus refusing to recognize that the women were living together as a couple" (Block, 1999, A6). When the city-region hosted the Outgames in 2006, Laval's heterosexual water polo team "blighted" the whole event by making "homophobic locker-room jokes" (Lamey, 2006, A2). There are also police reports of homophobic harassment, such as the 2014 case of a Longueuil household receiving threatening anonymous letters (stating "we don't want people of your type in our neighbourhood") for hanging a rainbow flag on their apartment balcony (Dumont, 2014, 12). As in Toronto, the geographical association between LGBTQ2S subjects and crime was perpetuated by pathologizing and often sensationalist coverage of police surveillance and court reporting. Whether it be an inmate coerced into a homosexual relationship by the assistant superintendent of the Pinel Institute for the Criminally Insane (Morin, 1987, A8) or an organized crime case involving a transgender man working as a prison guard in the local detention centre who "terrorized" co-workers into quitting (Kalogerakis, 1999, A4), such accounts are rife with salacious details intended to pique public interest.

Crime, law, and discrimination came together in the early 1990s print media representations of suburban police harassment and negligence regarding homotransphobic violence. The murders of 14 gay men between 1987 and 1993 combined with police brutality exhibited after a raid of an underground queer party called "Sex Garage" in 1990 intensified the queer anti-violence movement resulting in a Quebec Human Rights Commission inquiry in 1993 (Crawford & Herland, 2014). Montreal suburbs were implicated in the "gay serial killings" that led activists to accuse the police of necropolitical discrimination when they remained unsolved (Haritaworn, 2019). A home in the "tranquil" Chomedey region of Laval was the site of a thirteenth murder in 1993. It was exemplary of a pattern that police were reluctant to admit: middle-class, closeted suburban men were being "beaten and stabbed after returning home with someone they had befriended" in a gay village bar (Collister & Noel, 1993, A1). In 1988, it was also closeted suburban gay men who were subject to police harassment and entrapment in Longueuil's riverside Parc Marie Victorin, a site that the police chief told the press "has been used by gays as a favorite locale for the past few years" ("Longueuil Police", 1988, G4). Montreal gay activists, however, used press coverage of the 1988 crackdown on "gays on the South Shore" to identify an urban-suburban contradiction: while Longueuil police were conducting mass raids and entrapping gay men in Parc Marie Victorin, Montreal's "Station 33 sends patrols to protect gays from hostile young straight people in the Village" ("Longueuil Police", 1988, G4). Into the 1990s, Longueuil police campaigns to entrap and arrest men in Parc Marie Victorin provided spectacular media stories with 30 men being arrested in 1989 and another 13 in 1991 ("Longueuil police", 1991). Montreal activists argued that "police forces in smaller communities" needed more training to raise awareness that these suburban men were "faceless homosexuals" who were "married to women" and in professional jobs where they must hide their sexuality (Shepherd, 1996, E1). Reporting on the spectre of the faceless suburban homosexual having public sex in this park continued for the next two decades. As late as 2014, the sensationalist *Journal de Montréal* ran the headline "A park that has become a place for gay meet-ups" is "haunted by men seeking homosexual experiences in nature", seemingly caring "little for the people around them" (Theriault, 2014, 40).

In comparison to Toronto and Vancouver, the religious sphere was the most dominant in Montreal, a historic majority Catholic region. For the protestant United Church of Canada, these include a story of a Laval minister vigorously opposing the 1988 national proposal to ordain "practicing homosexuals" (Shepherd, 1988, A3) and sensational coverage of a gay West Island minister being sanctioned for posing nude on the cover of an "underground", "hard-core" "homosexual magazine" in 1993 (Kapica, 1993, A10). A suburban Catholic priest publicly opposed a 2003 Vatican statement urging Catholics to express their opposition to same-sex unions and marriages (Leclerc, 2008, B6). Characterizing this position as "discriminatory, hurtful and offensive", Raymond Gravel's open letter in *La Presse* urged the Church to "stop bashing gay and lesbian marriages and start blessing them" and baptising their children instead (Kalbfuss, 2003, A1; Parkes & Shepherd, 2005).

The second highest volume of articles ($n = 27$) was generated by the debates surrounding the ordination of Pastor Darryl Macdonald by a Presbyterian church in the western inner suburb and former municipality of Lachine. An interim minister at St. Andrew's, Macdonald was invited by the congregation to be ordained there in 1995, but in the interest of living "openly" and "honestly", he informed church elders of his two-year, same-sex relationship. The national body ultimately rejected his ordination because a minority within the Presbyterian College of Montreal objected unless he became celibate (Ross, 1996). Without an ecclesiastical policy forbidding gay clergy, opponents relied on the "defence of the Christian tradition" and flamed fears of internal division and loss of membership. In 1999 when his licence was revoked for defiantly continuing his service, the congregation opted to leave the Presbyterian Church in Canada. Their action was described as heroic rebellion: "Breakaway parish to ordain gay pastor" (Shepherd, 1999, L14), and "Renegade church ordains gay minister" ("Renegade church", 1999, A8). Print media coverage of this story contrasted gay rights activism with suburban conservatism, juxtaposing the misplaced gay priest against a suburban landscape and congregation.

As the site of a decade-long struggle to fight homophobic abuse in the domestic realm, Pointe-Claire received the most press coverage (n = 78) for a single story. In the early 2000s, a middle-class white gay couple, Roger Thibault and Theo Wouters, filed police reports and later a complaint with the Quebec Human Rights Commission, because their neighbours were harassing them by vandalizing their property and threatening physical harm.[8] Initial press coverage focused on the couple's suburban respectability, describing Wouters and Thibault as "non-ghetto gays", who prioritized homemaking and their careers (Lynch, 1987). In this "peaceful", "sleepy", and "tranquil" idealized suburb (Boisvert, 2001, E1), the couple lived a quiet life of 28-year monogamous domesticity and discretion (Hamilton, 2001, A5). The harassment they supposedly experienced was also stereotypically suburban with neighbours driving golf balls into their yard, spying on them through a hole in their shrubs, attempting to run them over in their driveway (Thanh Ha, 2001, A3), and calling them "fucking faggots" who should "move back to the gay village" (Thanh Ha, 2001, A10). They combatted this suburban harassment by organizing a "Saying no to violence march" ("Gay couple's neighbours", 2001, D10) to assert that "Gays and lesbians have the right to live in a family neighbourhood. We're considered families too" (Peritz, 2001, A6). Demonstrators from downtown Montreal marched through the suburb, passing the couple's home, and ending with a demonstration in a local park. *The Gazette* announced that "Gay pride came to suburbia yesterday in a parade through the land of Volvos and barbecues", pointing out that "as gay pride marches go, this was one tailored for the neighbourhood": it was "appropriately low-key and tranquil" with only a few "flamboyant, leather-clad folks from Montreal's gay village" (Peritz, 2001, A6). Summarized by the headlines of *Le Devoir*, Pointe-Claire's "Saying no to violence march" was "a suburban version of a gay pride march" (Dufour, 2001, A1).

Print media representations of LGBTQ2S identities and themes in the suburban contexts of Vancouver, Toronto, and Montreal offer many examples of city-regional and local variation. Through discourse analysis, however, specific queerburban identities, themes, and institutional spheres emerge as particularly suburban. In print media public discourse, queerburbanites are represented as gay, usually white, middle-class men, who are suburban teachers, politicians, activists, and residents. Yet this cisgender settler dominance of the public sphere only selectively surfaces other queerburban identities at the intersections of age, social class, religion, and ethnicity. In this public realm, queerburban subjects are variably criminalized, subjected to violence, misrepresented, and/or lauded for heroically fighting back against homotransphobic discrimination. Prior to and following the federal incorporation of sexual orientation into the Human Rights Act in 1996 and encompassing the election of the first "out" politicians and the granting of same-sex marriage rights in 2005, the predominant themes were discrimination and equality, kinship and family, and law and crime. Suburban LGBTQ2S subjects were both the perpetrators and victims of crime, but there were also debates about their criminalization that have continued into the recent past, suggesting ongoing police scrutiny and misrecognition. The church, the school, the home, the neighbourhood, and the voting booth are all institutions central to suburban life, and thus they are contested sites of queerburbia. While it remains unclear the extent to which print media representations influence public perception of suburban places, such narration discursively reveals "the real character of the place" or, more specifically, "the quality of place (i.e., what is there, who is there, and what is going on there)" (Nelson, 2015, 193).

Conclusion

This chapter has provided a preliminary analysis of Canada's queerburban places using some of the only quantifiable secondary data sources available on LGBTQ2S populations in Canada. Although abstracted from everyday life and limited by the underrepresentation of the full range of LGBTQ2S identities, same-sex couple data from the 2016 national census was mapped, offering city-regional portraits of urban-suburban patterns. Toronto was the most centralized of the three city-regions, with the highest concentration of male same-sex couples living downtown. Montreal exhibited moderate decentralization, with female same-sex couples (especially those with children) concentrated in peripheral municipalities. Likely due to the high cost of housing, the Vancouver city-region was also decentralized and the only city-region with a concentration in a particular peripheral municipality, New Westminster. While birthplace, age, and income added some nuance, the narrow definition of the population as "same-sex couples" is reductionist, leaving much to be explored regarding the range of LGBTQ2S identities involved in queerburban place-making.

Print media analysis from 1985 to 2014 added a public discursive layer to queerburban place-making revealing broader societal attitudes towards queerburbanites, general geographical patterns, and longitudinal portraits of thematic and institutional representations of queerburbia over time. These representations generally depicted queerburbanites as "out-of-place", interlopers in suburban institutional spheres dominated by education, politics, the public realm, and religion. City-regional print media examples contextualized the spatialities of queerburban life, but they also tended towards the spectacular, reinforcing the observation that when LGBTQ2S people appear in suburbia, it is neither neutral nor unremarkable but rather a signifier of queer out-of-placeness. Despite the many rights gains over the period, heterocisnormative socio-spatial orders continued to associate queers with metronormative discourse: their presence in suburbs generated conflict and surprise, problems to be solved and threats to be mitigated. These are, however, mere representations, revealing little of the place-making practices of civic actors in serving and including LGBTQ2S citizens, the unmaking resisting and organizing practices of activists and para-public organizations, or the remaking experiences of queerburbanites themselves.

Notes

1 Same-sex (cisgender) couple households are the unit of measure for this research, alternately described as same-sex households and same-sex couples. Using the complete 2016 data set, the year in which this project began, we describe the geographic distribution and characteristics of these households rather than offering a change-over-time analysis that integrates the 2021 census (released in 2022 at the close of this research).
2 Following North American suburban studies scholarship, the central city is defined by its pre-1946 housing stock while inner suburbs are postwar creations (1946 to 1971) and outer suburbs are a product of the post-1970s housing boom that continues to transform the urban-rural fringe into exurbs (Preston & Ray, 2020).
3 From the initial database of 1,721 articles, duplications were deleted, along with articles naming correctional or other government facilities in specific suburbs (rather than LGBTQ2S residents), discussing pedophilic crimes (where sexual orientation was irrelevant), and reporting on MP Svend Robinson's jewelry theft (rather than his political activities).
4 These are Canada's two "national" LGBTQ2S community-based archives along with regional community (e.g., Archives lesbiennes de Québec and the B.C. Gay and Lesbian Archives in Vancouver) and university (e.g., The Transgender Archives in Victoria and the Archives of Lesbian Oral Testimony at Simon Fraser) archives.
5 Relevant keyword searches were conducted for each municipality in English (gay, homosexual, lesbian, bisexual, transsexual, transgender, Two-Spirit, queer, sexual-orientation, same-sex, homophobia, homophobic, transphobic, transphobia) and French (gay, gai, homosexuel, lesbienne, bisexuel, transsexuel, travesti, transgenre, bispirituel, queer, orientation-sexuelle, même-sexe, homophobie, homophobe, transphobie).
6 Scanned archival materials were rendered readable and searchable using Adobe Acrobat Pro's OCR Text Recognition function.
7 Print media articles were catalogued by census metropolitan area (CMA), date, and municipality. In NVivo, they were coded by article type, key theme, the issue of contention, the spatial scale, LGBTQ2S identities, and institutional sphere.
8 This complicated story began with complaints to the police, followed by two criminal cases of harassment and physical assault, a 2008 defamation suit, and 2010 Quebec Human Rights Commission tribunal that found one neighbour guilty of harassment. It concluded with an out-of-court settlement based on admissions of baseless accusations (Kramberger, 2013).

3 Including

Introduction

Municipal social inclusion is a governance tool for enhancing the participation of marginalized populations in civic life (Fincher et al., 2016). Urban plans, policies, and strategies can embed the "social logics" of inclusion in place (Fincher & Iveson, 2008), but how this is done, and on whose behalf, constructs uneven civic terrains of power. Consequently, certain discourses and subjects become hegemonic, rendering forms of place-based action possible for some people while foreclosing it for "others". An unevenness of access to, and influence on, governance institutions produces overt exclusions and partial inclusions resulting from recognition and misrecognition (Fraser, 2000; Young, 2000). For Hines (2013), governance systems determine LGBTQ2S (mis)recognition by linguistically setting the terms of social inclusion using categories that bring minoritized groups into civic being even if it means adopting terms that are external to the group's frame of reference. Within governance systems, urban planning and policy stakeholders define inclusion/exclusion rhetoric, guiding processes of institutional recognition/misrecognition through municipal plans, policies, and strategies. These political institutional trajectories of purposive action – "structures exercising state authority in accordance with embedded values" – shape not only society but also place (Tuohy, 1992, xvii).

Much like laws and rules, civic plans, policies, and strategies are local technologies of power that guide governance priorities in place (Cooper & Monro, 2003). Within contemporary neoliberal municipal governance, "diversity" and "inclusion" are strategically promoted (Valverde, 2012) yet end up being "empty signifiers" of "everything" and "nothing" (Morrish & O'Mara, 2011, 974). Able to absorb multiple interpretations supporting coordinated action (Hillier & Abbruzzese, 2023), these empty signifiers "capitalize on and gain from a context of differentiation" (Jones, 2011, 163) while simultaneously erasing the most disempowered and non-commodifiable bundles of difference. Grassroots calls for a more "diverse city" are frequently co-opted through symbolic associations between "diversity" and "tolerance" that occlude distaste for discrimination (Valverde, 2012). Municipal leaders strive for social inclusion via the civic celebration of "diversity", but as "a simple plurality, a

DOI: 10.4324/9781032632544-3

harmonious juxtapositioning" (Massey, 2007, 88) rather than an appreciation of the "problems of difference, among a host of axes of difference" (Ettlinger, 2010, 54). These disconnections between inclusion and diversity as institutionalized discourses manifest in place, intensifying everyday lived frictions.

The "diversity turn" has integrated LGBTQ+ constituencies into the municipal social inclusion frameworks of many large metropolitan areas, but often in central areas rather than peripheries (Cravens, 2015; Murray, 2015). Historically, suburbs have been interpreted as places where "narratives of conservatism overwhelm those of inclusion" (Niedt, 2013, 1), particularly LGBTQ+ social movement politics. Given the morphological prioritization of housing for heterosexual nuclear families, suburban governments emphasize land development through infrastructural techno-production (Grant, 2009), neglecting the social planning necessary to support their demographic diversification. As independent municipalities with different political opportunity structures, suburban governments, as this chapter demonstrates, variably intensify place for sexual and gender minorities through their policy discourses of social inclusion.

This chapter answers questions about the LGBTQ2S-supportive political opportunity structures available in the plans, policies, and strategies of peripheral municipalities and how they are leveraged (or not) to "intensify" (Seamon, 2018) place possibilities for sexual and gender minorities. Political opportunity structures are the "consistent – but not necessarily formal, permanent, or national – dimensions" of political contexts that afford emplaced "opportunities" for political mobilization (Tarrow, 1998, 8). Such structures manifest as the division of government responsibilities, funding allocations, advisory committee formations, and in plans, policies, and strategies. As this chapter argues, when civic social inclusion governance frameworks inadequately address sexual and gender differences, the ensuing "place intensification" dynamics "enfeeble or squelch the life of the place", unmaking queerburbia (Seamon, 2018, 138). In what follows, urban planning and governance scholarship on LGBTQ+ social inclusion is reviewed. Next, social inclusion is introduced via a global inclusive cities agenda and linked to Canadian municipal policy frameworks. After considering how LGBTQ2S social inclusion has impacted peripheral municipalities in the Toronto and Montreal city-regions, the chapter provides a detailed analysis of Burnaby, New Westminster, and Surrey in the Vancouver city-region.

LGBTQ+ social inclusion in suburban municipal governance

In the 1990s, nation states and local administrations in contemporary liberal democracies initiated public policies to counter exclusion in the service of social justice and community cohesion (Cameron, 2007; Flint & Robinson, 2008; Parker & Karner, 2010; Askins, 2015). As neoliberal state restructuring reoriented government priorities to synergize with private investors, business interests, and the euphemistic "creative class" (Florida, 2002), the needs, rights, and responsibility for marginalized social groups receded from political view.

The inherent homocisnormativity, whiteness, and classism of Florida's (2002, 2005) creative class thesis have been deconstructed by Marxist, feminist, and queer critics (Muller Myrdahl, 2011; Oswin, 2012; Parker, 2017; Peck, 2005). When aligned with the "creative turn", urban governance deepens social inequalities because ideologies of competition, consumption, and individualism hold sway (Parker, 2017). Civic leaders tend to commodify "difference" in their place branding, opportunistically leveraging diversity, but rarely creating measures to enhance social inclusion. Municipal "social landscapes" become fodder for marketing campaigns that emphasize "diversity" at the expense of emplaced "community needs and problems" (Paganoni, 2012, 15). Creative city diversity branding thus cannot replace social inclusion policy as it undermines support for social groups by forcing competition for marketing attention and limited funding.

When social inclusion dialogues with its opposite, social exclusion, more policy nuance is possible. A multi-layered process through which individuals or social groups become ever-more socially disengaged from the places where they live (e.g., via citizenship rights, labour markets, public spaces, or civil society), social exclusion emerged in Europe and the UK garnering global policy traction with neoliberalism's rising inequalities (Gerometta et al., 2005). It influenced analyses of social disadvantage and directed attention to the "problems and deficits of those labelled 'excluded'" (Cameron, 2007, 397). Just as it identified barriers to participation, it also informed policy frameworks that pathologized the "excluded" rather than enhancing their inclusion. Problematically, this inclusion/exclusion binary distinguishes between "social being[s] characterized by normality, morality, responsibility, independence and competitiveness" and those marked as different, redundant, pathological, or immoral (Cameron, 2007, 401). Such norms-setting happens within the realm of a "straight state" that embeds straightness in place (Canaday, 2009) wherein unnamed heterocisnorms "permeate local government's internal culture, relations with citizens, and regulatory practices, as well as its policies and conduct" (Cooper & Monro, 2003, 248).

The heterocisnormativity of municipal governments has been enacted through the policing of homosex in public spaces (Catungal & McCann, 2010), the regulation of LGBTQ+ nightlife and commercial sex premises (Brown & Knopp, 2016), the gentrification of gayborhoods (Doan, 2015), and adoption of homonormative safety protocols (Hanhardt, 2013). Municipal improvement associations in gayborhoods (Nash & Gorman-Murray, 2015; Gorman-Murray & Nash, 2017) combined with sponsored Pride celebrations have also disciplined queer and trans bodies through "homo-entrepreneurialism" (Kanai & Kenttamaa-Squires, 2015). These neoliberal marketing pivots have driven LGBTQ+ activists to directly engage the local state whether with respect to HIV/AIDS service provision (Brown, 1997) or the promotion of "equal opportunity" agendas such as civil partnerships (Cooper, 2006). It is at the interface between LGBTQ+ activism and municipal governments that place is unmade and remade in ways that can forward their social inclusion (Cooper &

Monro, 2003). Electoral processes (Bailey, 1999; Brown et al., 2005) and anti-discrimination ordinances (Chapman, 2011; Fejes, 2008), for example, have provided local opportunities to unstraighten municipalities.

Differences in legislative and planning authority and policy priorities mean that LGBTQ+ social inclusion is fragmented across national, provincial or state, and municipal scales (Gorman-Murray, 2011). Since municipal LGBTQ+ recognition is generally "incommensurable" with other spatial scales (Hubbard, 2013), governance hierarchies exhibit a "scalar dissonance" of LGBTQ+ recognition (Gorman-Murray, 2011) that is compounded by the "scalar arrhythmia" of its diffusion (Bain & Podmore, 2022). Across purportedly "progressive" polycentric Canadian city-regions, the municipal scale masks spatio-temporal differences in the timing and types of LGBTQ+ social inclusion policy adoptions, especially between central cities and their peripheries. Not only do peripheral municipalities vary in the intensity of their LGBTQ2S-supportive social inclusions but their place constellations also produce different outcomes that are cross-cut by complicated stakeholder coalitions.

The inclusive cities agenda and Canadian municipal governance LGBTQ+ toolkits

From the Habitat Agenda (United Nations, 1996) to Agenda 2030's (United Nations, 2015) Sustainable Development Goals (SDGs), the United Nations has promoted "inclusive cities" as central to sustainable human settlements. The redistributive policies of the UN Human Settlement Programme's Global Campaign on Urban Governance (popularized as "The Inclusive City Campaign") promised to eliminate global poverty in developing (Escorcia Hernández et al., 2023; Liu et al., 2020) and developed economies (Cavicchia, 2021; De Filippi et al., 2019). This multidimensional humanitarian understanding of inclusive cities is characterized by: equitable access to public services, employment, and social networks; participation in decision-making; and environmental health and intergenerational resource sustainability (Liang et al., 2022; Ruble et al., 2005). It gathered traction due to its policy malleability and synergies with urban neoliberal competitive branding regimes (e.g., as sustainable cities, creative cities, smart cities, ecological cities, knowledge cities, and resilient cities) (Alsayel et al., 2022; Schraven et al., 2021) that determine which populations are targeted for inclusion (commonly women, children, elderly, refugees, disabled, and the urban poor). Sexual and gender minorities are seldom listed because they are considered "incommensurate subordinates, rather than people … who are systematically marginalised" (Duplan, 2023, 141).

The United Nations Educational, Scientific, and Cultural Organization's (UNESCO) launch of the International Coalition of Sustainable and Inclusive Cities (ICCAR) in 2004 inspired the formation of the Canadian Coalition of Municipalities against Racism and Discrimination (CCMARD) which became the Coalition of Inclusive Municipalities (CIM) in 2019. On behalf of the Canadian Commission for UNESCO, CIM (2019) is the first national coalition to prepare an LGBTQ2+ inclusion toolkit. Its chapters include: descriptions of the population, key terminology, and a pronoun overview; summaries of ongoing challenges (e.g., communication, crime and policing, human resources, infrastructure, recreation, and health) and prioritized populations (e.g., youth and seniors); an interpretation of municipal "quick wins", "do's and don'ts", and strategies; and appendices of historical rights timelines, a glossary of key terms, templates, and resources (e.g., gender-neutral/inclusive washrooms and change rooms, correspondence, survey questions, commitment statements, email signature samples, and name tags). It asserts that municipalities are uniquely positioned to promote LGBTQ2+ inclusion because "they are closely connected to residents" as service providers and employers (CIM, 2019, 6).

Canadian municipal government's concern for social inclusion policy depends upon their size, centrality, and autonomy from provincial governments. Their general responsibilities for urban planning, economic development, civic infrastructure, programming, cultural activities, and services implicate social inclusion, especially if they manage housing and transportation systems (Papillon, 2002). In Canada, a country of immigrant settlers, ethno-cultural diversity, and multiculturalism policies drive urban social sustainability agendas to mitigate the widening "fault lines of social isolation and spatial segregation in increasingly divided city-regions that concentrate patterns of advantage and disadvantage" (Bourne & Hulchanski, 2020, 11). Prioritizing economic development while integrating high volumes of newcomers, the peripheral municipalities of Canada's largest city-regions are where these tensions emerge, cross-cutting policy fields of education, health, housing, urban planning, and job creation (Papillon, 2002). Whether suburbs put social inclusion on the agenda, however, depends on upper-tier legislation, participation in regional management coalitions, the size of planning departments, municipal openness to non-profit partnerships, and citizen coalitions raising inclusion/exclusion concerns. Suburban LGBTQ2S social inclusion agendas are especially variable, requiring further research into their discursive embedding in municipal policy. Toronto and Montreal provide examples of these uneven peripheral municipal and regional governance inclusions followed by a detailed discourse analysis from the Vancouver city-region.

LGBTQ2S social inclusion in Toronto's regional and peripheral municipalities

In Toronto's regional municipalities, social inclusion policy momentum is nascent. For example, the Region of Peel's *Diversity, equity, and inclusion strategy* is unfolding with surveys and focus groups to determine gaps and barriers in the delivery of programmes and services (Peel Region, 2023). Yet back in 2009, four years after Canada legalized same-sex marriage, the Associated Youth Services of Peel undertook an assessment of LGBTTTIQQ (Lesbian, Gay, Bisexual, Transgender, Transsexual, Two-Spirit, Intersex,

Queer, and Questioning) needs and service gaps. A 2007 environmental scan of local service providers demonstrated that only one-third offered queer- or trans-specific programming and that many mainstream institutions lacked anti-homophobia/anti-transphobia policies and/or queer and trans competency staff training. The report distinctively references "social exclusion", underscoring the compounding of intersecting marginalities of sexual and gender identity, ethnicity, racialization, poverty, and homelessness (Access Alliance Multicultural Health & Community Services, 2009, 8). It also uniquely identified the "lack of visibility" and "isolation" of LGBTQ people in Peel as a cause of "loneliness, depression, anxiety, substance abuse, suicidal ideation and mental health issues" (Insauga.com, 2019, para. 6). However, it was not until 2019 that eight community organizations[1] created Rainbow Salad, a virtual information resource for the region promoting LGBTQ events, programmes, and services. Now joined by 16 community service organizations including the Brampton and Mississauga public libraries, Rainbow Salad (2023, para. 1) champions "meaningful 2SLGBTQ+ inclusion by taking a community-based and community-centered approach to equity" to combat "social isolation, lack of belonging, physical and emotional distress, and an overall lack of visibility within the landscape of the region".

The Peel District School Board has been another regional leader on sexual and gender minority inclusion through its *Equity and inclusive education policy #54*, *Safe schools policy #48*, and *Flag raising policy #68*. In 2018, board trustees revised the district's flag raising policy to permit the Pride flag and the Mississaugas of the New Credit First Nation flag to be raised. This policy update demonstrated "commitment to genuine inclusion" to help "all students to feel safe and welcome" (Peel District School Board, 2024). It originated with a Pride delegation of LGBTQ+ students and community organizations collaborating on safety, anti-bullying, and inclusiveness who spoke from personal experience about the significance of the rainbow flag. The decision, however, generated community controversy. The Chair of the school board received oppositional emails and telephone calls from religious groups asking "But, what about me?" requesting similar opportunities for visibility and recognition (19 June 2018).

The Durham Region opted to undertake an LGBTQ service competency and community asset mapping exercise in 2011 to uncover gaps and barriers and identify organizations, businesses, and groups who could be considered "allies" (Community Development Council Durham, 2011, 15). Describing Durham as lacking the "knowledge and culturally competent support services for sexual and gender minorities" and its LGBTQ community as "diverse and complex" (Community Development Council Durham, 2011, 4), the report advocated for queer- and trans-positive service provider training to improve attitudes, knowledge, and skills, and the development of policies and procedures for the creation of inclusive and positive spaces. It is distinctive for distinguishing between being LGBTQ *"friendly"* (no specialized services or products but striving to include everyone while serving and reflecting the

LGBTQ community), "*positive*" (specialized services, products, referrals, and a welcoming environment for the LGBTQ community), and "*focused*" (reflecting the LGBTQ community and reaching out with particular services, products, and programmes). Durham's LGBTQ assessment had two outcomes: a regional guidebook launched alongside *Prideline Durham* (Community Development Council Durham, 2011), a crisis and information telephone line operated by the Distress Centre in Durham (2021).

The Regional Municipality of York, in contrast, rarely specifies inclusions for LGBTQ2S populations. Working with the Municipal Diversity and Inclusion Group[2] and the Community Partnership Council, its *Inclusion charter for York Region* was adopted in 2018. While "diversity" was in its title, the term was later dropped because, while "a source of strength, vitality, and economic opportunity", diversity "highlights our differences" (Council of the Regional Municipality of York, 2018, 6). Inclusion, on the other hand, aligned with the region's neoliberal place brand because it promotes "an environment where differences are embraced and everyone feels included" so as to "attract and retain the best talent, promote innovation, and provide an excellent customer experience" (Council of the Regional Municipality of York, 2018, 6). Never mentioning LGBTQ2S communities, this charter became a best-practice model for constituent municipalities promoting "social inclusion" and "equality for all" while permitting each one to "customize" and "endorse their own version" (City of Markham, 2018, 5). How regional governance bodies frame social inclusion filters down to their composite peripheral municipalities.

City of Mississauga: A youth-friendly place

Within the Peel Region, the City of Mississauga has leveraged planning and policy frameworks to address significant demographic change over two decades. Its five-pillar *Strategic plan* (2009c, 2) – to move, belong, connect, prosper, and become green – envisions civic change to benefit "students and older adults, entrepreneurs and employees, new immigrants and families" such that they might "fulfill their future promise and realize their goals". The pillar most closely associated with social inclusion is "prosper", articulated in the plan as "Ensuring Youth, Older Adults, and New Immigrants Thrive". It strives to nurture diverse newcomer cultures, support ageing-in-place, and attract and retain youth, yet neglects sexual and gender minorities as an urban planning group.

Mississauga (City of Mississauga, 2009a, 1) especially strives to be "youth-friendly" – to make its "policies, programs, services, and facilities more inclusive and responsive to the needs and interests of youth". Neither its *Youth plan* (2009a) nor its follow-up report, however, considers that some youth might identify as LGBTQ+. The plan profiles youth discussion of the youth-friendliness of places (e.g., schools, colleges and universities, shopping malls, movie theatres, community centres, and public parks), barriers produced by inadequate, costly, and distant programming, limited public transit, racism, ageism, bullying, and police-harassment, and desires to become involved in civic decision-making. These findings resonate with LGBTQ+ young people (e.g., the lack of safety, mobility, programming, and social spaces), yet Mississauga youth are only disaggregated in terms of "race, ethnicity, socio-economic status, academic achievement, or interest" (City of Mississauga, 2009a, 19).

Responsible for parks and forestry, libraries, fire services, recreation, environment, and culture, Mississauga's Community Services Department seldom considers its LGBTQ2S citizens. Its *Cultural plan* (City of Mississauga, 2009b, 60), for example, accentuates festivalization to generate "platforms for a wide range of cultural and heritage expression and establish a sense of pride for communities". Such "pride" is not "Pride", although, in 2013, the city hosted its first Pride Week in Peel event in city hall's Celebration Square. While it does mention the necessity of year-round programming for "youth and diverse populations" in "'slow' periods of the year", there is no discussion of LGBTQ+ communities in the *Cultural Plan*, not even for Pride month (City of Mississauga, 2009b, 60). Instead, it is the arm's-length Art Gallery of Mississauga (AGM) in informal collaboration with public librarian insider activists and allies who are tasked with civically raising awareness of queer culture through exhibitions and displays.

The LGBTQ+ community is explicitly mentioned in Mississauga's *Master plan for recreation* (City of Mississauga, 2014), developed under the leadership of an "out" gay male Director of Recreation and Community Services (2012–2015). "LGBTQ" is listed with "marginalized and underrepresented groups" alongside people with low incomes, disabilities, and women. The plan mentions that "participation in family or couples-oriented programming may be avoided by members of the LGBTQ community due to perceived (or real) stigma or feeling uncomfortable" (City of Mississauga, 2014, 18). Using City of Toronto public health statistics (2001), it estimates that 10 percent of the general population is LGBT but also acknowledges the nascency of its outreach endeavours. The LGBTQ community is framed in "market" terms as "virtually untapped" and potentially "represent[ing] up to 70,000 residents" (City of Mississauga, 2014, 106), while opportunities for its inclusion are envisioned through the creation of "Safe and Positive Spaces" and the provision of "barrier-free recreation".

Outside of culture and recreation, no municipal department, committee, or sub-committee specifically advocates for LGBTQ+ residents. Instead, queerburbanites must self-cohere as a community group to make targeted requests that synergize with municipal political opportunity structures. Mississauga has developed a "Workforce diversity and inclusion strategy" (City of Mississauga, 2017) based on employee consultation with different equity-seeking groups (e.g., women, racial minorities, Indigenous people, and people with disabilities), but LGBTQ2S constituencies are not identified (only mentioned in descriptions of issues raised during focus groups). Despite being tasked with the removal of "barriers for all residents" since 2015, Mississauga's Diversity and Inclusion Advisory Committee (DIAC) focused on promoting "broader understandings and engagement between residents and ethno-cultural communities" (City of Mississauga, 2015, 1). It was replaced in 2022 by the Combating Racism, Discrimination and Hatred Advisory Committee (City of Mississauga, 2022a) which prioritized anti-racism and adopted a critical definition of marginalization: "the act of being relegated to the 'margins' of society (away from the centre) – it is both a condition and a process that prevents individuals and groups from full participation in social, economic, and political life enjoyed by the wider society" (City of Mississauga, 2022b, 2). While the Committee's terms of reference do cite sexual orientation, gender identity, and gender expression, LGBTQ2S constituencies are only mentioned in reference to recreation services access and tokenistic Pride month programming.

City of Brampton: Growing a mosaic

Elsewhere in the Peel Region, Brampton's *Living the mosaic: Brampton 2040 vision* (2018b) is a community-inspired "peoples" official plan with seven aspirational target statements of 250-year transformative re-invention. Inspired by Vancouver's city-centric policies, it envisions Brampton as a transit-oriented, sustainable place with a compact downtown and neighbourhood condominium and commercial employment centres that offer physical and social connection, community, productivity, and artistic expression (Keil & Üçoğlu, 2021). The fifth vision, "Social matters and housing", considers social inclusion under the sub-heading "Brampton will be a rich mosaic of cultures and lifestyles, coexisting with social responsibility, respect, enjoyment, and justice" (City of Brampton, 2018b, 73). In the primary action towards a social policy framework, sexual and gender minorities are mentioned for the first and only time in this 100-page document: "Inclusionary provisions will be made for social, ethnic, cultural, sexual orientation, and gender diversity" (City of Brampton, 2018b, 73). No roadmap is provided.

Brampton, however, lacks a specific equity and inclusion plan. In 2015, the city established a short-lived 20-member Inclusion and Equity Committee that met bimonthly to advise City Council. Only one committee member represented an LGBTQ+ perspective. Mandated to build awareness about the city's diverse communities and increase access to civic services and programmes (City of Brampton, 2024a), over its three-year existence, the committee was "plagued by attendance issues", making it hard to achieve quorum, conduct official business, and advise the council in a sustained and meaningful way (Frisque, 2018, para. 1). As a result, intersections of community diversity remain largely unaddressed. In Brampton's public facing communication, it is only in a *2017 city satisfaction survey* that "LGBTQ*" is mentioned in an overview of survey respondents, indicating that 5 percent identified as LGBTQ and less than 1 percent identified as having a gender "other" than male or female (City of Brampton, 2018a). The minimal inclusion agenda of this municipality is far from addressing the intersectional concerns that determine access to political power and resources.

City of Markham: Integrating newcomers

Further east, the City of Markham has realized the Region of York's *Inclusion charter* with civic commitments to respect, dignity, and equity in service delivery, the embrace of "diverse perspectives and constructive conversations", and a commitment to proactively addressing discrimination (City of Markham, 2018, 8). Its promotion of inclusion was forwarded through the *Markham diversity action plan: Everyone welcome* (2008) and, more recently, through its targeted combatting of anti-Black racism. The *Diversity action plan* (2008, 25) acknowledges that "Markham's new communities face modern versions of old problems: political representation, social recognition, and prejudice" and singles out income disparity as the most damaging to social cohesion. It spotlights the need to "work harder to recruit seniors, newcomers and visible minorities, youth, and persons with disabilities to serve on boards, agencies and commissions, and to advise on municipal services" that "everyone is equally entitled to" (City of Markham, 2008, 26). This diversification strategy ignores sexual and gender minorities, prioritizing race relations, disability, seniors, and youth as these groups have greater visibility in the national census data that drive urban planning inclusion targets. Without an advisory committee to champion LGBTQ+ needs and with no city council LGBTQ+ community deputations, inclusion policies are forwarded in terms of ethno-racialized newcomer integration. The singular mention of civic "accomplishments" in "leadership in diversity" with respect to "LGBT" community groups refers to Markham's creation of "gender welcoming, accessible washrooms" in city hall and newly built civic facilities (City of Markham, 2018, 8). Beyond washrooms and flags, LGBTQ+ inclusion is sidelined by the prioritization of newcomers and anti-racism initiatives.

Town of Ajax: LGBTQ2S inclusion through anti-racism

For the Town of Ajax in the Durham Region, social inclusion refers primarily to anti-racism policies, practices, and strategies. Unlike Markham, however, there are overlaps with LGBTQ2S inclusion here due to the work of one insider activist (who later led the Recreation Division of Mississauga's Community Services Department). With the establishment in 1997 of its Race Relations Committee to raise awareness about anti-Black racism, Ajax was an early social justice advocate. This committee was later renamed Diversity and Community Engagement (DACE) with a broader awareness-raising remit to respond to changing demographics in a formerly white, working-class community, while fostering inclusion for a greater diversity of marginalized social groups (e.g., Indigenous people, racialized newcomers, youth, people with disabilities, women, people of low-income, and LGBTQ2S). Composed of a steering group of two city councillors and 15 community leaders and residents, it was charged with developing a plan to engage disenfranchised populations in discussions about access to services and (sub)urban development processes.

With its 51 recommendations to ensure that "all residents have access to employment opportunities, programs and services", Ajax's *Diversity and community engagement plan* (Town of Ajax, 2010) was unanimously adopted by City Council in 2010 and celebrated with its first Pride flag raising. It would be another five years, however, before civic leaders budgeted for the hiring of Ajax's first Diversity and Community Engagement Coordinator, the only such position in the Durham Region focused on ensuring that "everyone living in the community is included and feels welcome" (Gilligan, 2015, para. 1). In the second phase of the plan (2017–2022), this Coordinator oversaw the development of four overarching policy themes of potential interest to the LGBTQ2S community: the town as an employer; programme and service priorities; community engagement and neighbourhood strengthening; and youth engagement. In employment terms, the human resources department has prioritized, since 2016, staff training on gender identity and expression in the workplace (Town of Ajax, 2017). Its workplaces and facilities are now governed by policies in line with the Canadian and Ontario Human Rights Codes to foster "a safe, inclusive, equitable, and welcoming space for everyone regardless of their gender" (Town of Ajax, 2017, 1).

In Ajax's first 88-page *Diversity and community engagement plan*, the "LGBT" community is mentioned 15 times (mostly as targeted bank clients), but also in lists with other marginalized groups. These listed groups are prioritized for civic engagement ("stronger connections"), assessment (for "programs, service and engagement needs"), and collaboration ("to become a more inclusive Ajax") (Town of Ajax, 2010, 9). In the focus group feedback summary, Ajax is described as "supportive" of the LGBT community but that it could do more in terms of service and programme provision. Critical mention occurs in a statement of a guiding principle: "[t]he complex issues of racism and relations with LGBT communities play out as blatant or subtle injustices and must be candidly addressed" (Town of Ajax 2010, 33). Such naming of the LGBT community in the plan received additional advocacy from the gay-identifying Director of Recreation and Culture from 2007 to 2012 who collaborated with the Durham Board of Education and the local PFLAG chapter to host the first LGBT Prom in 2009 as a Durham Pride event. Despite having the rainbowized subtitled moniker "fostering a culture of equity and inclusion" on the cover, however, the updated plan for 2017–2022 contains no mention of LGBT. While a key employer priority is expanding Ajax's Positive Spaces initiative from the central public library to other recreation and seniors' centres (Town of Ajax, 2017), LGBT positive space production has been limited to hosting Durham Pride in 2018. A second initiative was the 2022 mayoral unveiling of a new rainbow crosswalk at the entrance of the Audley Recreation Centre. It was civically framed as a "powerful symbol of support and acceptance – a landmark that reaffirms the importance of Pride Month and recognizes the joy and value of love, diversity of identity and authentic self-expression in all forms" (Town of Ajax, 2022). Until the 2022 Christian evangelical student disruption of a Pride flag-raising ceremony for IDAHOBIT outside of an Ajax elementary school, however, no civic official had publicly spoken about LGBTQ2S issues ("DDSB investigating", 2022). LGBTQ2S social inclusion in Ajax evolves through collaborative allyship configurations, rendering it an inclusive employer and an occasional institutional ally to para-public and select community organizations.

Across the case study peripheral municipalities of the Toronto city-region, social inclusion addresses ethno-cultural cohesion, combating racism, and supporting youth. Within such frameworks, LGBTQ2S inclusions are predominantly generic, manifesting as an acronym in a list of minoritized groups. They are sporadically mentioned in minor documents but never make it into strategic social inclusion policy directives or urban planning master plans that create more substantive political opportunity structures. When municipal committees are formed to address "diversity and inclusion", they have broad mandates that are occasionally refined to target racism and marginalization, but seldom appreciate the intersections of age, ethnicity, and racialization with sexuality and gender non-normativity. The programming of recreation divisions may be the singular civic path to greater LGBTQ2S inclusions, but this as yet unrealized opportunity for social change is limited to the sphere of leisure.

LGBTQ2S social inclusion in the Montreal city-region

At the centre of the Montreal city-region, the Ville de Montréal has witnessed international moments of LGBTQ2S recognition and inclusion. During the International Conference on LGBT Human Rights that launched the World Outgames in 2006, its central borough of Ville Marie was the first government body in the world to adopt "The Declaration of Montreal on Lesbian, Gay, Bisexual, and Transgender Human Rights" demanding that the United Nations address global rights abuses (Podmore, 2015). Today, regional leadership on LGBTQ2S municipal inclusion comes from the para-public organization, the Conseil québécois LGBT. Founded in 1993 to represent and unite Quebec's LGBTQIA2+[3] organizations and advocate for the respect of their legal, social, and institutional rights, the Conseil now has over 70 organizational members with mandates that include social services, rights lobbying, health, and education (Conseil québécois LGBT, 2024, 6). After organizing a round table of regional organizations in 2017, it collaborated on municipal LGBTQ2S awareness-raising and a 2021 province-wide survey of LGBTQIA2+ municipal social inclusion practices summarized in its "*Guide pratique pour des municipalités québécoises inclusives des personnes LGBTQIA2+*" (Conseil québécois LGBT, 2024). For the Conseil (2024, 18), municipalities are important jurisdictions of LGBTQIA2+ inclusion because they assume "many responsibilities that directly impact the population" and "have many means to concretely promote these groups". Enhanced municipal LGBTQIA2+ inclusion in regional small towns and suburbs can potentially halt "the great gay migration" to Montreal and even attract "sexually diverse and gender plural families who wish to leave large urban centres" (Conseil québécois LGBT, 2024, 18).

The Conseil's guide summarizes many potential inclusion practices: increasing access to municipal services; creating a supportive work environment; encouraging the election of "out" officials; enhancing citizen participation; partnering with organizations; redesigning municipal facilities; changing administrative forms; adopting non-gendered dress codes; promoting LGBT organizations and events; and engaging municipal services (e.g., administration, communications, human resources, social development, urban planning, libraries, recreation centres, public transportation, and police). It makes ten recommendations for LGBTQIA2+ municipal inclusion: designate a responsible official; train employees; connect with local organizations; raise the flag for IDAHOBIT; develop "best practices" with police services; provide gender-neutral facilities; develop policy supporting employees undergoing gender reassignment; adopt an anti-harassment policy; perform an LGBTQIA2+ inclusion analysis; and include sexuality and gender diversity issues in strategic plans (Conseil québécois LGBT, 2024, 40). Tailored to municipal administrative needs, the guide's appendices also offer a glossary, visual lists of various rainbow flags, and a calendar of annual events.

Some of the peripheral municipalities in the Montreal city-region participated in the Conseil's study, but it was smaller regional cities and villages that provided examples of LGBTQIA2+ inclusion. The research team understood that Montreal's smaller suburbs might not have a social development mandate, but it was the absence of LGBTQIA2+ social inclusion innovation in its larger suburbs that was surprising: "When you have a municipality of 7000 inhabitants, you are not there. I can understand that, but when you are large like Longueuil or Laval?" (19 August 2021). Only ten references to homosexuality, homophobia, sexual orientation or LGBT identities were found in policy documents collected for all 97 CMA municipalities. Flags were the most commonly mentioned (e.g., as flag raisings for IDAHOBIT and half-masting in remembrance of the victims of the Orlando nightclub shooting in 2016). Sexual orientation appeared in listed statements of non-discrimination in funding guidelines, police policy, and regulations on visual displays. While few West Island municipalities referenced LGBTQ2S inclusions, there has been a recent increase in the public-facing documentation of Laval and Longueuil.

Independent West Island Municipalities: Age-friendly inclusions

The independent municipalities of Montreal's West Island are part of the Communauté métropolitaine de Montréal (MMC), a regional body dealing with urban planning, environment, economic and social development, heritage, and transportation. The Island's English and French school boards and police services are also regional. Since social development is MMC's primary concern and Quebec municipalities have not historically been responsible for social planning, these small peripheral municipalities make few mentions of LGBTQ2S inclusion.

Across this cluster of eight municipalities, social inclusion was rarely a dimension of town planning and service provision. Kirkland, Dorval, and

Sainte-Anne-de-Bellevue referenced social inclusion practices in their plans of action for seniors (as part of participation in Municipalité Amie des Ainés (MADA), a provincial certification programme for municipal age-friendliness) and residents with disabilities. With a certified seniors' action plan, Sainte-Anne-de-Bellevue (2023) interprets social inclusion as the promotion of intergenerational activities and the municipal recognition of the International Day of Older Persons. In their urban planning master plans, both Dorval and Dollard-des-Ormeaux considered how social sustainability could enhance community life (City of Dorval, 2023). In Dorval, social inclusion was not used as a policy framework, but its sentiment was conveyed through its Youth Advisory Committee and Committee for the Integration of Individuals with Disabilities.

Despite more robust approaches to social inclusion, Baie-D'Urfé, Pointe-Claire, and Beaconsfield still did not mention LGBTQ2S inclusions. The wealthy suburb of Baie-D'Urfé launched its *Strategic vision, 2022–2026* (Town of Baie-D'Urfé, 2022, 14) with consultations that identified "community life" as a strategic priority, encompassing the integration of "cultural and linguistic diversity", a "strong sense of belonging", and "citizen involvement by residents of all ages" (Town of Baie-D'Urfé, 2022, 18). With its MADA Committee, Beaconsfield strives to be age-friendly by adapting its structures and services, but its only inclusion mention is in regard to recreational programming expansion beyond young families and retirees (Ville de Beaconsfield, 2024). Pointe-Claire city council (City of Pointe-Claire, 2020) made a rare diversity statement in response to Black Lives Matter protests of the police-killing of George Floyd:

> Regardless of our background, our beliefs, our gender, our sexual orientation, our skin colour, our language or our means, we are all equal and must be able to enjoy the same rights. Pointe-Claire is a tightly knit multiethnic community that stands together, respectful and proud of its diversity. Our values of inclusion and justice that guide us must reflect daily in all facets of our community life, whether at home, at work or in our various activities.

This is the only reference to LGBTQ2S populations and sexual orientation. Pointe-Claire's extensive strategic plan prioritizes "Supportive, caring and welcoming" as strategic actions, but its inclusions focus on "attracting and retaining families" through the creation of affordable housing options and optimizing cultural infrastructure, parks, and play areas (City of Pointe-Claire, 2023, 28). Its civic intention to maintain "safe, compassionate living environments" neglects ethno-cultural diversity, social inequality, and the heterocisnormativity of queerburbia.

The only references to the region's growing socio-economic and ethno-racialized inequalities and heterocisnorms come from the para-public sector. The social development plan of the Table de Quartier Sud de l'Ouest-de-l'Île (TQSOI), a non-profit organisation to improve citizens' quality of life, for example, highlights housing precarity, poverty reduction, and food insecurity, providing a completely different portrait of social inclusion issues in these relatively affluent suburbs (Table de quartier sud de l'Ouest-de-l'Île, 2023). The English-language Lester B. Pearson School Board and its Task Force on Equality and Inclusivity developed an anti-discrimination action plan that considered multiple forms of racism (anti-Indigenous, Black, immigrant, and refugees) and discrimination based on gender, sexuality, and ability (Task Force on Equity, Diversity and Inclusion, 2021). Having investigated multiple dimensions of intersectional student exclusions, the task force recommended practices and policies to combat ethno-racialized, gender normative-based, and sexuality-based bullying. Yet none of these issues have ever been addressed by social inclusion policies in the West Island peripheral municipalities where these youth live.

Ville de Laval: Queering human nature for trans youth

In the 2010s, Laval initiated plans for a sustainable future with *ÉvoluCité* (Ville de Laval, 2011b), prioritizing public domain investment, including planning collaboration with citizens and community partners. Following citizen consultation and with the slogan *Urbaine de nature* (Ville de Laval, 2015a, 2015b), it cast its strategic planning vision to mobility, heritage, and ecological management. With the monikers "One city with 14 flavours" signalling the diversity of its neighbourhoods and "nature is ever present" conveying the interplay between its natural, agricultural, and built landscape (Ville de Laval 2015a, 13, 14), diversity is prioritized over social inclusion but both are trumped by environmental concerns. Social inclusion is mentioned twice, in relation to mobility enhancement for an ageing population and access to sports, leisure, and cultural activities for vulnerable populations, specifically new immigrants. Community life is framed as constrained by obsolescence, renovation deficits, and "underinvestment in community infrastructure" (e.g., arenas, indoor pools, community centres, libraries, and cultural institutions) (Ville de Laval, 2015b, 39). Laval's main challenge is identified as building a "dynamic and inclusive community", meaning one that is safe and attractive, promotes family well-being, harmoniously integrates immigrants, mitigates risks of poverty, and retains seniors and youth (Ville de Laval, 2015a, 70).

In *Urbaine de nature*, social inclusion is associated with one of six interpretations of nature; "human nature" refers to the provision of neighbourhood infrastructure to integrate families and other social groups (Ville de Laval, 2015b, 19, 20). Following the provincial government's abolition of Conférences régionales des élus (CRÉ) and Centres locaux de développement (CLD) in 2015, civic responsibility for social development and health was assumed by the Service du développement économique and the Centres intégré de santé et des services sociaux (CISSS). Social inclusion became the responsibility of Laval's Social Development Division of the Cultural, Leisure, Sport and Social Development Department. Their first report (Ville de Laval, 2016b) on solidarity and social inclusion does not mention sexual orientation, homophobia, gender diversity, or LGBTQ2S populations. Its target groups are immigrants, women, youth, seniors, families, and the homeless and the main themes are housing, food security, and the social economy. This department, however, does circulate the term "sexual orientation" in public-facing documents in reference to non-discrimination and intersectional Gender-Based Plus (GBA+) Analysis. Its funding guides and reports occasionally refer to LGBT people, especially youth, and their representative community groups. Funding programmes targeting disadvantaged youth mention the development of "inclusion services for LGBT people, specifically trans people" (Ville de Laval, 2020a, 13). Its "Youth Action Plan" stresses "respect for and appreciation of sexual and gender diversity and gender equality" and the encouragement of projects "aimed at raising awareness and valuing sexual and gender diversity" (Ville de Laval, 2021a, 25).

Amongst Laval's civic leaders, LGBTQ2S inclusion is championed by two city councillors, feminist and former president of the Centre des femmes de Laval, Cecilia Macedo, and Alexandre Warnet elected in 2021. Warnet (23 June 2022) argues that his election was an important milestone for LGBTQ2S inclusion in Laval: "I identify as homosexual, the only one on the council. This, in itself, represents a glass ceiling that I have personally broken". In 2024, following an IDAHOBIT proclamation, the mayor designated Councillor Warnet as the elected official responsible for LGBT issues: "It is important to have a city, to have a living environment that is free from prejudice, where everyone can thrive … We have never had an elected official responsible for this dossier in Laval" (Ville de Laval, 2024). Queerburban Laval is manifest in Figure 3.1 where civic and community leadership poses in city hall chambers to celebrate this important coalitional moment.

Ville de Longueuil: LGBTQ2S inclusions in a place of multiple exclusions

South of the Island of Montreal, the Ville de Longueuil's (2020) strategic plan uses the slogan "essential, innovative and inclusive". Citizen services prioritize: participation; sustainable economic revitalization; fair taxation; and strong leadership. The first alludes to the efficient administration of citizen requests, accelerated digitalization, and increased engagement. It references the generic "citizen", with no analysis of exclusions nor mention of any specific target populations such as seniors, youth, or LGBTQ2S communities. The city's official plan makes no references to social inclusion, defining diversity through the built environment and modification of municipal facilities. The excuse of ageing infrastructure and concession of single-room family washrooms and change rooms in city facilities justify indifference to the Coalition of Inclusive Municipalities (2019) *LGBTQ2+ Inclusiveness* toolkit recommendations.

Municipal concerns with social cohesion and inclusion operate through upper-tier governance, specifically the regional Agglomération de Longueuil

Figure 3.1 Councillor Warnet posing with Councillor Macedo, the current president of the Centre des femmes de Laval and the Director General of Sphère, 2024

(*Source:* Martin Alarie).

where inclusion is facilitated through the Social Development Office in conjunction with the Social Development Commission and specific citizen advisory committees (mainly intercultural, seniors, and youth) (Head of Social Development, 5 July 2022). Action plans referring to social inclusion for seniors, families and children, immigrants, homeless, disabled people, precarious households, and at-risk youth date back to 2013 shortly after the Social Development Office was created and it joined CCMARD and MADA. Longueuil prioritizes anti-racism, social cohesion, and newcomer integration as specified in its new *Action plan to address racism and discrimination and promote inclusion 2024–2026* (Ville de Longueuil, 2024). The plan interprets Longueuil as embedded in a global "interconnected ecosystem" where state secularism, the COVID-19 pandemic, increased xenophobia and hate speech, and even cyber racism are considered drivers of intolerance.

Longueuil strives to be a city that respects the fundamental rights of all citizens; it makes no reference to LGBTQ+ acronyms and sexual orientation. Sexual and gender diversity only appear in definitions of diversity and intersectionality provided by local organizations. It is in the *Corporation de développement communautaire de l'agglomération de Longueuil* (2021), composed of cross-sectoral social justice-oriented community organizations, where variations on the acronym LGBTQ+ and the term rainbow (rather than sexual orientation and gender and sexual diversity) describe partnerships, training sessions, and funding programmes. Struggles to integrate LGBTQ+ inclusion, however, remain due to a prioritization of ethno-cultural diversity and immigration. A regional community organizer (28 November 2022) explains: "We all have a rhetoric of inclusion, but the practical actions, understanding that someone has a different orientation or a different way of understanding their identity, we're not there … We continue to operate according to dated parameters".

From 2021 onwards, a new mayor and City Council implemented governance initiatives to enhance LGBTQ2S inclusion. Planning and social development staff view this political change as pivotal to inserting "inclusion at the heart of the electoral platform" (Head of the Office of Social Development, 5 July 2022). New files were created where there were "blind spots", including on the status of women, public safety, and intercultural community relations (Laramée, 2022). In 2022, Councillor Rolande Balma was named as the official responsible for LGBTQ+ community relations following the consensus of the Executive Committee: "in an effort to represent our society properly, we believe that the creation of a file dedicated to LGBTQ+ relations is becoming a must" (Laramée, 2022). The councillor is mandated to meet with organizations addressing LGBTQ+ realities, initiate culture, recreation, and social development projects, and possibly obtain a mandate for an LGBTQ2S-specific commission (Dostie, 2022). In 2023, city councillors supported the first raising of the Pride flag for IDAHOBIT and initiated the process of creating a local action plan on sexual and gender diversity inclusion. Responsible for creating this LGBTQ2S "portrait", the Community Development Advisor (25 January 2023) details the ongoing challenges of undertaking a GBA+ analysis in the absence of adequate demographic data: when there are "no statistics it is difficult to put in place an action plan" because "we end up forgetting a group of the community". In Longueuil, there remains no civic LGBTQ2S inclusion strategy.

In the Montreal city-region, LGBTQ2S inclusions rarely enter civic discourse unless through para-public community organizations, the appointment of elected officials, and youth programming. In the independent municipalities of the West Island, LGBTQ2S inclusion and highly politicized social differences like racism and economic inequality are never addressed outside school board task forces and community group reports. In Laval, LGBTQ2S inclusion is equated with sexual and gender diverse youth whose community groups require funding but are never accounted for in civic action plans. In Longueuil, where tensions around secularism, language, immigration, and poverty intersect, LGBTQ2S inclusion is sidelined by administrative procedures that require empirical evidence to forward an action plan. In sum, it is a few community organizations and elected officials who are the primary shapers of local political opportunity structures that can enable LGBTQ2S inclusion.

Articulating LGBTQ2S inclusion in Vancouver's peripheral municipalities

The plans, policies, and strategies of peripheral municipalities are political opportunity structures that afford civic leaders procedural and interpretive authority in place. As administrative tools of the state, they produce unique social inclusion policy landscapes with their own "codebooks" of shared meaning that assert a "political language" for stakeholder dialogue that "*is* in important ways political reality itself" (Fischer, 2003, 56). As discussion of the Toronto and Montreal city-regions has illustrated, language codes for representing sexual and gender minorities within social inclusion governance frameworks are limited. Instead, suburban social inclusion remains largely a heterocisnormative concept that materializes the life cycles of reproductive futurity. Such continued sidelining of intersectional understandings of inclusion forecloses political opportunity structures and contributes to the de-intensification of LGBTQ2S capacities to generatively remake place.

To illustrate how queerburban linguistic foreclosure operates, this section draws upon an LGBTQ2S social inclusion database constructed for Burnaby, New Westminster, and Surrey in the Vancouver city-region. A place-specific and temporally sequenced database of records (council and committee minutes, departmental reports, and plans, policies, and strategies) was derived from keyword searches of civic websites (prior to May 2020) for public-facing communication references to LGBTQ2S people and topics. Linguistic discourse and content analysis distilled place-specific language of LGBTQ2S positioning within social inclusion frameworks (Bain & Podmore, 2021a). Inductive qualitative document clustering produced three analytical categories: *proclamations* (ceremonial acknowledgements of LGBTQ2S social inclusion), *policy inclusions* (naming of LGBTQ2S concerns in civic policies, plans, and reports), and *procedural mentions* (passing references to LGBTQ2S identities in official documents). In decreasing order of social inclusion potential, "policy inclusions" are the most overt and integrative; "proclamations" are the most symbolic; and "procedural mentions" are the most ephemeral. When temporally ordered, the proclamations, policy inclusions, and procedural mentions show the unique timing, sequencing, pace, and duration of LGBTQ2S-supportive policy development (Bain & Podmore, 2022; Podmore & Bain, 2023). Although most Canadian LGBTQ2S equalities legislations precede this timeline, the 2016 federal law to prohibit discrimination based on gender identity and expression impacted municipalities due to implications for educational systems, employment policy, and infrastructure (Figure 3.2).

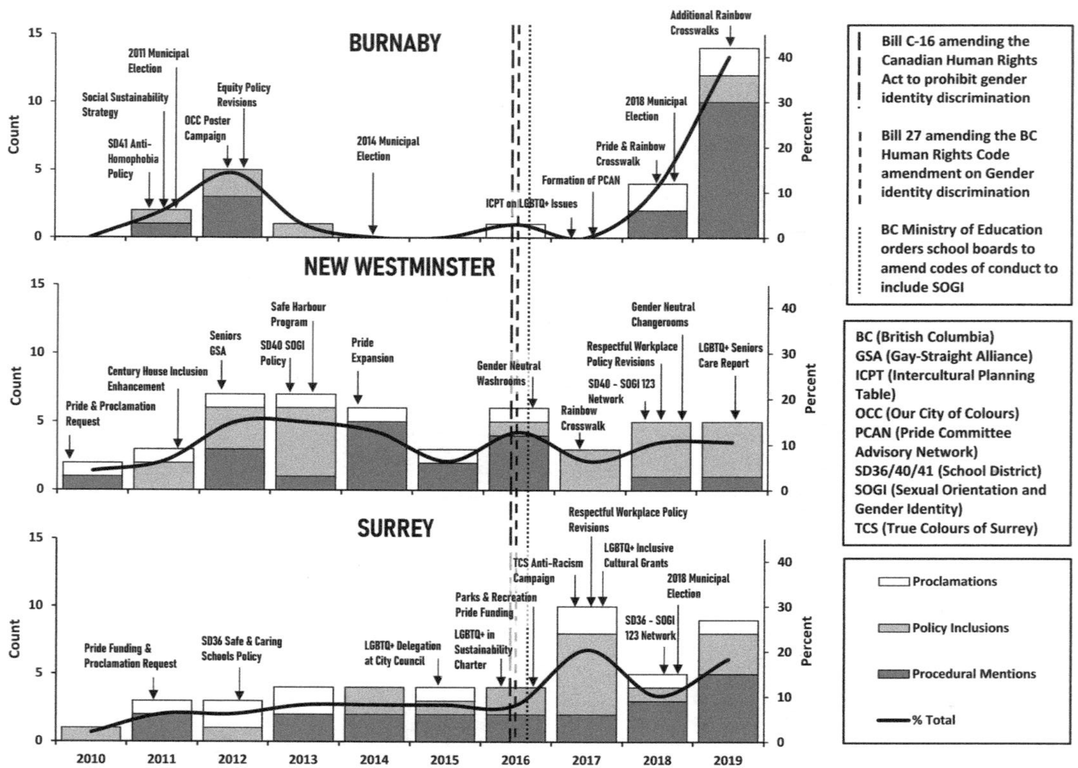

Figure 3.2 Timelines of LGBTQ2S-supportive municipal policies in Burnaby, New Westminster, and Surrey, 2010–2019 (*Source:* Alison Bain and Julie Podmore).

City council minutes articulate governance by conveying who may speak and about what, who references the LGBTQ2S community, and the extent of civic action. LGBTQ2S issues come before these suburban municipal councils as requests for proclamations, funding, and festivalization. In Burnaby and Surrey, proclamations account for 31 and 56 percent, respectively, of LGBTQ2S council content while New Westminster's annual Pride proclamation has been independent of the council since its inauguration in 2010. LGBTQ2S topics are found in proclamations for International Day of Pink (Burnaby), IDAHOBIT (Burnaby), and HIV-AIDS Awareness Week (Surrey). The language of these proclamations is formulaic (see Chapter 4), yet place-specific – an annual regurgitation of the political positioning of the municipality that also juggles other interest groups and awareness events (e.g., domestic violence, homelessness, bullying, National Seniors' Day, and International Women's Day). Funding requests mostly support annual Pride festivities, but they also allow activists to frame their organization's work and articulate needs. In such mundane instances, the language of social inclusion shifts from community needs to municipal festivalization, a move most apparent in New Westminster where 73 percent of the Council's records are Pride-related. However, even in Surrey where there is greater thematic variety, Pride still constitutes 52 percent of all LGBTQ2S council minute content.

Municipal committees and departments push councils for social change. Reference to LGBTQ2S concerns within social issue-related committees depends upon the reception of advice and delegations, but here there is also potential for social issue siloing. In New Westminster's Seniors' and Youth Advisory Committees and Surrey's Multicultural Committee, for example, LGBTQ2S populations are referenced in passing remarks, interjections, or points of information usually when overlapping with other social groups. Departments, however, have a more evidence-based role as they research requests, document needs, verify suitability, and raise awareness of social concerns. In all three peripheral municipalities, it was the planning, development, engineering, or culture and recreation departments with the greatest number of references to LGBTQ2S populations. Finance, human resources, and corporate services also fund community groups and initiate LGBTQ2S-competency training.

At the macro-scale of official community plans, LGBTQ2S subjects are generally listed amongst marginalized populations. This was the case in New Westminster and Surrey, but not in Burnaby which defines its service clientele through "family citizenship". New Westminster's (2017) *Age-friendly strategy* is exceptional as it profiles a retired gay male couple's relocation to the municipality. LGBTQ2S constituents are more likely to be found in the plans and strategies of para-public partners because of their more robust equity, diversity, and inclusion training. In Surrey, for example, para-public plans regarding policing, safety, homelessness, refugees, or Indigenous populations all mentioned LGBTQ2S subjects. In its municipal plans, however, references are as infrequent as in council minutes, reinforcing queerburban marginality.

Across the three peripheral municipalities, the social groups prioritized in social inclusion plans were those people experiencing homelessness (a national crisis) and life course groups (children, youth, and seniors). Of the 7971 references to social groups (1995–2020), only 7 percent refer to LGBTQ2S identities, constituting 1 percent less than seniors. Figure 3.3 illustrates the municipal variability of LGBTQ2S references with Burnaby at 6 percent, New Westminster at 38 percent, and Surrey at 4 percent. In New Westminster, LGBTQ2S people are both the largest social category and the target subjects of social inclusion. In Burnaby and Surrey, however, sexual and gender minorities are infrequently mentioned and thus incidental to social inclusion.

Social inclusion-related themes rarely co-occurred with LGBTQ2S content[4] indicating disconnections between the two. Instead, as an analysis of the top 20 social inclusion words revealed, the most common LGBTQ2S theme is "Pride"[5] suggesting a link to diversity festivalization and aestheticization. In Burnaby, "Pride" and iterations of "rainbow crosswalk" have higher co-occurrence than with the LGBTQ2S acronym itself. New Westminster was the only municipality with over four LGBTQ2S keywords in the top 20 and where the acronym ranked above "Pride". In Surrey, there are frequent references to Pride, but LGBTQ2S populations are not in the top 20 keywords because other "at-risk" social groups (e.g., unhoused people, Indigenous people, children and youth, immigrants, and refugees) predominate. In all three municipalities, LGBTQ2S social inclusion discourses prioritize Pride and festivalization over LGBTQ2S identities, gender diversity, and sexual and gender discrimination. However, important differences in their vernacular vocabularies of LGBTQ2S social inclusion stemming from municipal diversity place-brands are revealed in the following linguistic portraits.

City of Burnaby: "Socially sustainable" yet discordant

Reconciling a working-class history with contemporary ethno-cultural diversity, Burnaby civic leaders have turned their award-winning 2011 *Social sustainability strategy* into a place-brand (Bain & Podmore, 2021a, 2025). Social sustainability involves "individuals and the community working together to meet their needs, realize their potential, and prosper in a healthy environment" (City of Burnaby, 2011, 2). It is also realized through social policy adoption (e.g., Burnaby Multiculturalism Policy, Civic Youth Strategy, and Healthy Community Policy). Burnaby's vocabulary of social sustainability accentuates a "sense of belonging" and "social cohesion", the elimination of "barriers", "social bias", and "discrimination", and "linguistic", "ethno-cultural", and "faith groups". To "prepare the city for the future", civic leaders celebrate Burnaby's "increasing diversity" while "maximiz[ing] the potential of all" to "build a stronger community" that is based on "inclusion", liveability, and "resilience" (City of Burnaby, 2011, 38).

While Burnaby appears committed to social inclusion and the harmonious celebration of intercultural diversity, it is also a discordant city. As Chapter 4

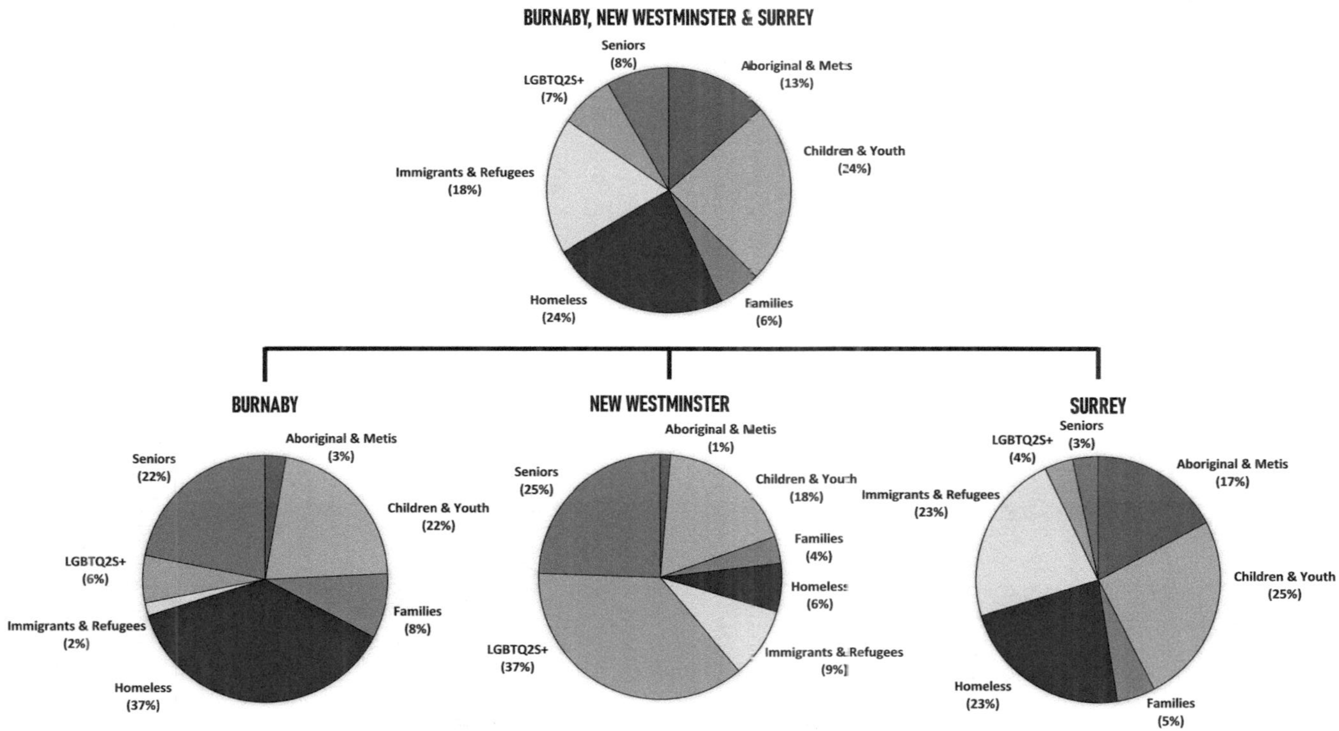

Figure 3.3 Percentage reference to social inclusion target populations in Burnaby, New Westminster, and Surrey, 1995–2020 (*Source:* Alison Bain and Julie Podmore).

elaborates, the place intensity of the municipality's social landscape includes tensions wrought by its juxtapositions of labour politics, religious lobby groups, progressive school boards, and redevelopment pressures. These contentious dynamics have slowed LGBTQ2S inclusion and limited it to youth inclusion and rainbowization. Its LGBTQ2S vocabulary has the term "questioning" reflecting a prioritization of LGBTQ2S youth. An occasional critical stance on heterosexism is reflected with the keywords "homophobia", "biphobia", and "transphobia", but these appear only after a 2018 municipal regime change. City hall's extended silence on LGBTQ2S inclusion delayed more systemic institutional changes and built a dependency upon para-public organizations to mobilize on behalf of sexual and gender minorities. Distinctive to Burnaby's LGBTQ2S inclusion is the aestheticized "visibility" of the "rainbow crosswalk", infrastructure that was extensively, rapidly, and reparatively installed in 2018–2019 (see Chapter 4). Rainbow crosswalks now visibly embed Burnaby's LGBTQ2S-friendly status in its suburban landscape but constituency discord over more substantive LGBTQ2S social inclusion long-hampered policy development.

City of New Westminster: "Compassionate" yet festivalized

Like Burnaby, New Westminster is a former working-class city but also a regional leader in compassionate social inclusion. Its community plan, *Our city, 2041* (City of New Westminster, 2017, 27), envisions a "healthy, inclusive and thriving community where people feel connected with each other". Prioritizing social and physical connection for a "socially minded" citizenry, civic care is conveyed through the "compassionate city" framing in the 2019 New West Hospice Society partnership proposal recommended to the Community and Social Issues Committee by the Social Planner (City of New Westminster, 2019). Compassionate communities "publicly encourage, facilitate, support and celebrate care for one another during life's most testing moments and experiences" (City of New Westminster, 2019, 42). Civic leaders empathize with populations seldom publicly validated (e.g., people living with intellectual and developmental disabilities, chronic and terminal illness, in recovery, dealing with renoviction, and low-income seniors). Thus, New Westminster's language of social inclusion includes keywords like "social capital", "equity", and "empowerment" to convey civic concern with social exclusions. "Acceptance", "awareness", and "tolerance" mingle with unique references to "exclusion", "stigma", and being "disassociated from society". This reckoning with socio-economic minorities can be seen in New Westminster's vision as a "safe city", a place of "shelter", "liveability", "meaning", and "well-being" for all its residents.

LGBTQ2S inclusion is interwoven within a complex and contradictory vocabulary sometimes adopting a critical stance on heteronormativity and referring to distinct local activities: a Seniors' Gay-Straight Alliance; the Sexual Orientation and Gender Diversity (SOGI) school board policies; and

LGBTQ2S "vigils". Other distinctive terms describe gender diversity (e.g., "gender expression", "gender identit(ies)", "gender reassignment", "gender-free", "in transition", "trans-inclusive", and "trans-competent") and prioritize civic "queer competency". Gender diversity also manifests in reference to "universal access" to restrooms and changing rooms in New Westminster's newly built and renovated recreational facilities. However, compassionate attention to diversity in this peripheral municipality is undermined by economic redevelopment strategies that capitalize on LGBTQ2S "difference" through white-Anglo "Royal City" heritage festivalization. New Westminster's queerburbanites are associated with the commercial revitalization of Columbia Street through Business Improvement Association programming alignments with Pride's rainbow iconography. Its annual family-friendly Pride festivities selectively employ the rainbow to rework the colonial and working-class heritage of its streetscape.

City of Surrey: "Inclusive" yet preferential

Across the Fraser River, Surrey uses its *Sustainability charter 2.0* (City of Surrey, 2016c) to brand itself as "the inclusive city". Its social sustainability brand aggregates complex social exclusions (e.g., homelessness, youth disenfranchisement, addictions, and poverty) to the neglect of LGBTQ2S inclusions. "Other concerns" supersede LGBTQ2S inclusion, because the local political opportunity structure relies on problem/deficit frames to balance solutions and garner electoral support from ethno-religious constituencies. Although significantly larger in size than either Burnaby or New Westminster with numerically more, and seemingly comprehensive, social inclusion policies, Surrey's civic leaders deflected explicit LGBTQ2S recognition until 2024 when city hall hosted an inaugural Pride gala night.

The City of Surrey's (2016c, 16–17) *Sustainability charter 2.0* envisions "a thriving, green, inclusive city" that is "sustainable", "resilient", and "welcoming" to specific marginalized social groups (e.g., immigrants, newcomers, youth, and Indigenous people). "Youth engagement" and "youth friendliness" are central while there are explicit commitments to poverty and homelessness reduction for "at-risk" and "vulnerable" populations. Aspirations to eradicate exclusion inform the multifaceted interpretation of Surrey's social inclusion landscape described using terms like "social infrastructure", "social innovation", "social equity", and "social responsibility". Surrey's action verbs are distinct: "outreach" and "advocacy" augment "awareness", increasing the "capacity", "connectedness", and "participation" of marginalized residents. For Indigenous peoples, "restorative justice" and "reconciliation" are concepts unique to Surrey's public vocabulary, as is a more general articulation of "support" for disadvantaged communities. Its civic language also leverages "human rights" and "multiculturalism" to build "cultural awareness" and "cultural competency".

While Surrey strives to be "culturally inclusive" and "culturally appropriate", cross-cultural awareness does not extend to LGBTQ2S inclusions. It prioritizes "safe space(s)" or "safe environment(s)" and is concerned with violence (particularly gang conflict and youth bullying) that accompanies "discrimination", "prejudice", and "stereotypes". Surrey's vernacular vocabulary uses critical terms about marginalization and exclusion (e.g., "neglect" and "segregation") and power relations (e.g., "anti-oppression" and "social justice"), but these are only applied to discrete social groups and seldom extend to LGBTQ2S people. Here, "Pride" is more commonplace than the LGBTQ2S acronym. Frequent mentions are made of Pink Shirt Day (to eradicate bullying), identifying LGBTQ2S youth among its victimized groups. Surrey's place-specific vernacular includes "allies" and "anti-homophobia" and the singular reference to the Rainbow Refugee Committee, a community group advocating for "people seeking refugee protection because of persecution based on sexual orientation, gender identity, or HIV status" (City of Surrey, 2014c, 22). Larger than either Burnaby or New Westminster, Surrey has more social issues (e.g., homelessness, youth disenfranchisement, addictions, poverty, and ethno-cultural segmentation); more plans, policies, and social inclusion initiatives; and a more sophisticated understanding of social justice. Despite its history of local queerburban activism, civic leaders tend to "obscure" LGBTQ2S social inclusions in official documentation because it is an "electoral liability" (Cooper & Monro, 2003).

In the vernacular social inclusion vocabularies of Burnaby, New Westminster, and Surrey lie distinctive queerburban place-brands and political opportunity structures. By universalizing social sustainability, Burnaby's LGBTQ2S inclusions are limited to young people, prioritize rainbow-washing, and over-rely upon para-public organizations to do the work of creating a language of inclusion. New Westminster deploys compassionate inclusion as civic queer competency to support a rich vocabulary of gender diversity within its neoliberal place-rebranding. Surrey's more social justice inflected language of inclusion permits the articulation of complex social problems that vie alongside LGBTQ2S activism for civic attention. These are all detailed linguistic examples of "place intensification" (Seamon, 2018, 138). They demonstrate how municipal social inclusion governance frameworks create discursive political opportunity structures that can enhance or squelch queerburbia through their dynamics.

Conclusion

Across the peripheries of Canada's three largest city-regions, LGBTQ2S inclusion is variably enacted in response to place-specific political opportunity structures. As emplaced technologies of power, municipal social inclusion frameworks should enable the emergence of queerburbia, but as these dramatic discursive silences and linguistic divergences suggest, the forwarding of LGBTQ2S inclusion in peripheral municipalities is sporadic, limited, and context dependent. The social inclusion frameworks that are in place heterocisnormatively guide civic priorities creating political opportunity structures that mostly foreclose possibilities for sexual and gender minorities. In rapidly diversifying peripheral landscapes, social inclusion tensions and juxtapositions are discursively embedded in municipal plans, policies, and strategies.

Municipal political opportunity structures can be informed by best-practice toolkits such as CIM's (2019) *LGBTQ2+ inclusiveness* and the Conseil québécois LGBT's (2024) *Guide pratique pour des municipalités*. By bringing together "best practices" from Canadian municipalities large and small, urban and suburban, these toolkits provide examples of inclusive possibilities. As "professional visions" (Goodwin, 1994, 606), they address civic leaders by providing them with "forms of knowledge" and "visions for the way the world should be" while simultaneously flattening nuance and suggesting that "the tools to solve complex problems lie within the confines of the kit" (Wong et al., 2023, 145–3). Their "quick wins" set expectations that have yet to be realized in queerburbia (e.g., safe washrooms and change rooms, gender-inclusive language audits, or LGBTQ+-specific civic commitment statements). While these toolkits demonstrate that LGBTQ2S inclusion is on municipal political agendas, the more detailed discussions of suburban municipalities in the Toronto, Montreal, and Vancouver city-regions highlight scalar incommensurabilities and metropolitan unevenness.

In the Toronto and Montreal city-regions, LGBTQ2S inclusion in peripheral municipalities is lightly inflected in administrative regimes that prioritize anti-racism, newcomer integration, and youth support, but rarely substantially manifest as political opportunity structures. Given the centrality of the family to the definition of suburban constituencies, LGBTQ2S inclusions are usually considered in youth programmes, plans, and strategies and materialized in related institutions such as youth centres, libraries, and schools. Awareness of the needs of LGBTQ2S youth (and sometimes seniors) is articulated by community organizations and para-public agencies who witness their intersectional exclusions. Few have even begun to aspire to the best practices outlined in existing toolkits. Across the Vancouver and Toronto city-regions, there are fragmented examples of the gender-neutral redesign of municipal facilities, supportive employer policies, support for community organizations, and the recognition of LGBTQ2S events, but access to services and political power remains limited. In Montreal, the delegation of duties to elected officials and city council backing of advancements on LGBTQ2S inclusions have the potential to bring change, but these too are as yet unrealized.

These city-regions provide examples of multi-tier governance whereby regional bodies encourage LGBTQ2S inclusion and municipalities respond by developing their own interpretations that are embedded in pre-existing symbolic place-brands and vernacular planning and policy vocabularies. The queerburban vernacular vocabularies distilled in the Vancouver city-region convey a degree of civic ambivalence towards LGBTQ2S social inclusion amidst suburban diversity where anti-racism, immigration status, and youthfulness take precedence. As the examples from this city-region demonstrate, integration into inclusion frameworks depends on synergistic alignments with civic priorities for festivalization, safety, welcoming newcomers, integrating seniors, and anti-discrimination initiatives. While political opportunity structures shape place, they have not necessarily made any one peripheral municipality any more LGBTQ2S-inclusive than another.

Notes

1 Rainbow Salad includes the Peel HIV/AIDS Network, Associated Youth Services of Peel, Caledon Community Services, Volunteer MBC, East Mississauga Community Health Centre, WellFort Community Health Services, Family Services of Peel, and Punjabi Community Health Services.

2 It was formed in 2016 from municipal, school board, hospital, police, and environmental organizational partners to develop the charter through community consultation to respond positively to rapidly changing demographics.

3 The acronym LGBTQIA2+ includes intersex and asexual alongside the more commonly listed identities.

4 The top 20 keywords were distilled from database textual segments referring to LGBTQ2S populations and themes. Linguistic discourse analysis identified, extracted, and interpreted references to LGBTQ2S populations in co-occurring relation to social inclusion language. The content analysis identified, selected, and quantified keywords (n = 20,384) that were then condensed by grouping words with similar roots and meanings and isolating LGBTQ2S topics, events, and identities. This process generated 20 clusters of social inclusion terms.

5 The top 20 key LGBTQ2S words for all three municipalities in descending order are: communities; children/youth; homeless/ness; Indigenous; safe(ty); event(s)/festival(s); immigrant(s); diversity; senior(s); inclusion; refugee(s); Pride; famil(ies); access(ibility); support; at-risk; LGBTQ+; sustainability; vulnerability; and discrimination.

4 Serving

Introduction

Municipal place-making is an instrumental, universal, and material process of "place realization" (Seamon, 2018, 91). LGBTQ2S-friendly municipal place-making involves leveraging municipal infrastructure, resources, and community services to "progressively actualize" selective local possibilities for sexual and gender minority inclusion in place (Seamon, 2018, 91). LGBTQ2S-friendly municipalities function to varying degrees as upperground civic institutions whose actions and practices of LGBTQ2S allyship shape access to political opportunity structures and resource landscapes for sexual and gender minorities (Seamon, 2018, 134). As discussed in Chapter 3, political opportunity structures are the external political contexts that can encourage or discourage collective action and shape the resources made available to citizens (Tarrow, 1998). Municipal resource landscapes are the economic, material, emotional, and symbolic collective goods that are available to support municipal constituencies (Nicholls, 2009). In the service of social inclusion, these resource landscapes condition which stakeholders and local assets can be mobilized to meet the needs of marginalized groups.

This chapter examines the role of peripheral municipalities as institutional allies in LGBTQ2S suburban place-making with particular attention to the political opportunity structures and resource landscapes they generate to serve sexual and gender minorities. Allyship involves reflexively acknowledging and questioning institutional structures of power and privilege and confronting institutional biases and hierarchies (Russell & Bohan, 2016). Allies are generally members of advantaged groups who espouse egalitarian ideals and advocate for marginalized populations, working to dismantle the systems of oppression that afford themselves greater privilege and power (Radke et al., 2020). Where individual allyship can be interpreted as performances of friendship that are crosscut by currents of concern, kindness, care, and understanding (Sumerau et al., 2021), institutional allyship "focuses on the role of institutions, specifically organisational entities, in advancing the interests of the marginalised" (Dobele et al., 2022, 188). It concerns settings where individuals meet institutional dynamics to advance first- and second-order collective social

DOI: 10.4324/9781032632544-4

change by integrating equity outcomes into all facets of political decision-making including strategic agendas, yet without reinforcing the power structures that are targeted for dismantling. First-order changes are superficial "performances" that pay lip service to equity and focus more on words than deeds while second-order changes are deeper and more responsive, involving strategic institutional actions to foreground the voices of disadvantaged groups (Pettinicchio, 2012).

This chapter analyzes first-order civic place-making practices across the case study peripheral municipalities, appreciating its limits and the necessity of partnership with para-public institutions to realize more substantive and critical second-order social inclusion changes (see Chapter 5). It argues that when suburban municipalities seek to signal service of LGBTQ2S constituencies and their inclusion, their institutional allyship revolves primarily around rainbowization as place-making, a limited form of first-order "rainbow-washing" (Misgav, 2015). Such civic actions of rainbowization inculcate standardization and "automatic environmental ordering" (Seamon, 2018, 135), enabling peripheral municipalities to perform LGBTQ2S progressiveness through aestheticized and routinized practices without reconfiguring political power structures and redistributing resources. The chapter investigates the four most common civic first-order LGBTQ2S-supportive services found in peripheral municipalities: mayoral proclamations; rainbow and trans flag raisings; painting rainbow crosswalks[1]; and illuminating civic plazas. It concludes by considering why such suburban civic allyship is limited to first-order changes that generate minimal LGBTQ2S-friendly "place realization" (Seamon, 2018, 91).

First-order municipal LGBTQ2S place-making practices

Despite negative connotations associated with the bureaucratic logics of "institutions" and "institutionalization" that often rely upon administrative automation rather than social innovation (Tarrow, 1998), municipalities provide access to politico-procedural legitimation and civic materiality within the public realm, both elemental components of articulating LGBTQ2S social inclusion through place-making. In their symbolic enactment of first-order LGBTQ2S institutional allyship, peripheral municipal leaders have performatively leveraged the rainbow to signal equity within, and the progressiveness of, their respective jurisdictions. This section outlines first-order institutional allyship starting with repetitive and rote annual verbal proclamations and Pride flag raisings through to permanent installations of rainbow crosswalks at prominent intersections, and luminescent projections onto exterior façades of city halls and plazas. It asserts that while annual declarations, flag raisings, and illuminations are more ephemeral civic rituals, the permanence and publicity of rainbow crosswalks have a year-round impact through the embodied movements of pedestrians and drivers. In augmenting LGBTQ2S visibility and awareness, they elicit heteroactivist vandalism, compelling civic leaders to take a stance against prejudice and hate. Ultimately, however, first-order institutional allyship remains procedurally (council meetings) and structurally (flag poles, light fixtures, and street intersections) confined, leaving the sustained second-order changes required to unmake suburbia's heterocisnormativity to para-public institutional partners.

Mayoral LGBTQ2S proclamations

City halls spatialize the relationship between citizens and municipal authority. As "instance[s] of civic materialism" (Chattopadhyay & White, 2014, 5), they are contentious public spaces whence competing demands and regulatory regimes of surveillance condition access and representation. Nowhere is this more apparent than in the civic rituals of municipal proclamations and city hall flag raisings – two widely used civic place-making practices to signal LGBTQ2S inclusion. Mayors, the honorary figureheads of municipal regimes, officiate these place-making practices, leading ceremonial flag raisings and issuing proclamations on behalf of their councils to promote "good relations, particularly across ethnic and cultural cleavages" (Cooper, 2018, 121).

Within contemporary, neoliberal diversity regimes, Pride proclamations have become annual rituals used to signal official LGBTQ2S social inclusion and articulate municipal positions vis-à-vis nationalized LGBTQ2S equalities. The first municipal Pride proclamation and flag raising was in San Francisco in 1978 for the ninth anniversary of the 1969 Stonewall Riots in New York City when the insider-activist Harvey Milk raised a rainbow flag commissioned from Gilbert Baker with an impassioned speech to the crowd gathered for Gay Freedom Day (Tracey, 2022). Since then, these rainbow rituals have been adopted by civic leaders within and beyond the Anglosphere to declare local Pride events (e.g., Pride weeks, festivals, and parades) or pronounce municipal alignment with national or international dates of recognition (e.g., International Day against Homophobia, Biphobia, and Transphobia (IDAHOBIT) or International Day of Pink). Appreciated for their temporary conferral of special municipal status, LGBTQ2S community groups view these micro-symbolic practices as forms of localized recognition. For municipalities, however, proclamations are temporary yet expedient forms of inclusion that can be procedurally muted – lists of proclamations can be presented at the beginning of city council meetings to be adopted and minuted, often without being read aloud. In some municipalities, therefore, this ritual has become normalized because it provides the bare minimum and is relatively inconsequential compared with more substantive civic acts.

Pride proclamations in the case studies of the Vancouver city-region began in 2010 with a citizen request in New Westminster. The municipality welcomed the opportunity to expand its inclusion practices and adopted a distinctive date in the summer Pride calendar. During the "Open Delegations" portion of the council meeting on 25 January 2010, "Vance McFadyen, resident of New Westminster, requested that 7 August 2010 be declared New Westminster Pride Day to promote trust, understanding and a means of the gay community connecting for the greater good of all". The council proclaimed that 7 August would be "Gay Pride Day" (City of New Westminster, 2010, 3). New Westminster rapidly adopted the annual Pride proclamation ritual in conjunction with a flag raising ceremony during its Pride week. Mayoral proclamations have, therefore, remained external to the city council and never been passively listed in the minutes. Rather, the mayor and councillors attend the flag raising with community activists and federal and provincial politicians in a public park adjacent to city hall.

A year later, nearby Surrey made its first official Pride proclamation. While proclamations recognizing sexual minorities were previously made – including National AIDS Awareness Week as of 1995 – the first specific council minute reference to an LGBTQ2S proclamation was in 2011 to mark Surrey Pride Weekend, from 8 to 10 July. The mayor issued the following proclamation:

> WHEREAS the City of Surrey is proud to be the home of ethnically, culturally and racially diverse people who contribute to our strength and vitality; and WHEREAS the concept of an annual pride celebration in Surrey originated in 1999 and today there are hundreds of Pride celebrations, festivals, and Pride-related special events held in Canada throughout the year. This event is targeted to the thousands of gay, lesbian, bisexual, transgender individuals and families including the straight community living within Surrey and the Fraser Valley area; and WHEREAS held at Holland Park, this event will bring together people; providing an excellent opportunity for businesses, organizations and entertainers to reach out and proudly support diversity; and WHEREAS OIS Rainbow Cultural Society and its many tireless volunteers provide education, support and a safe environment for people of alternative lifestyles and encompass all lifestyles in a celebration of life; NOW, THEREFORE, BE IT RESOLVED that I, Dianne L. Watts, do hereby declare July 8–10, 2011 as "SURREY PRIDE WEEKEND" in the City of Surrey.
>
> (City of Surrey, 2011, 15)

Although the proclamation does not include sexuality and gender in its definition of municipal diversity and makes specific reference to "the straight community", it does cite the municipality's long history of LGBTQ2S activism and thus plays a politically educative role. Despite a series of rejected proclamation requests, when finally accepted, it was local activists who wrote the text, shaping the language of civic recognition. In the face of overt homonegativity amongst Surrey constituencies, the mayoral proclamation attempts to neutralize its political message by stressing the civic entertainment and commercial value of the Pride event. It presents sexual and gender minorities to Surrey residents as a social group with an "alternative lifestyle" (City of Surrey, 2011, 15). Despite a mayoral change, the same text was regurgitated in subsequent years. These proclamations were minuted until 2017 when they become ordinary, referenced in a list of community proclamations (e.g., National Blood

Donor Week, Amateur Radio Week, World Refugee Day, and National Aboriginal Day) (City of Surrey, 2017, 11). Due to the competing political interests of its large religious and ethno-cultural constituencies, Surrey has been slow to recognize any "special interest group" proclamations.

Burnaby was much slower to adopt the practice of Pride proclamations. Its first LGBTQ2S proclamation came in 2016 when it marked the International Day of Pink, a campaign to take a "stand against bullying, discrimination, homophobia and transphobia" by recognizing "the dignity and worth of all". Like many suburban policies that support heterocisnormative family social reproduction, its first foray into LGBTQ2S proclamations centred on youth and the more universal concern with bullying. More overt support for LGBTQ2S communities came in 2018 with a proclamation marking IDAHOBIT:

> WHEREAS Burnaby is a community that strives to be a welcoming, diverse, inclusive and safe community for all its citizens; and WHEREAS intolerance of members of the LBGTQ2S+ communities is still a common aspect of society; and WHEREAS Burnaby is a community where the contributions of all residents are acknowledged and appreciated; and WHEREAS The United Nations annually recognizes May 17 as a day to call for equal rights for all people regardless of sexual orientation or gender expression; now, therefore, I, Derek R. Corrigan, Mayor of Burnaby, do hereby proclaim 2018 May 17 as The International Day Against Homophobia and Transphobia.
>
> (City of Burnaby, 2018)

The "WHEREAS" clause is used throughout to justify recognition and legitimate support for a specific community group and to emplace the city within an international discourse of human rights. LGBTQ2S inclusion is also neutralized by articulating that Burnaby is a "welcoming, diverse, inclusive and safe" city for all citizens rather than one that panders to "special interests". A similar place-making discourse was used one month later when Burnaby's Mayor issued the City's first Pride proclamation that began by mentioning the "equality and freedom" of all citizens before listing multiple axes of social difference neglecting gender identity and expression and leaving sexual orientation for the penultimate reference alongside disability. This oversight was rectified following a 2019 mayoral regime change when the proclamation stated: "All people regardless of gender identity or sexual orientation have the right to be treated on the basis of their intrinsic value as human beings" (City of Burnaby, 2019c). Gender identity and sexual orientation are the central focus while intersex and asexual are added to the municipality's acronym, "LGBTQIA2+". In subsequent years, Burnaby's mayoral proclamations moved out of city hall's council chambers and minutes onto stages as a ritual performance to open its now annual Pride festival.

Across the country on the peripheries of Toronto, LGBTQ2S proclamations frequently accompany Pride flag raisings and have been organized within many regional municipalities to align with Pride festivities. LGBTQ2S community groups have lobbied political leaders via letters and more formal council recognition request forms. A key difference between Toronto and Vancouver is the role played by second-tier regional municipalities that create pressure on member municipalities and provide a degree of regional coherence. For example, every year since 2006 the York Region chapter of PFLAG on behalf of Stand Proud York Region extends invitations to member municipalities to join in regional recognition of Pride. In 2015, the York Regional Municipality's Pride Week was launched with five official municipal flag raising ceremonies that also included the first by a major police service in Canada (York Regional Police headquartered in Aurora). Such inter-organizational relations and parapublic institutional allyship are, as Chapter 5 details, central to LGBTQ2S place-making, but so too is intermunicipal competition within regions. For example, in 2016 the York Region's chapter of PFLAG Canada sent its annual letter to Markham requesting participation in regional place-making by raising the Pride flag and proclaiming 11 to 19 June as Pride Week, stating that the proclamation "would be a valued acknowledgement of the lives and contributions of people of all sexual orientations and gender identities living in York Region" (PFLAG Canada–York Region, 2016, 1). As regional municipal civic leaders are attentive to how their counterparts enact LGBTQ2S place-making, the letter also stressed that "For more than a decade, York Region's Pride Week has been recognized by the Regional Municipality of York and proclaimed by most of its municipalities" (PFLAG Canada–York Region, 2016, 1).

Pride proclamations are often the starting point of LGBTQ2S civic recognition and a key discursive component of official place-making practices. A momentary public declaration of LGBTQ2S support, they are largely a first-order form of symbolic recognition. Nevertheless, they offer reluctant municipalities acknowledgment opportunities in the face of political pressure from homotransnegative constituencies rendering even these minimal acts highly political. While municipalities may make hundreds of proclamations annually, only some are accompanied by flag raisings, which are anything but quotidian acts. Since city hall flag poles are governed by strict protocols, municipal flag raisings add another layer of ceremonial officialdom to proclamations with their visual place-making symbolism.

Pride flag raisings at city hall

Historically, flags publicly symbolize "imagined communities" (Anderson, 1983, 1) and materialize (often banal) forms of nationalism (Billig, 1995). In municipalities, flags territorialize civic identity, representing a city's history and aspirations, marking centres of institutional power, and officially making place. According to Canadian federal protocols, most contemporary city halls fly three flags (federal, provincial, and municipal) on separate poles, lowering them to half-mast to signal mourning. They are governed by the Order of Precedence in accordance with national etiquette necessitating consistent and hierarchically appropriate displays of flags from the national to the provincial to the city flag (Canadian Heritage, 2022). Due to protocol rigidity and the sanctity of the national flag, the order is seldom disrupted making flag protocol management a highly politicized component of municipal diversity recognition. Some municipalities have therefore opted for a dedicated community "diversity" pole where ethno-cultural, Indigenous, and Pride flags can be flown. The epitome of Canada's multicultural aspirations, diversity flag poles permit superficial signaling of minority recognition by shifting the boundaries between public and private expressions of difference. However, whose private expression of difference is publicly articulated, is politically contested, especially in suburban areas where homotransnegative publics have treated rainbow flags at city halls as a special interest infrastructural surplus that is not a public "good" but rather a usurper of government space.[2]

Given these protocol and political complexities, requests to fly Pride flags at city halls can be interpreted by municipalities as potentially disruptive of the multi-scalar governance order and its inculcation of heterocisnormativity. For activists, however, the rainbow flying over city hall is a powerful transnational symbol of sexual and gender diversity recognition in civic life. While the rainbow flag was originally used to signal an individual public declaration of "coming out" (Bitterman, 2021), it is now a global political symbol that is used to proclaim LGBTQ2S identification, combat homotransphobia, and designate specific geographical areas as safe(r) (Wolowic et al., 2017). Designed over 50 years ago by artist Gilbert Baker for San Francisco Supervisor Harvey Milk, by the 1990s the rainbow became an alternative to the politicized Nazi-era pink triangle that was reclaimed by queer activists (Victoria & Albert Museum, 2023). Each of its eight colours represents the politics of LGBTQ+ life-affirming hope: pink (sex); red (life); orange (healing); yellow (sunlight); green (nature); turquoise (art); royal blue (harmony); and purple (spirit) (Bitterman, 2021). The six-colour flag became the most common version after hot pink and turquoise were phased out for manufacturing and design reasons (Human Rights Campaign, 2023) until 2018 when non-binary artist Daniel Quasar added triangular lines to represent trans and non-binary (white, light blue, and pink) and QTBIPOC (brown and black) to create the "Progress Pride Flag" (Victoria & Albert Museum, 2023). Variations include the addition of a yellow triangle with a purple circle to represent intersex people which is sometimes overlaid with a Black Power fist or feathers representing Two-Spirit Indigenous identities. A federalizing force, iterations of the rainbow flag communicate complex webs of social norms, languages, and fictions that provide visible connection in place to transnational communities (Klapeer & Laskar, 2018). While seemingly static, these "multifocal and multivalent" boundary objects facilitate "interactions, translations, and coherence, but also stir conflicts over meaning across diverse social worlds" (Laskar et al., 2017, 194).

Within the Vancouver city-region, the rainbow flag has generated significant municipal conflict over LGBTQ2S recognition in Surrey. Its civic leaders have long denied local activist requests to fly the flag on city hall's official flagpoles. In a "rhetorical move" conditioning public expression of inclusion (Brodyn & Ghaziani, 2018), Surrey's mayor rejected the first activist request via email in 2005. An activist explained to the press, "the policy at city hall is only to have whatever the regular flags are … if they put ours up, then they'd have to do it for everybody" which they feared "would open a floodgate of requests to have flags flown over city hall" (Colley, 2005, 1). Nearly a decade later, a request to the Council by the 15th Annual Pride celebration delegation was again rejected on the grounds that "only Government flags are flown at City Hall" with the caveat that the City Clerk would investigate "what other regions and cities have done in terms of flag protocol" (City of Surrey, 2014b, 2). The report later recommended that the Council adopt Corporate Policy R026 which "ensure[s] that all flags at City Hall and other City operated municipal facilities are flown and displayed in a consistent manner" (City of Surrey, 2014a, 2) in accordance with the British Columbia Office of Protocol and the federal Department of Canadian Heritage. The City Clerk noted that as for the "recent request to fly a symbolic flag on one of the City's official City flagpoles", Surrey should not fly flags that "identify people belonging to a group" (City of Surrey, 2014a, 1) and be restricted to representing municipal, provincial, and national communities because they symbolize unity (and not division) "without distinction of race, language, belief or opinion" (City of Surrey, 2014a, 1). The report concluded that "Flying our Canadian, Provincial and City flags on the official City flag poles fully represent[s] embracing diversity and inclusiveness in our City" (City of Surrey, 2014a, 2).

After repeated requests from Pride organizers, Surrey would finally fly the rainbow flag at city hall for the first time in 2022. Until that point, only two non-state flags had ever been flown: first, in 2010 when Surrey was an Olympic Venue City and flew the Winter Olympic flag; second, in 2016 when the rainbow flag commemorated the lives lost in the Orlando, Florida gay nightclub mass shooting. Shockingly, it took extreme homotransphobic violence and an impassioned plea from Surrey's first openly lesbian city councillor, Vera LeFranc, for the City Council to consider flying the rainbow flag (from 13 June until the Surrey Pride Festival on 26 June 2016) and lowering their official flags to half-mast. The mayor described the flag-raising as a demonstration of "solidarity, sympathy and support for all those who lost their lives under these extraordinary and trying circumstances" (City of Surrey, 2016b, 1). This civic response did offer a degree of political absolution for a historic lack of municipal solidarity, but subsequent requests by LGBTQ2S activists were still denied, an inconsistent outcome according to journalists who ridiculed arguments that a fourth community flagpole was cost-prohibitive (Booth, 2014). When civic leaders finally conceded in 2022, a small flagpole was erected above the entrance to city hall where rainbow and trans flags now fly with those of other communities.

In Toronto's peripheral municipalities, it has also taken significant LGBTQ2S activism and labour to get the rainbow and trans flags flown. A Peel Pride co-organizer (3 July 2018) describes it as "political schmoozing and playing politics" – "you have to get to the people" to make a flag raising happen (Figure 4.1). The first flag raising in the Regional Municipality of Peel occurred in 2015 at a police station on Mississauga Road after activists engaged the Peel District School Board and agreed to buy the CDN$40 flag. Like Surrey, the City of Brampton follows strict flag protocols for its nearly 100 civic flag poles (located at transit buildings, fire stations, recreation centres, and public libraries). Where three flag poles are available, the National Flag of Canada, the Ontario Provincial Flag, and the City of Brampton flag fly, but if a facility has only one flagpole, then the national flag takes precedence. In order to collectively "celebrate … the strength of Brampton's diversity, equality and inclusivity, the City offers the opportunity for communities to raise nation[al] and community flags on the city hall designated Community Flag Pole, enhancing awareness of special occasions and activities" (City of Brampton, 2024b). There are detailed guidelines for community flag raisings, requiring that they be no more than an hour in length, include remarks from community organizers and the mayor or another city council representative, and possibly offer a brief cultural performance. The city provides a podium and microphone, chairs, and tables, but no flag. Community organizers must obtain a 3-foot flag with metal grommets and deliver it one week in advance to the City Clerk.

Figure 4.1 A community organization photo-opportunity following the Pride and Trans flag raising at Brampton's city hall, 2018

(*Source:* Alison Bain).

In 2018, the former Chair of Peel Pride purchased a trans flag online and delivered the flag along with protocol recommendations. This "intense hand holding" was what enabled the trans flag to be flown at Brampton's city hall (Community organizer, 20 June 2018). Each year, these flags fly only if the request is submitted on time and conforms to guidelines. In 2023, no such formal request was noted, but its online registry of community flag raisings documents a range of other approved ethno-cultural group sponsors and flag-raising events (e.g., Hindu Heritage Month in November, International Day of Elder Persons and Islamic Heritage Month in October, World Peace and Franco-Ontarian Days in September). The timing of flag raisings also matters. In Brampton, these are scheduled to align with council meetings because "more people are around, giving photo ops that can be posted on websites" and circulated in print and social media (Pride Peel co-organizer, 26 June 2018).

In neighbouring Mississauga, the mayor refused for decades to make a proclamation or to fly the rainbow flag which local LGBTQ2S advocates describe as a "ban" (Peel Pride Co-organizer, 3 July 2018) derived from homotransphobia but also a misrecognition of diversity that involved "berating communities for not integrating" (QTBIPOC activist, 10 July 2018). It was only when a mayoral regime change brought a PFLAG mother and ally, Mayor Bonnie Crombie, into office that the rainbow flag flew in Celebration Square in 2016 with eight people in attendance. As in many municipalities, the flag raising is a daytime mediatized event, attended by civic officials (e.g., the Mayor, councillors, police officers, and fire and emergency service officials) and community organizers but few participants due to work and school schedules. In 2022, Mississauga passed a mayoral-initiated motion (in part motivated by the burning of a Pride flag at a local high school in 2021) to annually recognize June as Pride month. They pledged to "honour and celebrate LGBTQ2S+ communities" by purchasing their own Progress Pride flags to fly at city facilities including one showcased on the main city flagpole in Celebration Square (Cornwell, 2022). Mississauga also makes use of its digital infrastructure to fly virtual flags on Celebration Square's screens, extending the flagpole's reach by broadcasting and amplifying its representations on a scheduled rotation. Despite the short duration, personal screens on cell phones and computers permit repeated viewing and circulation of the event for a wider social media audience.

In the Montreal city-region, Pride flag raisings respond to a different event. Since no peripheral municipality outside the City of Montreal hosts its own Pride celebrations, its suburbs signal LGBTQ2S inclusion by recognizing *La journée internationale contre l'homophobie, la transphobie et la biphobie* (IDAHOBIT) on May 17 (Fondation Émergence, 2020). The Francophone-initiated IDAHOBIT still prompts civic rainbowization, but its goal is to promote awareness, sensitize, and prevent homophobia (as well as biphobia and transphobia[3]) on a global scale. Launched in 2003 as a Quebec "national" event by *Fondation Émergence* and adopted by organizations in France in 2005, it went international in 2007 (Director, 19 August 2021). Louis-Georges Tin, a French-Martiniquian activist and anti-racist scholar, chose 17 May to commemorate the 1990 decision by the World Health Organization to declassify homosexuality as a mental illness (Tin, 2008),[4] a date that is now recognized in 60 countries around the world. In Canada, Fondation Émergence supplies organizations (e.g., government offices, community centres, and businesses) with supporting materials (e.g., flags, pamphlets, digital imagery, and video capsules) (Director, 19 August 2021). Each year they thematically raise awareness about violence, discrimination, and phobias in different facets of life (e.g., sports, parenting, or mental health) that can impede LGBTQ+ rights.[5] Raising municipal awareness about the importance of civic support for LGBTQ2S communities was the goal of the Canadian IDAHOBIT campaign in 2021. City councils were invited to adopt a resolution, proclaim their recognition of the event, and demonstrate their adherence to its principles. More than 287 Canadian municipalities participated, including three case study suburbs (Laval, Longueuil, and New Westminster) (Fondation Émergence, 2022, 15). Although only 260 of the 1100 municipalities in Quebec raised the rainbow flag, many of them were on the Montreal city-region's periphery, including Laval and Longueuil but not the independent West Island municipalities.

Laval's recognition of IDAHOBIT was gradual. Municipal records indicate that its city council acknowledged 17 May for the first time in 2016 as an event that "highlights the problems related to the reality of older adults living in these situations" (Ville de Laval, 2016b, 4). While this minuted remark by a city councillor inserted LGBTQ2S rights into city hall, it also demonstrated a lack of awareness of intersectional complexities, including the growing needs of younger rather than older adults. In 2022, Laval participated in the IDAHOBIT campaign, adopting its resolution, illuminating city hall, and raising the rainbow flag for local press and social media platforms, although due to lack of personnel and dedicated leadership, it neglected to sign the actual declaration (Librarian, 15 July 2022). "Out" city councillor Alexandre Warnet (23 June 2022) championed the IDAHOBIT campaign because it is "an important day that allows us to reflect upon what we still need to accomplish as a society, to reassure that all of us can live freely, no matter what our sexual orientation or our gender identity". For Laval, IDAHOBIT is an opportunity to show that it "is a profoundly inclusive city and that diversity is our pride"; take "a stand against prejudice and discrimination"; and "reiterate its values of inclusion, equity and solidarity" (Ville de Laval, 2022, para. 1). In 2023, Laval again raised the rainbow flag for 17 May against the backdrop of its defunct city hall using the designated community pole (Figure 4.2). In contrast to Longueuil, where the progress flag flew on an extra improvised flagpole, Laval's fourth flagpole is permanent and equal to the federal, provincial, and municipal poles.

From the city-regions of Vancouver, Toronto, and Montreal, city hall rainbow and trans flag raisings for Pride and IDAHOBIT are mediated ceremonial performances by local political dignitaries who, with their embodied presence and speeches beside civic flagpoles, emplace municipal support for

Figure 4.2 Laval and Longueuil raise Progress Pride flags for IDAHOBIT, 2023 (*Source:* Julie Podmore).

LGBTQ2S citizenry. They are attended by a coterie of municipal service providers, community organizations, and local residents but remain oriented to a social and print media audience. In the face of discriminatory rhetoric and actions, elected civic leaders use these honourary place-making opportunities to assure LGBTQ2S communities of their ongoing political commitments to "stand with you and for your right to acceptance and equality" (Collier, 2023, para. 1). Pride proclamations and flag raisings have become standardized elements of first-order LGBTQ2S municipal institutional allyship that are increasingly accompanied by more permanent visual symbols of support in suburban landscapes via rainbow benches (e.g., Scholars' Green in Mississauga or Garden Square, Brampton), gazebos (e.g., Ajax), or, more ubiquitously, rainbow crosswalks.

Rainbow crosswalks

Rainbow crosswalks are a globalized, but relatively recent, phenomenon. The first was a temporary public art installation at the National Taiwan Normal University in Taipei in 2008 to promote a gender equality campaign ("Street cred", 2020). Los Angeles, California, and Tel Aviv, Israel followed in 2012 with temporary rainbow crosswalks to celebrate Pride. In 2013, the mayor of Sydney, Australia was inspired to champion a permanent rainbow crosswalk in the historic gay district at Oxford Street for the 35th anniversary of the city's

Mardi Gras parade. While ultimately removed, it initiated a global race to be the first city to paint and maintain a permanent rainbow crosswalk. In 2013, the City of Vancouver made its 2012 gay village Davie Street rainbow crosswalk permanent, becoming Canada's first. Toronto followed in 2014, installing two on Church Street in the gay village to celebrate the hosting of World Pride, and Montreal installed one at the entrance of its Village gai in 2017.[6] In Vancouver, an "out" gay city councillor remarked that this rainbow crosswalk "mark[ed] the strongest support the city has ever shown towards the GLBTQ community"; it also enabled civic leaders to proclaim Vancouver as a "supportive" and "inclusive city" where "everyone is welcome" (Takeuchi, 2013, para. 4). LGBTQ2S inclusion aspirations have since diffused to the city-region's periphery where municipalities have competed to paint rainbow crosswalks, with New Westminster installing the first outside of downtown Vancouver in 2015, Surrey in 2017, and Burnaby in 2018. Within Canada, the Vancouver city-region's rainbow crosswalk diffusion has been exceptional and the area now has the greatest national concentration. Even provincially, "[i]t feels like everywhere – from the heart of Davie Village to the tiny town of Princeton (population: 2,828) – has embraced the rainbow" (Wells, 2018, para. 1). In contrast, suburban rainbow crosswalks in the Toronto city-region are much rarer and more recent, often involving the participation of para-public sector actors. In the Montreal city-region, suburban rainbow crosswalks are almost non-existent except for in the small municipality of Varennes and a recent addition in Saint-Jêrome. This section, therefore, focuses on how rainbow crosswalks are brought into being in the Vancouver case studies with some concluding comparison to Toronto municipalities.

As "public goods" that legally prioritize and safeguard pedestrian travel and safety on roadways, crosswalks are part of an intricate traffic geography reducing the negative impacts of automobilization (Loukaitou-Sideris et al., 2007). Their public visibility, generic social safety role, and the ability to increase access to urban life have made them an ideal infrastructure to metaphorically bridge LGBTQ2S representation and municipal inclusion. These very same attributes, however, are also referenced by homotransnegative critics, who question the replacement of generic (white and yellow) street markings with vibrant rainbow colours. In the language of road safety, red, orange, and green signal temporary disruptions to the road pattern, but these are rarely deployed permanently or together. In addition to the technicalities of installation (e.g., finding materials and appropriate revenue sources), concerns about "universal access" to pedestrian infrastructure (especially for citizens with mobility and sensory impairments) are frequently raised as issues. Municipalities are also concerned that rainbow crosswalks provoke homotransnegative vandalism with spray painted swastikas and Bible verses, tire skid marks, or crudely tossed white paint. Ultimately, rainbow crosswalks remain contentious because they politicize walking. To step upon a national flag is considered a desecration of the body politic; a rainbow crosswalk invites all pedestrians to celebrate LGBTQ2S lives while crossing a street.

There are many examples of rainbowized municipal infrastructure (e.g., street banners, benches, garbage bins, stairs and ramps, lamp posts, planter boxes, utility boxes, and benches), but crosswalks have become a civic staple for signalling progressiveness, inclusivity, and LGBTQ2S service. An affordable surface treatment of already-existing road infrastructure, they require minimal taxpayer investment or upkeep. With appropriate paint choice and selection of intersections, rainbow crosswalks offer municipalities an inexpensive but visually and physically impactful place-making tool to be integrated into bureaucratic processes of financing, planning, and maintaining civic physical assets. As an LGBTQ2S community activist remarks, "Who doesn't like a rainbow?" (Peel Pride co-organizer, 3 July 2018). The rainbow's malleability, combined with its child-like, hopeful, playful, and caring symbolism, allows for a civic stretching of the rhetoric of diversity while also obfuscating a direct LGBTQ2S reference. Rainbows morph into more palatable civic symbols of ethno-cultural, family, intergenerational, or linguistic diversity, representing everyone yet no one.

In 2015, New Westminster became the second municipality in the Vancouver city-region to create a rainbow crosswalk. For civic leaders, there was a synergy between the crosswalk and their compassionate city place-brand. As the mayor told the local press, "New Westminster is an incredibly inclusive community. We do a lot to really deal with marginalized individuals. That's the type of city we want to be and be known for" (Verenca, 2015, para. 5). With its large Pride street fair and rebranding as the region's second gay village, the rainbow crosswalk also dovetailed with downtown urban redevelopment plans to aestheticize its main commercial artery, Columbia Street. The rainbow crosswalk further cemented the Royal City Pride Society's role in the local BIA's calendar of street festivities animating the downtown core. Prominently located, it connects what was once the hub of drag performance at the Heritage Grill with a gay bathhouse and a mural celebrating the municipality's heritage, both adjacent to the old Columbia Theatre where the voices of drag show contestants echo. Despite such civic support, New Westminster's rainbow crosswalk was vandalized with white paint upon installation by an elderly resident who was later arrested on charges of mischief. Soon restored, it inspired the 2018 rainbow crosswalk installation in adjacent Surrey.

Surrey's mayor Linda Hepner was commuting to a regional TransLink meeting when she saw New Westminster's crosswalk and thought that she liked its aesthetic and the message it delivered about diversity and inclusion. Once installed, she justified its CDN$8500 expense by referring to these civic values: "When you look at [the] 102 languages spoken here in our city, that rainbow [cross]walk has gone way beyond just being an original LGBTQ [cross]walk … it represents now a message of inclusiveness" (Reid, 2018, para. 3). This argument permitted her to politically own the installation of a rainbow crosswalk in Surrey and use her authority to become an ally to queerburban activists, expediting its installation for Pride celebrations without council approval (Reid, 2018). Although opponents questioned her unilateral decision, Hepner asserted that as a message of inclusion, the rainbow crosswalk fell under her mayoral purview. The day after the crosswalk was laid, it was vandalized with white paint, leading the Royal Canadian Mounted Police (RCMP) to investigate the incident as a hate crime. Since there had been no celebratory unveiling, queerburban activists recounted that they were actually "thrilled" with press coverage of this "desecration" because finally "we were in the media, on the news" which means that "we're on the map" and we are not "going to shut up and go away" (FG2 Surrey, 8 June 2019). Neither civically celebrated nor integrated into pedestrianization or urban redevelopment plans, this crosswalk was a singular instance of civic service in recognition of historic LGBTQ2S activism celebrated by Surrey's queerburban activists during their 2018 Pride march. Long excluded from hosting their Pride events in the civic plaza, the march over the crosswalk is as much a protest as a celebration of a singular gain in what amounts to unreliable institutional allyship.

In contrast with Surrey, civic allyship in Burnaby was non-existent until 2018, when the first rainbow crosswalk was painted by the municipality in response to para-public institutional lobbying for the city's first Pride event. As civic compensation for this peripheral municipality's late service to its LGBTQ2S citizenry (Bain & Podmore, 2022, 2025), this crosswalk was painted with deck paint in four hours by the city's social planner with help from the Engineering Department. Hidden on Jubilee Avenue, a "redundant street" with little traffic and in proximity to youth and recreation centres, Burnaby's first rainbow crosswalk was an "interim measure" that was rushed with no road crew dedicated to its service (Social planner, 9 July 2019). Hours after installation, the crosswalk was vandalized with a spray-painted green question mark and then swiftly repainted and guarded by 24-hour security until Pride was launched the next day. During the event, the Mayor, flanked by Indigenous elders and political dignitaries, unveiled and metaphorically interpreted the crosswalk as a multicultural place-maker (Figure 4.3):

> In this city, we believe that it is essential for all of us to cross the bridge of our differences … in a community with over a 100 languages, that brings people together from all over the world, this diversity is our strength and this crosswalk is an important statement that all of us believe in each other (Field notes, 11 August 2018).

This road infrastructure also emerged through city hall's backdoors as an exceptional instance of institutional allyship politically brokered outside of council chambers.

With a fall election looming and weakness in LGBTQ2S inclusion and housing justice, Burnaby's first rainbow crosswalk had served as limited compensation for one area of the regime's inaction but it was not enough to hold power at city hall (Bain & Podmore, 2025). Thus, leading up to the fall 2018 election, "there was a lot of willingness to act on things" leading Burnaby's social planner (9 July 2019) as the primary city hall "insider activist" to

Figure 4.3 Inauguration of Burnaby's rainbow crosswalk, 2018
(*Source:* Julie Podmore).

champion the rainbow crosswalk as part of preparations for Burnaby's first Pride events. Over two months, the city flew its first rainbow flag, installed a rainbow crosswalk, and sponsored Pride festivities. When the City Council's approval of the crosswalk plan was publicized, however, complaints "blew up" from "the antis" opposing this infrastructural change (Programme Manager, 5 July 2018) leading community groups to leverage a networked response: "By Monday morning", we didn't "have to worry anymore. There had been such a huge response from everybody … saying, we need to be more inclusive" (Burnaby Intercultural Planning Table Coordinator, 16 August 2018). Following Burnaby Pride, the single, temporary crosswalk became permanent, accelerating civic infrastructural rainbowization.

A citizen request to Burnaby's Sustainable City Advisory Committee for six additional rainbow crosswalks further motivated staff to determine new rainbow place-making intersections and funding sources. Urban planners began consulting citizen advisory groups and departments to determine crosswalk locations. The social planner (9 July 2019) took these recommendations to the Department of Engineering to evaluate questions of visibility and long-term maintenance, cautioning "[i]t is worse to have a rainbow crosswalk that looks terrible – you want it to look good". The city's engineers proposed four "high community profile [intersections] close to civic facilities where higher pedestrian activity can be anticipated" (City of Burnaby, 2019d, 2). These "pedestrian gateways to community services" align with the City's *Corporate strategic plan* and its broader goals to "encourage and welcome all community members and create a sense of belonging" while celebrating diversity (City of Burnaby, 2019d, 2). Engineers rejected more visible high-traffic intersections because they were less "hospitable for pedestrians" and "would increase maintenance costs due to increased tire wear across the painted markings" (City of Burnaby, 2019d, 2). To garner support, the council's motion aligned the crosswalks with "pedestrian travel" as a "valued aspect of life" and a "favorite recreational activity" of all Burnaby residents (City of Burnaby, 2019b, 21) only mentioning Pride in reference to LGBTQ2S constituencies. Upon a directive of the new Mayor, the crosswalks were funded with CDN$35,000 of revenues from the casino fund to sidestep concerns about the "impact on the Burnaby taxpayers" (City of Burnaby, 2019a, 17). The Finance Department stipulated that the crosswalks would all be painted with a durable product, namely methyl methacrylate valued for its "excellent brilliance of colour, good traction, and a long life expectancy" (City of Burnaby, 2020, 42). Even in its paint choice, Burnaby was committed to the rainbow as a place-making tool of sexual and gender minority inclusion and wanted to ensure the crosswalks could withstand high use and vandalism.

As institutional allies, therefore, Burnaby's politicians gave the rainbow crosswalk to all citizens disassociating it from LGBTQ2S bodies and making it resonate with suburban technocratic governance. Burnaby's city council used LGBTQ population estimates from the Canadian Community Health Survey and the BC Centre for Excellence in HIV/AIDS to bolster its rationale. Given estimates of the City of Vancouver's sexual and gender minority population at 40,000+ ("in an environment where not everyone is willing or able to declare"), the council speculated that "proportionally comparable populations of LGBTQ are believed to exist in neighbouring metro communities" (City of Burnaby, 2019b, 21). Appreciating these statistics and the place-making power of the rainbow crosswalk to convey LGBTQ2S inclusion, the Mayor's Office requested that a sixth be installed in front of city hall and Burnaby Central High School, a location described as "well used" with symbolic value "consistent with the goal of signalling to the public the City's goal of inclusivity" (City of Burnaby, 2019d, 21). Although these rainbow crosswalks aestheticized inclusion in civic infrastructure, it was perceived by activists as an attempt at political absolution: it "doesn't mean that the city is supportive, an ally, or educated"; instead, it serves as "a visual representation of municipal support around something that is controversial" (Coordinator, Burnaby Youth Hub, 11 July 2019). Burnaby has excessively used the rainbow to reinforce its place-brand, leveraging it for legitimacy as socially sustainable and diverse while minimizing political controversy. Such first-order institutional allyship is feasible because of the banal malleability of the rainbow as a proxy for LGBTQ2S inclusion.

In the Toronto city-region, rainbow crosswalk installation on the periphery is piecemeal, incidental, and recent, with city councils "trying to play catch-up" and relying on para-public institutions for support (Insauga, 2023, para. 1).

The Regional Municipality of York initiated crosswalk rainbowization in 2020 with their own installation in the Town of Aurora and the unveiling of a York Regional Police crosswalk at the Community Safety Village in Markham in celebration of Pride. A year prior, the Regional Municipality of Durham installed its first rainbow crosswalk in front of its corporate regional headquarters in Whitby shared by Durham Regional Police Service and Durham Regional Transit. The Town of Ajax followed in 2022 with a rainbow crosswalk in front of the Audley Recreation Centre. As in Burnaby, Mayor Shaun Collier's speech universalized the crosswalk as a symbol of "the incredible importance of love, acceptance, and equality – everyone is welcome" (Town of Ajax, 2022, para. 6), but also encouraged citizens to learn more about the particular histories and everyday discriminatory challenges of LGBTQ2S communities "so we can take part in shaping a better tomorrow" (Town of Ajax, 2022, para. 6). The mayor presented the crosswalk as an opportunity to share in a positive, celebratory, and educative place-making experience that strives to civically chart a discrimination-free future.

The struggle to spread rainbow crosswalks throughout the Regional Municipality of Peel is, like Surrey, more fraught. In 2018, the Peel Pride Committee was "looking for ways to kick it up a notch" and began to "plant ideas" about the regional municipality's "responsibility for our community" (Peel Pride co-organizer, 3 July 2018). The Committee requested CDN$25,000 from the regional municipality to fund a rainbow crosswalk in each member municipality (Brampton, Mississauga, and the Town of Caledon). Their request for institutional allyship was approved in a 2019 resolution but was later rescinded in 2020 and replaced with the following motion:

> Whereas, the Region of Peel recognizes Pride Month in July and is a supporter of 2SLGBTQ+ communities;
>
> And whereas, symbols of Pride are a reflection of the diversity in the Region of Peel;
>
> Therefore be it resolved, that the Region of Peel raise the rainbow flag annually for the month of July in recognition of Peel Pride;
>
> And further, that the Region of Peel allocate funds on an annual basis, up to a maximum of $25,000, between the Cities of Brampton and Mississauga and the Town of Caledon with each receiving up to one third of the funds;
>
> And further, that the funds be applied toward visible demonstrations of Pride across the Region as determined through consultation with the communities and staff of each respective municipality;
>
> And further, that visible demonstrations may include but are not limited to flags, murals and crosswalks;
>
> And further, that funds will not be used for entertainment or event purposes.
>
> (Region of Peel, 2020, 2)

This resolution interprets the "symbols of Pride" as indicators of "diversity", but it contains no discussion of how to forward a municipal social inclusion agenda nor a required deadline for a rainbow crosswalk. This lack of clarity was compounded by a gap in the funding allocated (CDN$8333) per municipality and the actual costs (approximately CDN$50,000) giving member municipalities an escape option (Region of Peel, 2020, 3). In Brampton, for example, where the city council approved funding for a rainbow crosswalk in 2022, "the project sat in limbo" with "no direction … given to City staff" (Insauga.com, 2023, para. 3). While the municipality has since installed two rainbow benches in Garden Square and Pride flag decals in Vivian Lane behind it, it is still "looking for more opportunities to work with organizations to add more visible demonstrations of Pride as symbols of support" (Insauga.com, 2023, para. 9). Brampton has politically and financially committed to rainbow crosswalk infrastructure, but the extended timelines and delivery delays suggest a civic ambivalence about implementing visible representation of LGBTQ2S inclusion.

For suburban municipalities, rainbow crosswalks materialize LGBTQ2S social inclusion by visibly underscoring civic diversity place-brands and progressive distinction. They exist on a spectrum of absence to abundance. Their almost complete absence in Montreal's peripheral municipalities is largely cultural, while in Toronto their uneven distribution can be attributed to multi-tier governance structures that enable member municipalities to deflect responsibility for rainbowizing crosswalks. Toronto's suburban super-diversity also structures local socio-political priorities, making the more permanent modalities of LGBTQ2S social inclusion such as rainbow crosswalks more difficult to achieve and sustain across election cycles. In the Vancouver city-region, crosswalk rainbowization requires political maneuvering, but the region's rainbow crosswalk excess is further driven by intermunicipal competition for an inclusive place-brand. Rainbow crosswalks provide an especially expedient performance of LGBTQ2S inclusion, a highly performative, visible transformation of municipal infrastructure with multiple referents. Unlike other first-order forms of institutional allyship, rainbow crosswalks require budgeting for maintenance costs, interdepartmental decision-making, and recuperation in the face of (sometimes repeated) vandalism. Seemingly aesthetic and trivial instances of first-order allyship, rainbow crosswalks have a material permanence that borders on requiring second-order levels of civic commitment. One final, and more recent, iteration of first-order institutional allyship is civic plaza illumination.

Civic plaza illumination

Where the materiality of rainbow flags and crosswalks may elicit peripheral municipal homotransnegative responses around community representation, the ephemerality and surplus luminescence of rainbow civic plaza lighting as aestheticized infrastructure in the service of LGBTQ2S inclusion can be more politically palatable, difficult to vandalize, and adaptable to multiple publics. Artificial lighting is appreciated by urbanists for the experimental and experiential design possibilities it affords cities (Ebbensgaard, 2020). For municipalities, illumination is a "sacrosanct" way to shape public nightscapes (Edensor, 2015). When applied to civic spaces, like the plazas and buildings of city hall, lighting becomes a place-making tool to enhance place-branding and public realm usage.

In Surrey, lighting became central to civic place-making in the 2014 redevelopment of its 210,000-square-foot civic complex intended to transform "the culture and ethos" of Surrey "from 'suburb' to thriving business and urban center" (Architizer.com, 2021, para. 1). Surrey's new civic plaza was designed for concerts, community festivals, and holiday celebrations, but only some community groups have received official plaza hosting permits (e.g., Surrey Filipino Heritage Festival, African Heritage Festival of Music and Dance, Surrey Latin Festival, Surrey Tree Lighting Festival, and Party for the Planet). Prior to 2022, the LGBTQ2S community was missing from this list of community groups with access. Instead, they made do with mounting history displays in city hall's atrium and informally beginning Pride festivities with a small march from the plaza to a nearby park and, in recent years, a shopping mall (see Chapter 6). Mounted along the walls of the plaza are black luminaires and integrated at grade into the stonework is stainless steel encased underfoot illumination. Both lighting technologies accentuate the plaza's architectural features and create a celebratory atmosphere.

The plaza's lighting is curated as part of Surrey's official annual event calendar. This calendar specifies municipal cultural governance priorities, standardizing civic place-making actions in recognition of diversity. A staff resource for the Parks, Recreation, and Culture and Human Resources departments, it is an inclusion tool to expand "awareness and education of the diversity of Surrey" (City of Surrey, 2016a, Appendix, 1). Beyond national holidays, the 2020 calendar listed 32 events, nearly half of which were ethno-cultural or religious, and the rest dedicated to awareness raising for social issues (e.g., autism, disability, mental health, and the environment). For these events, the atmospherics of artificial lighting dramatize the "hard" plaza surfaces with increasingly complex colour schemes and patterns: red (Lunar New Year, Canada Day, and Remembrance Day); pink (Pink Shirt Day); orange (Orange Shirt Day); yellow (Vaisakhi, and Easter); green (Earth Day, Nowroz, and Mental Health); dark blue (Autism Awareness); purple (Epilepsy Awareness, and Rakhi); rainbow (LGBTQ+ Pride); white, blue, and pink (Diwali); and red and white (Canada Day, and Christmas Day) (Podmore & Bain, 2023). The calendar expands annually to include new celebrations signalling growing local awareness of rights recognitions, national celebrations, and religious events. Such municipal lighting solutions provide civic aesthetic representation to, and initiate public dialogue about, causes, populations, or events.

Although all the colours of the rainbow are available for Surrey's plaza celebrations, it was not until 2018 that rainbow colours were combined to civically mark Pride week from 21 to 25 June. Such illumination allows civic leaders to offer first-order institutional allyship while simultaneously avoiding the queer "excesses" of Pride festivities (e.g., face paints, feathers, sequins, satins, flags, banners, and slogans). In its diffusivity, plaza rainbow lighting is partial compensation for political refusal to emplace queer and trans bodies in civic space offering officials the capacity to "redistribute the field of the sensible in ways that neutralise opportunities for contesting this very distribution" (Ebbensgaard, 2020, 1973). Rather than being a tool of resource redistribution to address everyday material and service inequalities and discriminations, the dramatic evanescence of rainbow lighting is playful and celebratory, avoiding homotransnegative defacements. Its use of a similar colour palette and lumination treatment as that of other community celebrations mutes rainbowization controversies. In a peripheral municipality where the only municipal rainbow crosswalk has been repeatedly vandalized and the rainbow flag has only been flown twice, such luminary rainbow aesthetics are an ephemeral expression of first-order institutional allyship.

Other suburban jurisdictions also make use of rainbow lighting to serve their LGBTQ2S citizens during the "bombardment of Pride season" (Peel Pride co-organizer, 3 July 2018). In 2023, Mississauga recognized Pride month by hosting "Lighting the clock tower" in its civic plaza, Celebration Square. For this Pride event, the distinctive, yellow-bricked postmodern city hall and clock tower are illuminated at night, one colour at a time with a programmed sequence of rainbow colours throughout the month unless another cause is scheduled. Much like Surrey's illumination calendar, this peripheral municipality's Community Recognition Programme allows Canadian registered charities and non-profit community group events that "create a spirit of community, celebrate Mississauga's diverse cultures and contribute to enriching the quality of life" in the city (City of Mississauga, 2024, para. 1). Protocol dictates that community lighting requests must be made weeks in advance and only one recognition per community (either a flag raising or lighting) is permitted annually. Despite Mississauga's significant ethno-cultural diversity, combinations of two and three coloured lighting are only used to celebrate Indigenous and Franco-Ontarian cultural identities, in the service of settler Canadian traditional holidays, and to raise medical awareness. Rationed in this way, the rainbow is a technical impossibility because it relies upon a colour spectrum that cannot coexist in the current system.

As a testament to the multiple meanings of the rainbow, in 2020, the Ville de Laval used the multicolour lighting system to illuminate the facade of its city hall with the rainbow. It was listed on a local blog report on Montreal places using the rainbow to signal support for the population at the beginning of the COVID-19 pandemic (Hanrahan, 2020). On Facebook, it hashtagged the photo #ÇAVABIENALLER ("it's going to be okay") – a reassuring slogan used in Francophone contexts during the COVID-19 pandemic lockdowns – and captioned it with the following: "Our city hall is bathed in the colours of the rainbow. A symbol of hope and support for our essential service workers"

(Ville de Laval, 2020b). Since the pandemic accelerated the integration of the rainbow into Laval's lighting scheme, it was symbolically available to serve sexual and gender minorities after a municipal regime change in 2021. Following the City Council's resolution to recognize 17 May and raise the rainbow flag, the rainbow colour lighting scheme was redeployed to illuminate Laval's city hall in August 2022 to celebrate central Montreal's Pride week. The hyper-metropolitan character of Montreal's LGBTQ2S geographies is again underscored by Laval's performance of its satellite status.

The most recent addition to the rainbow place-making repertoire, rainbow lighting, provides civic leaders with a more ephemeral, yet extensive, means to signal LGBTQ2S inclusion. These nocturnal circles of illumination that vertically bathe seats of political power in the rainbow are visible from afar. For peripheral municipalities, rainbow lighting, therefore, scales up LGBTQ2S visibility while simultaneously limiting its duration and curtailing vandalism. Yet the very ephemerality and versatility of civic lighting is its strength as an official place-making inclusion tool because, with the flick of a switch, it is gone, or with the changing of light filters, it can represent something else to another public.

Conclusion

First-order LGBTQ2S institutional allyship has gradually gathered civic momentum and diversified its rainbowizing repertoire of place-making strategies. Initiated verbally within council chambers as minuted mayoral proclamations, LGBTQ2S municipal place-making has moved outside of city halls, first to flag poles and then to proximate street intersections, and to civic plazas. From the pages of printed text to the surfaces of fabric, roadways, and buildings, the rainbow – as a symbol of LGBTQ2S social inclusion – has incrementally manifested in more durable ways in queerburbia. Nevertheless, an oscillation persists between predominantly temporary and permanent visibility. Civic rituals to mark Pride month and 17 May (IDAHOBIT) have become increasingly familiar, while through rainbow crosswalks, more permanent and visible articulations invite public debate about LGBTQ2S inclusion throughout the year.

If the rainbow was originally intended to code spaces as safe and inclusive for LGBTQ2S people, peripheral municipal rainbowization is misleading for four reasons. First, the slippery significations of the rainbow make it palatable to multiple publics, neutralizing its political implications for sexual and gender minorities through universalized meanings. Second, while rainbowization politicizes municipal infrastructure, its visibility often surfaces pre-existing tensions between local and regional governments, civic ambivalence about visibly representing LGBTQ2S inclusion, and conflict with homotransnegative constituencies. Third, while queerburban constituencies, their allies, and some civic leaders may view rainbowization as representing LGBTQ2S inclusion, its bureaucratization remains performative, reiterating the rainbow's impermanence. Finally, uncertain civic motivations for the rainbow's infrastructural application – progressive place-branding, urban redevelopment, and diversity discourses – make the rainbow as a place-making tool an unreliable indicator of LGBTQ2S safety, support, and inclusion. For these reasons, peripheral civic rainbowization remains primarily a form of first-order allyship upon which municipalities rely to serve LGBTQ2S constituencies while blocking demand for more redistributive social inclusion policies. Thus, in suburbs, it is para-public institutions that are the political agents pushing for second-order allyship.

Notes

1 Crosswalk is the North American term for marked pedestrian crossings elsewhere referred to as zebra crossings.
2 A controversial question raised in a mayoral candidate's debate in Ajax in 2010 is summative, "What makes the gays any more deserving of their flag hanging than anyone else?" (Vyhnak, 2010, GT2).
3 The Fondation refuses to integrate the word lesbophobia.
4 IDAHOBIT became a global symbol during the Montreal human rights conference that accompanied the World Outgames and the Montreal Declaration in 2006. It was later recognized along with the declaration by the United Nations and the European Union.
5 For example, in 2022 "*Le montre de 54 secondes*" highlighted the frequency of violence experienced by LGBTQ+ people around the world and in 2023 "*Les LGBTQphobies sont des peurs irrationnelles*" situated LGBTQphobias within the framework of other phobias to point out its irrationality.
6 Montreal's gay village BIA created impermanent rainbow crosswalks around the western entry to the pedestrianized portion of Sainte-Catherine Street but all have since been removed and replaced with street art.

5 Resisting

Introduction

Resisting the limitations of peripheral municipal political opportunity structures and non-medicalized resource landscapes for sexual and gender minorities, some suburban para-public institutions perceive this gap and become LGBTQ2S allies. Para-public institutions are organizational entities invested with arm's-length public authority (be it municipal, regional, provincial, or national) to provide infrastructure that is partially funded or managed by the government (e.g., public schools, public libraries, police forces, hospitals, medical and social services, and community centres). Although LGBTQ2S service delivery is seldom a primary responsibility, suburban para-public organizations still initiate it due to the "insider-activism" of LGBTQ2S-identified employees (Browne & Bakshi, 2013), following consultations that inductively reveal gaps in LGBTQ2S service needs (Cooper & Monroe, 2003), or as a result of awareness-raising from within multi-scalar organizations. In nuclear family-centric suburbia, where institutional structures centre on reproduction, para-public resistance to LGBTQ2S exclusions often prioritizes young people and their promise of a different future.

This chapter attends to the inter-organizational relations and second-order para-public institutional allyship that creates the "coalitional moments" and place-release necessary to queerly remake suburbs. Following queer Chicana feminist scholar Karma Chávez (2013, 9 & 8), "coalitional moments" are understood as spatio-temporal turning points that integrate queer politics into the public sphere by re-envisioning and possibly reconstructing "rhetorical imaginaries". "Place release" occurs when para-public allies produce these turning points, enabling minority groups to embed in place via "all manner of happenstance" which extends "from the momentary and prosaic to the revelatory and life-changing" (Seamon, 2018, 122). LGBTQ2S activist resistance takes sustained planning and interactive organizing and is unlikely to "just happen", but para-public institutions do create "unexpected happenings" (Seamon, 2018, 91) within place that can temporarily "release" it from hetero-cisnormativity's disciplining order. While Seamon (2018, 124 & 126) interprets place release as "spontaneous and unsolicited" and seldom intentional, he does

DOI: 10.4324/9781032632544-5

acknowledge that advocacy can indirectly initiate it. In queerburbia, para-public institutions produce place release through their internal and inter-organizational relations, stepping in for, putting pressure on, and filling the gaps in political opportunity structures and resource landscapes. While organizations invariably have different goals, tactics, and deliberative logics, their cooperative dynamics augment and add cohesion to queerburban resource landscapes, forwarding first- but also second-order social changes.

In the face of the attenuated political opportunity structures of municipalities, para-public organizations augment suburbia's limited and fractured resource landscapes through critical allyship with LGBTQ2S communities leading to enactments of second-order change. For individuals, critical allyship is a self-reflexive process fraught with missteps that necessitates understanding one's own role in upholding systems of oppression as well as working in solidarity with (and not for) and under the leadership of marginalized people (Nixon, 2019). It involves unlearning privilege, experiencing discomfort, and working in solidarity with marginalized groups. For institutions, critical allyship involves radical reconfigurations of leadership opportunities, governance priorities and processes, and worldviews. It may be augmented when multiple organizations competitively propel social change, or, on a micro-scale, when "insider activists" build consensus for collective action. Institutional activists are "insiders with access to resources and power – who proactively take up causes that overlap with those of grassroots challengers" (Pettinicchio, 2012, 499). Such institutional resource mobilization and extension of political opportunity structures for allyship are nevertheless shaped by an insider/outsider dichotomy whereby outsiders temporarily leverage insiders to address their claims and insiders mobilize resources, power, and policy. Foregrounding the lived experiences of queerburbanites, para-public insider activists can elicit intense and sustained institutional change and emplace it in programming initiatives, buildings, and inter-organizational networks.

In what follows, this chapter examines how critical institutional allyship and insider-activism bring about second-order para-public place-making. It focuses on selected non-medical institutions, including secondary schools, colleges and universities, public libraries, municipal and national police divisions and detachments, and churches that support queerburban resistance with resources and programming, while strengthening inter-organizational relations. Across the range of para-public institutions serving queerburbanites, only those with a sustained central leadership role facilitating place-based LGBTQ2S inclusion in the three case-study city-regions are explored.[1]

Secondary schools, colleges, and universities

As state-regulated and adult-led educational institutions, public schools, colleges, and universities teach young people how to participate in society. Over decades, the activism and allyship of teacher, faculty, and student unions,[2] equity, diversity, and inclusion officers,[3] sexual and gender diversity offices,[4] as well as individual students and educators, have remade high schools and post-secondary institutions as more LGBTQ2S-inclusive places through positive space campaigning, student club formation, and event hosting.

Positive space campaigns originated in Canada in 1995 at the University of Toronto's downtown campus with an ad hoc group of faculty, students, and staff who addressed how to increase visibility for lesbians and gays (University of Toronto, 2023). They developed stickers for office doors and windows to publicly declare spaces as lesbian- and gay-positive as well as educational flyers and resources. Formalizing as the Positive Space Committee, they eventually expanded to its suburban Scarborough and Mississauga campuses. Two decades later, a Sexual and Gender Diversity Officer (16 August 2018) explains that the positive-space campaign's inverted rainbow triangle has retained "currency on campus … and has been borrowed by community partners locally and globally". Inspired by this campus activism, Canadian public schools, colleges, and universities adopted LGBTQ2S-positive space campaigns to combat discrimination using "safe space" designation (e.g., policies, codes of conduct, guidelines for participation, or positive-space stickers or posters) to unmake the heterocisnormative codes of institutions with daily visual reminders (Sharp & Nilan, 2017).[5] During Pride month or for IDAHOBIT, educational institutions also generate first-order change by issuing statements and raising rainbow and trans flags. For LGBTQ2S young people, these para-public institutional allies and their spaces offer an affirmative contrast to their hetercisonormative parental homes, particularly for under-18 youth. For some queerburbanites, therefore, educational institutions may offer safer micro-spaces of social affirmation.[6]

GSAs and Pride proms in secondary schools

Within suburban secondary schools, Gay-Straight Alliances or Gender and Sexual Alliances (GSAs) are organizational configurations intended to enhance LGBTQ2S belonging. Beginning in the United States in the 1970s,[7] they gained traction in Canada in the early 2000s. In Ontario, the government passed the Accepting Schools Act in 2011, following the death by suicide of the son of an Ottawa city councillor. It mandated stricter and clearer anti-bullying responses and required publicly funded schools to permit the establishment of GSAs. These extracurricular student "clubs", recognized by school authorities and hosted by an insider activist or ally teacher, serve LGBTQ2S youth and their straight allies by raising awareness about, and reducing, school-based homo-, bi-, and transphobic discrimination, victimization, and violence. Some GSAs are social spaces, while others sponsor events aligned with national and international human rights movements, educate the school community, and/or advocate for policy changes. GSAs also engage in micro place-remaking practices within their schools through decoration, participation, sociability, reconnaissance, and activism, while the national GSA network connects LGBTQ2S students to each other and community resources across school districts, provinces, and territories to host Pride proms and national conferences (Bain & Podmore, 2020a).

GSA teacher-sponsored classrooms are often colourfully inviting, adorned with LGBTQ2S life-affirming slogans and images of hope, resilience, and advocacy that rarely coalesce elsewhere in queerburbia (Figure 5.1). GSAs concentrate resources (e.g., books and films to borrow, events to attend, and social service and community organization contacts), their classrooms generating a surplus visibility of positive symbols and resources, that can extend into the transitional semi-public spaces of hallways. They unmake heterocisnormativity by promoting acceptance and respect for LGBTQ2S identities. Sponsorship of annual district events like Pride proms creates opportunities to remake school and community spaces as LGBTQ2S-positive via joyful demonstrations of same-sex dancing and affection. Connections to community organizations via guest talks, protests, and reconnaissance field trips to access central-city queer infrastructure enhance appreciation of LGBTQ2S community organizations that sustain queer histories and places beyond educational institutions. Such ongoing place remaking also has a history and geography that is rooted in lived experiences yet also welcomes the next generation of activists.

Pride proms are particularly resonant events because they challenge the institutionalization of heterocissexual dating norms. In the Canadian suburban context, youth activism to change school prom regulations is exemplified

Figure 5.1 The Sullivan Heights Secondary School GSA classroom, 2017 (*Source:* Alison Bain).

by the political organizing of queerburban teenager Marc Hall, whose story is told in *Prom fight: The Marc Hall story* (Peloso, 2002) and the docudrama *Prom queen: The Marc Hall story* (L'Ecuyer & MacLennan, 2004). In 2002, this openly gay 18-year-old Grade 12 student used pro-bono representation to legally challenge the Durham Catholic District School Board's (DCDSB) requirement for opposite-sex prom dates. A judiciary injunction required the school to hold the event and permit Hall to take his boyfriend, concluding that "the principal unjustly discriminated against the plaintiff" (McConnell, 2002, para. 13).[8] Across most Canadian school districts, students can now bring same-sex dates to the prom; however, Pride proms, as inaugurated in the school districts of the Durham region, Surrey, and Burnaby, offer LGBTQ2S students and their allies safe(r) and more socially inclusive alternatives.

In 2008, almost 10 years after Marc Hall's "prom fight" with the district's Catholic school board, the Regional Municipality of Durham became an early suburban Pride prom adopter. This civic regional Pride prom coalitional moment occurred because ally para-public institutions leveraged their inter-organizational networks to resist suburban heterocisnormativity. Upon the initiative of one of the most active PFLAG chapters in Canada and modelled on the Supporting Our Youth (SOY) Pride Prom initiative at the Sherbourne Health Centre in downtown Toronto, it was realized through the collaboration of local health (Carea Community Health Centre), education (Durham District School Board), policing (Durham Regional Police), and library para-public institutional service providers. A volunteer-driven effort in which "everybody brings something to the table", the prom's goal is to bridge queer-burban administrative and resource silos "in a challenging geography where youth were isolated by transit, and invisibilized" (Health Promoter, 12 July 2018). In its first year, the Durham Pride Prom attracted 80 youth on "the spectrum from trans masculine to trans feminine, non-binary" for an opportunity "to get to be who they are … and not be beaten up for it" and those numbers have continued to increase to over 100 (PFLAG Manager, 17 August 2018). Para-public institutional allies provide funds or donate in-kind sponsorship (e.g., photo booths, decorations, printing, DJing, and swag for prizes) or serve as duty officers to create a "celebratory, safe space, where young people can wear what they want, be affectionate with whom they want, see drag performance, and find role models in the local youth performers and celebrity host MCs" (Health Promoter, 12 July 2018). A free bus collects participants from neighbouring Oshawa, Whitby, Ajax, and Pickering and youth come from as far away as Port Perry, Peterborough, and Bowmanville.

Across the country in the Vancouver city-region, suburban secondary school Pride proms are more recent expressions of second-order institutional allyship initiated by school boards. In Surrey, the first Pride Prom was hosted by the Guildford Secondary School's GSA with the goal of creating an inter-school "lavender dance" that offered an "open and welcoming … safe space for queer kids and their allies" from Grades 8 to 12 (Teacher-sponsor, 23 June 2017). School district administrators responded by going into "crisis mode", emphasizing the necessity of going "by the book" (e.g., providing security, having registration lists, and requiring parental permissions) and prioritizing institutional liability so as not to incur negative media attention or legal complications (Teacher-sponsor, 23 June 2017). By reconstructing suburban imaginaries of high school inclusion, the Surrey school board's Pride prom became a coalitional moment of heterocisnormative place release receiving media attention and inspiring mimicry from nearby suburban school districts. The following year, a Burnaby GSA student initiated the province's second district-wide Pride Dance. Afterwards, ally teachers (26 June 2017) said they "want to supervise the dance again" and administrators proclaimed the event "the best school dance they've ever been to". In Burnaby and Surrey, adult teachers and administrators became LGBTQ2S institutional allies by witnessing student joy. Leaving high school to attend post-secondary institutions and transitioning into adulthood often marks the first sustained engagement with LGBTQ2S communities through college and university student clubs and collectives.

Post-secondary queer student clubs

Although peripheral municipal colleges and universities are often "commuter schools" where few students, staff, or faculty socialize after-hours because of lengthy journeys, queerburbanites may still identify them as "my school of refuge" (FG2 Burnaby, 9 June 2019). LGBTQ2S post-secondary student clubs, like Out@UTM, Douglas College's Pride Collective, or John Abbott's Queer Hub, are important nodes of suburban para-public resistance through student activism, described by one Equity and Inclusion officer (14 August 2018) as "bubble[s] in … communit[ies] with few resources". These small clubs and collectives are student-led, facilitating adulthood transitions and the gaining of greater autonomy and political voice for queerburbanites beyond the neighbourhood school catchment areas of their familial homes on campuses with city-regional reach. They also offer programming and co-facilitated events to share lived experiences, integrate students into multi-scalar community organizational networks that can support them, and foster coalitions of LGBTQ2S students and their allies queering post-secondary educational institutions through flag raisings, participation in local Pride events, and advocacy for all-gender washrooms and changing facilities.[9] At Sheridan College's Brampton and Oakville campuses, sexual and gender minority student union members collaborated with the equity and inclusion officer (14 August 2018) to identify gaps in service delivery and to propose the need for "an LGBTQ-positive space – a physical space with resources" on campus.

Such a space – Out on Campus – has operated for over 30 years at Simon Fraser University in Burnaby. This predominantly undergraduate-oriented advocacy group and resource centre is an official student club offering space, care, and respite. While only used by a few students each day, it provides queer-related fiction and non-fiction books, safe-sex resources, peer crisis counselling support, social and educational outreach, and a place to socialize with other

Figure 5.2 Simon Fraser University's Out on Campus meeting space, 2017 (*Source:* Alison Bain).

LGBTQ2S people over work and food (Figure 5.2). With their campus-wide event postering, activism to make washrooms accessible to trans people, and positive-space workshops, Out on Campus affirms its university mandate. Its two paid part-time staff (14 June 2017) describe it as "*the* queer space on campus" and as "a pocket of queerness in Burnaby". They underscore the socio-spatial necessity of concentrating queerness: "queer people need a space for their own survival … [where they] don't have to explain [their] marginalization to people who don't experience those things". Within suburban Burnaby, Out on Campus is where students intentionally co-create a "place for queerness to survive; this is the space for education, support, and safety" (Staff, 14 June 2017).

In Montreal's West Island, John Abbott College's Queer Hub (QH) is a suburban locus of the LGBTQ2S community in a transient two-year educational environment. As a student club, it organizes activities, only specifically marking LGBTQ2S inclusion with a drag show and a college IDAHOBIT flag-raising event. However, students have queerly remade QH's office-sized basement club room as a place "to be with their friends" where "they feel comfortable" (Co-President, 1 December 2023). From the moment the first executive member arrives, gets the key from security, and virtually declares the room open on the Discord server, students access this space, the last member returning the key in the evening. All students are welcome, including cis-straight allies, as long as they observe basic club protocols (e.g., no smoking, drinking, sex or public displays of affection, or garbage) and the club's rules of obtaining group permission to discuss emotional and political topics. For college students who have seldom participated in high school GSAs, QH is their first queer space. Outside of public schools, colleges, and universities, another key queerburban institutional resource is public libraries that, with their informational infrastructure and programming, quietly provide support and recognition.

Public libraries

As "adaptable" civic institutions interfacing with multiple publics, municipal libraries are more than physical buildings offering materials and services. Not just quiet places of educated middle-class contemplation (Freeman & Blomley, 2019) or learning commons of knowledge and skill circulation (Leorke et al., 2018), public libraries, like public schools, have social and educational "common good" service mandates (Pateman & Vincent, 2016). Unlike schools, these lower-intensity and more voluntary institutional meeting places are embedded in locales that reach an all-ages client base (Audunson, 2005). As local community centres and educational repositories, they aspire to represent and proactively engage diverse social groups without discrimination. With adaptive dexterity and significant autonomy, suburban public library systems provide important "queer affordances of care" to queerburbanites through their collections, displays and finding aids, and event programming (Bain, 2022).

For sexual and gender non-normative clients, many factors determine whether a library is perceived as queer-positive, but most especially the content, organization, and visibility of LGBTQ2S collections. When LGBTQ2S materials are lacking, hidden, or difficult to access, unavailability may be interpreted as exclusion. Librarian "censorship by omission", fears about collecting controversial materials, and institutional reliance on "public requests" can compound queerburban library resource deficiencies (Alexander & Miselis, 2007, 44). Cataloguing and shelf-labelling also influence perceptions of availability and visibility (Schrader, 2007). While LGBTQ2S content may be found in different library sections (e.g., fiction, health, parenting, erotica, and/or feminism), the adult and children's 306.76 section (the Dewey Decimal call number for "Sexual Orientation and Gender Identity") holds the largest concentration of non-fiction resources.

Whether Canadian suburban libraries serve as plentiful sources of LGBTQ2S information and visible beacons of inclusion depends upon client demand for materials and staff insider activism. In Laval's nine libraries, some branches have almost a complete absence of LGBTQ2S materials because "there is no interest" (i.e., no one requests or checks them out) (Librarian, 15 July 2022) while others signal allyship through their semi-permanent density. This outcome derives from a central library "floating" collection shared between branches via client demand and LGBTQ2S insider-activist librarians who work to enhance access and create a welcoming environment (Librarian, 15 July 2022). In Burnaby's four-branch library system, insider-activist librarians have built the LGBTQ2S collection by consulting the American Library Association's Rainbow Book List and monitoring the offerings of well-known LGBTQ2S publishers (Melili, 2023). According to its librarian (22 June 2017), the collection is comprehensive meaning that "it's been rare that I've seen something and been interested in it, and it wasn't here". Collections are also shaped by broader societal trends, community needs, and the availability of resources. A Surrey librarian (23 June 2017) noted an increase in trans materials for child and youth audiences, "more titles on parenting an LGBTQ child and on queer and trans healthcare than previously", and "a growth in young adult fiction about 'coming out'". In New Westminster, an insider-activist librarian (11 August 2017) compared her collection to the most recent rainbow lists and reported being "especially pleased at our young adult collection … I found 'best books' for intersex teens, for questioning teens, for trans teens, for younger kids and parent support, and excellent non-fiction". Having a comprehensive collection necessitates offering classics and recent publications, but also analyzing low circulation rates and using "protected status" rather than culling (25 June 2017). Attention-drawing materials are often read in the library or online in digital format rather than checked out in hard copy. A Burnaby librarian (22 June 2017) underscores: "The nice thing about reading eBooks of course is that … nobody else can tell what you're reading". This discretion is important because "it's still not an easy thing to be 'coming out', no matter how much more tolerant we are" (Librarian, 22 June 2017).

Public library collections in Toronto and Vancouver were predominantly in English with small minority language collections to reflect local ethno-cultural demographic patterns. However, these rarely included intersections with LGBTQ2S content (other than in film). The discursive politics of LGBTQ2S resistance, as mobilized through suburban public library para-public institutional allyship, therefore, is distinctly Anglo-American, bringing with it a specific understanding of identities and experiences. In contrast, Montreal's public libraries have predominantly French-language content, except in officially bilingual municipalities. In the small, bilingual Pointe-Claire library system on the West Island, the LGBTQ2S collection is divided across age and language categories that are organized, as its head librarian (14 July 2022) notes, in an "antiquated" French-language catalogue system that obscures materials. Such limitations are overcome by ally and insider-activist librarian prioritization of child and adolescent needs through annual displays for Pride and digital reading lists for other significant dates such as the Trans Day of Remembrance or Pink Shirt Day. Collections, therefore, form the backbone of LGBTQ2S library visibility, recognition, and allyship, which are enhanced by displays and programming.

Across the three city-regions, rainbowized visibility signals LGBTQ2S inclusion using ritualized Pride and IDAHOBIT displays, material and virtual readings lists and bookmarks, and positive-space sticker iconography to disrupt suburban heterocisnormativity. Library materials are incorporated into community celebration calendars to foster social inclusion and increase their visibility and circulation through annual displays. Often arranged on pedestals in central, high-traffic locations, LGBTQ2S Pride and IDAHOBIT displays are interiorized, micro-place-remaking initiatives that temporarily redistribute surplus visibility to queerburban patrons who may nevertheless struggle the remainder of the year to see themselves within collections and programming (Figure 5.3). From system to system and branch to branch, Pride and IDAHOBIT displays vary in size, material currency, use of rainbow semiotics, and location.

Figure 5.3 A teen “Read with Pride” book display at the Guildford branch of the Surrey Libraries, 2017

(*Source:* Alison Bain).

Most Pride displays are produced by insider activist and ally branch librarians: "We do a Pride display in June because we have a transgender and a gay librarian on staff who likes to do one. It will be in the centre of the library … and we'll have a big rainbow" (Librarian, 1 June 2018). However, not every branch in this or other city-regional public library systems will have such displays, as LGBTQ2S patrons are seldom officially named in institutional mission statements as designated "marginalized groups" in need of targeted initiatives. As one Mississauga branch manager (13 June 2018) explains:

> [w]e have never enshrined the words LGBTQ+ anywhere in our mission. When we talk about core programming groups we talk about the pillars of families, youth, newcomers, older adults, and adults. They don't fit into one of our core programming groups and we don't have anyone really telling us "Hey, you should really make sure to include LGBTQ+ representation, or have you tried running a Pride story time?"

Such policy inattention reproduces LGBTQ2S resource unevenness, especially in smaller branches where collections are limited, and community affirming displays are rare unless there is an insider activist on staff:

> If you're at a branch and there is someone who is an ally or part of the community, you're likely to see displays, book lists, decorations, anything like that. If you're at a branch that doesn't have anyone who is connected to the community, crickets!

Pride library display creation is therefore idiosyncratic as an indicator of para-public institutional allyship because it relies on individual insider activists or ally initiation. When LGBTQ2S advocates or materials are absent at a branch, there will be an eclectic mix of materials with limited quality control, if they are mounted at all. There are also many co-occurring community events competing for representation (e.g., June celebrates seniors, First Nations, and Quebec national identity). In Quebec, IDAHOBIT book displays are more common than Pride displays, particularly in predominantly French-language municipalities. In Laval, for example, a librarian (15 July 2022) explains that IDAHOBIT displays "recalibrate" its floating collections because every branch is mandated to have an LGBTQ2S literature display for the event, however paltry.

Pride and IDAHOBIT displays may include LGBTQ2S reading lists and bookmarks created for distribution on library shelves, but also circulated at community events like Pride or digitized and posted on library system websites. The materials they showcase can be lifelines for queerburbanites who cannot afford books and may struggle to find community, yet desire "a way to access queer people", even if only imaginary characters (Librarian, 21 November 2019). Books are "a portable community" that can be "brought wherever you are" (Librarian, 21 November 2019). Simple and functional, LGBTQ2S finding aids enable place-remaking visibility year-round, yet they too require the initiative of insider activists and allies, and labour for compiling and updating in line with local availability. Such work is seldom an individual's "official" responsibility, but rather is voluntarily undertaken (or not). Like Pride and IDAHOBIT displays, the finding aids of book lists and bookmarks are selectively available resources that are rarely official institutional requirements.

Similar to schools, peripheral municipal public library systems largely cater to children, young people, and their families. For queerburban youth whose lives are institutionally circumscribed by heterocisnormativity, libraries provide a free and safe(r) public gathering space outside of home and school, to socialize, study, find information, and participate in programming (Gamerel et al., 2014). To address queerburban youth disenfranchisement that contributes to diminished emotional well-being and belonging, insider-activist librarians and informal community actors have established book clubs, writing workshops, and story times. These initiatives emerging from individual lived experiences, skills, and social networks are rarely profiled. Like reading lists and displays, LGBTQ2S programming is done by individuals without management backing and seldom scaled up. Instead, particular library branches become known via word of mouth as places remade as sites of queerburban care.

At Surrey's Guildford library branch, for example, a Pride Alliance "book club" was established in 2017 by two insider-activist librarians who saw a need for queerburban youth friendship when GSAs dissolved at the school year's end. In the summer, it becomes "the queer hub of Surrey" (16 August 2017), offering queerburban youth a discreet micro-place to gather, talk, eat, craft, have guest speakers, watch movies, and make friends. Under the guise of discussing books, it teaches youth to "create a community for themselves, and become leaders" (Librarian, 16 August 2017). Youth colloquially refer to this queerburban meeting place as "the little gay room" (Youth participant, 2 July 2019), the one place in the library and wider community centre that they have remade together and locate feelings of validation, belonging, and agency. However small, this welcoming and free place, with its co-created ground rules of interaction and protection from prying patron and parental eyes, permits some queerburban youth to "authentically" express their convivial selves using their preferred pronouns, and socializing without fear of discrimination and harassment (Bain & Podmore, 2020b, 2021b). As word has spread, queerburban adults and seniors have requested a similar social space. Yet, as with many peripheral municipal public library services for queerburban communities, this library branch and others remain youth focused.

In contrast, suburban libraries in the Toronto city-region have experimented with the creation of adult suburban library clubs. A Brampton library branch, for example, partnered with a local AIDS Service Organization in 2018 to pilot an LGBTQ2S book club following insights gained on LGBTQ2S service needs during a Regional Diversity Roundtable. Since local library management opposed the initiative on the grounds of limited demand and concerns about funding the purchase of new LGBTQ2S books, an insider activist and an ally librarian persisted and joined forces with the library systems in Mississauga and Caledon to host a seasonal online Pride book club as a way to assume "collective responsibility" for "uphold[ing] safer spaces for all community members" (City of Mississauga, 2019). A longer running grassroots community micro place-remaking initiative is the bimonthly queer writing group Writing Rainbow launched in 2011 by a local resident (24 November 2019) at Ajax's main library branch. The goal was "to create a creative community of fellow LGBTQ+ people because it's very difficult to meet queer people in the suburbs". An LGBTQ2S-identifying branch programme coordinator in neighbouring Pickering found a free meeting space for the group which continues to mentor aspiring LGBTQ2S writers who learn about the volunteer-run programme through word of mouth.

LGBTQ2S library programming also includes Drag Queen Story Hours (DQSHs). Initiated in 2015 in San Francisco by author-activist Michelle Tea, DQSHs have spread across North American and British library systems as a mode of queer cultural mainstreaming. These early reading programmes involve drag performers in colourful makeup and clothes reading developmentally appropriate children's books that celebrate diversity, provide LGBTQ2S representation, and use inclusive language (Radis et al., 2022). More common in central cities, many suburban librarians remain cautious about organizing DQSHs. As Mississauga's library system manager for community outreach and programming explains, while they want to be an institutional ally, they are not ready to jump on "the current hot trend in programming" because of uncertainty about the "readiness" of the community and fears of "rushing in" only to "create a situation that isn't positive" (5 July 2018). This reticence speaks to the persistence of heterocisnormative constructions of suburban childhood and family (Bain et al., 2020). "The Child" remains abstracted as a figure of "innocence" needing "protection" from the knowledge of same-sex relationships and diverse genders to ensure national reproductive futurity (Edelman, 2004). Heteroactivist backlashes against DQSHs have occurred in Vancouver and Montreal suburbs with religiously conservative populations and growing support for right-wing populism.[10]

In municipal library branches across the peripheries of Montreal, Toronto, and Vancouver exist a range of small and free programming initiatives run and supported by insider-activist librarians and allies that make micro and temporary places for queerburban clients. These publics are often youth who lack alternatives outside of school-based GSAs and online worlds for finding LGBTQ2S communities. Another important, although many activists would argue inappropriate, para-public institutional ally for queerburban publics, especially youth, are municipal, regional, and national police forces.

Municipal, regional, and national police forces

Given the surveillant and hyper-visible position of police forces relative to LGBTQ2S and especially QTBIPOC communities (Weissinger & Mack, 2018), suburban police allyship comes with significant challenges. A "queer-friendly" police image is the contemporary goal (Russell, 2018), but police forces have been agents of homotransphobia, enacting raids on LGBTQ2S communal spaces (e.g., bars, bathhouses, and park cruising grounds) (Bain & Nash, 2007), entrapping men-who-have-sex-with-men in public restrooms (Bain et al., 2020), harassing trans people in the highly policed cabarets of red-light districts (Namaste, 2005), and stoking public panic about "gross indecency" (Dagenais, 2020).[11] While gay village community leaders have sought to build organizational bridges with the police, events such as the 2016 Pride protest by Black Lives Matter Toronto (Davis, 2021) and the serial gay village murders of South Asian and Middle Eastern gay men from Toronto's peripheries demonstrate uneven and necropolitical experiences with police (Haritaworn, 2019). Such traumas have not, however, similarly informed peripheral place-making dynamics and inter-organizational relations. While uniformed police officers are banned from Toronto and Vancouver's Prides, the same institutional distancing has not occurred in queerburbia.

Canada's largest cities[12] are policed by municipal forces, except in the case of the Vancouver city-region where the well-resourced national Royal Canadian Mounted Police (RCMP)[13] have suburban detachments. Given that there are no suburban Pride events in Montreal and few efforts are made to liaise with LGBTQ2S populations outside the Service de la Police de la Ville de Montréal (SPVM),[14] there is limited evidence of peripheral police allyship. In Toronto's and Vancouver's regional and municipal police forces, and RCMP detachments, para-public institutional allyship involves inter-organizational LGBTQ2S advisory committees and community round-tables. Such police allyship efforts can concretize LGBTQ2S equality initiatives leading to outreach activities, event attendance, and inclusion policies, but significant organizational unlearning is also required when police forces participate in queerburban place remaking.

Organization-wide competency in the language, identities, and histories of LGBTQ2S communities is central to this unlearning with the goal of prejudice reduction. Post-2014, many regional police forces have strategic plans for diversity and inclusion, specialists on staff, and internal support networks (ISNs) via WhatsApp offering mentorship, education, and awareness-raising. Some local detachments have established equity, diversity, and inclusion units that strive to build "diversity-competent members" and create supportive workplaces through discrimination, harassment, cultural awareness, and language training. Sometimes it is the settlement of police mistreatment human rights complaints that reactively triggers the development of mandatory, short e-learning modules (e.g., "human rights 101", "language matters", "positive space", or "transitioning in the workplace") and workshops (e.g., "transphobia and LGBTQ+ awareness"), while at other times it is inter-organizational round-table feedback from local school boards and social service agencies (e.g., the 2018 "We Belong" breakfast initiated in the Regional Municipality of Peel that inspired a similar one hosted by the York Regional Police). National training programmes are also offered such as the Toronto Police Service's 2020 online 2SLGBTQ+ awareness course via the Canadian Police Knowledge Network (CPKN) or the RCMP's 2018 "Trans 101" training delivered by transgender Officer Rielly Knock. While the impact of such national training programmes is difficult to evaluate, they present opportunities to change the social climate, foreground the leadership of sexual and gender minorities, and invite critical reflection on heterocisnormative privilege. With their wider reach, the CPKN and RCMP national training programmes can more effectively standardize and diffuse basic queer and trans competency, empowering small-town and suburban detachments to align with federal policy requirements.

In queerburban police divisions and RCMP detachments, queer competency is also informally delivered through local community consultations. In Ajax, an allied insider activist who facilitates the Colours 2SLGBTQI youth programme invited young people to the division to share their lived experiences with local officers. In Surrey, sporadic round-table discussions with community leaders inspired education workshops about police (mis)treatment. As a Surrey RCMP officer (11 July 2019) notes, it is often the transgender/non-binary community with whom police "have the least amount of experience" and thus require training "to make sure that we're staying up to date on what is happening in the community" (RCMP Officer, 11 July 2019). Local ad hoc workshops with community members also help to build queer competency: rather than "blaming and putting officers on the defensive", community dialogue offers a different vantage point and the opportunity to "find the root cause of mistrust" (RCMP Officer, 11 July 2019). Police officer attendance and outreach at community events, especially during Pride, is another important educational tool and practice of institutional allyship.

Police outreach intensifies during community events, often manifesting in queerburbia as iterations of rainbow-washing. In the Regional Municipalities of Peel, York, and Durham in Toronto and in Surrey, there are rainbow-wrapped police patrol cars (with other cruisers wrapped in hot pink year-round to raise awareness about anti-bullying). Costing an average of CDN$1500, these police cars affordably attempt to signal LGBTQ2S institutional allyship. Speciality cruisers are assigned to divisions and detachments and driven in local and more distant Pride parades, bringing colour and social media photo opportunities to queerburbia (Figure 5.4).[15] One police officer (14 September 2018) maintains that the rainbow treatment "pays dividends", while an inspector (13 August 2018) is sceptical because "its effects can't be measured", suggesting it would be better to "spend money impacting a life rather than having a photo op".

Many uniformed police and RCMP officers also choose to wear rainbow pins and pull off Velcro flash panels at ceremonial Pride events. New Westminster's small municipal police force follows the "Vancouver Model" to train officers in "diversity" and works closely with other municipal agencies to build LGBTQ2S inclusion. The first in Canada to adopt rainbow shoulder flash panels[16] to celebrate municipal Pride events, it reissued police shirts and

Figure 5.4 A Durham Regional Police car wrapped in LGBTQ2S supportive messages at York Region Pride, 2019

(*Source:* Alison Bain).

jackets with outer shoulder Velcro to enable officers to interchangeably adhere specialized event celebration flashers following an annual calendar. When providing security detail at Pride events, police flashers demonstrate institutional allyship, but when these are worn weekly and everywhere, they extend their impact. The Chief of Police (27 June 2018) explains: "If you walk down the street and the people at the Heritage Grill … see it … they're always very appreciative and say, 'thanks' or 'that looks great'", but if you take "a call in Queensboro at the Casino and someone will come up to you and say, 'Why is your shoulder flash like that?' and you say, 'Well, it's Pride week'", it offers moment of public education.

Beyond celebratory events, senior officers (13 August 2018) (particularly if they hold an equity and inclusion portfolio) may choose to defy bureaucratic rules governing rainbowization: "I wear my rainbow pin all the time, contrary to directive … I just keep wearing it and showing up and no one is telling me to take it off". Police officers (28 August 2018) serving in diversity units and/or in community policing as insider activists or allies may also wear rainbow pins to "identify them as safe" to sexual and gender minority citizens. At community events, some may be more playful in their attire, such as the Surrey RCMP insider-activist Corporal Elenore Sturko,[17] a former Media Relations Officer charged with the diversity relations portfolio. An "out" lesbian, Sturko was active in Surrey Pride until her election to British Columbia's Legislative Assembly. At Pride, she wore a rainbow boa over her RCMP red serge, liaised with community members, posed for photographs, and lip-synced on stage alongside drag performers. She supported Surrey's LGBTQ2S community with a speech about her family's story of reconciliation with Canada's Pink Purge. She asserts:

> Our LGBTQ community … [has] me. It does make a difference when you have someone who is in the public eye and willing to be out. It changes people's confidence in having a conversation when you know they have chosen a spokesperson who is part of the community.

Serving as a media relations officer in this suburban detachment, Sturko used her body, appearance, and story to model queer inclusion and reconciliation.

For other municipal police and RCMP officers, LGBTQ2S event attendance falls under the remit of community engagement and diversity portfolios, reorienting institutional allyship outwards. In the Regional Municipality of Durham where the Pride Durham Committee has voted to continue to involve uniformed police in the Pride parade, officers march on their own time, line the streets in specialty t-shirts, and run a recruitment booth, "grateful" for the opportunity to "show solidarity and respect for others" (Police inspector, 13 August 2018). Some have volunteered on the planning committee for the Durham Pride Prom since its inception in 2008 and served as duty officers at the event, generating only "positive feedback" (Health promoter, 12 July 2018). In Surrey, RCMP officers (27 June 2017) volunteer at LGBTQ2S events, like Surrey's Pride festival, as part of their community policing programme: "We're just here to show support for everybody. We're not here to police". Aware that their uniformed presence remains controversial and sometimes unwanted, they await LGBTQ2S community welcoming signals. In Toronto suburbs, there is an appreciation for police forces that invest in LGBTQ2S community outreach since they are "working harder than Toronto police at being respectful and checking in with the community" (Social worker, 22 August 2018).

A significant act of institutional allyship comes from the rainbow ritual of Pride and Trans flag-raisings which synergize with police ceremonial protocol. By adopting rainbow iconography, suburban police detachments offer official sites of recognition that counter homotransphobic and heteroactivist constituencies. With their accompanying media relations, photo opportunities, and guest speeches by Chiefs of Police and LGBTQ2S community leaders, police detachment flag raisings for Pride, IDAHOBIT, and the Trans Day of Remembrance may also signify LGBTQ2S allyship when municipal support is not forthcoming. Surrey's long denial, for example, of activist requests to raise the rainbow flag over city hall inspired the local RCMP detachment to host an ad hoc ceremony in 2019 at its main entrance (Figure 5.5). Media coverage was amplified when the right-wing group "Culture Guard"[18] protested, asserting that the RCMP should represent all Canadians without

Figure 5.5 Flag raising at the entrance to the Surrey RCMP detachment, 2019 (*Source:* Alison Bain).

pandering to special interest groups. Heteroactivists carried signs shaming rainbowization schemes, referencing Evangelical Christianity, and drowning out LGBTQ2S-supportive speeches with the national anthem. In alignment with their professional mandate to afford public safety, the RCMP was given an opportunity to deepen their institutional allyship by countering community-based homotransphobia.

Suburban police forces also promote local LGBTQ2S allyship through community safety programmes using rainbowized decals to signify ally businesses and institutions modelled on a 2015 initiative from Seattle, Washington[19] now taken up by select American and Canadian police forces. A shield-shaped rainbow decal identifies private and public sector allies who, completing a training programme, can offer LGBTQ2S people shelter from harassment, violence, and bullying and support individuals in reporting these incidents to the police. The City of Vancouver was one of the first Canadian municipalities to adopt the programme, followed closely by Surrey and the Regional Municipality of Peel on the initiative of their respective diversity units. In Surrey, the RCMP's "Safe Place" programme emerged from a desire "to expand services and show support for the queer community" and demonstrate that "we are willing to support our community" to reduce hate crimes (RCMP corporal, 11 July 2017). A similar "Safe Place" programme was unveiled in 2018 by the

Regional Municipality of Peel, Ontario's first, by the eight-member Equity and Inclusion Bureau team responsible for community engagement and consultation. This programme has helped to rebuild "trust with the LGBTQ+ community post-bathhouse raids in Toronto", countering LGBTQ2S concerns about "not feel[ing] seen and represented in the community through signage" and feeling "like ghosts" (Peel police officer, 14 September 2018). Within the programme's first six weeks, 100 Peel businesses and organizations registered, yet without an interactive map to guide queerburban citizens, it remains unclear which sites display the Safe Place decal. A website invites a commitment pledge to "providing training to your employees and/or volunteers and posting the decal on your organization's window to showcase your support", but the self-directed training merely involves reading an orientation letter, reviewing a PowerPoint presentation, and committing to employee knowledge-sharing prior to affixing the rainbow Safe Place sticker in a high-traffic and visible location. Nevertheless, in queerburbia, with its dearth of tangible LGBTQ2S inclusions, these various "Safe Place" programmes are important second-order changes that turn the policing gaze outward to build future multi-organizational partnerships in the service of community safety.

Safety programmes like Safe Place involve community outreach and relationship brokering, a key expression of institutional allyship that can lead to facilitated dialogue around ongoing LGBTQ2S discrimination, safety concerns, and service delivery. For example, the Police Chief's LGBTQ+ Advisory Committee of the Peel Regional Police (complementing five others for Muslim, Chinese, Black, Sikh, and Youth constituencies) was established in 2016. Its quarterly in-person discussions bring local non-profit community health and social groups (e.g., Serving with Pride (SWP),[20] Peel HIV/AIDS Network (PHAN), Mississauga Community Health Network, and Peel Pride) into collaborative dialogue with the police. In the Regional Municipality of York, the "out" lesbian Police Superintendent (22 August 2018) established the Rainbow Network Table, a community consultation group to address LGBTQ2S York resident needs and enable inter-organizational "relationship brokering". Its quarterly meetings allow community social service agencies with LGBTQ2S programming mandates (e.g., public school board, AIDS Committee of the York Region, 360 Kids, York Pride, and York PFLAG) to discuss issues and establish the trust necessary for future collaborations. For LGBTQ2S organizations, such community-police alliances may raise the profile of their concerns, create funding opportunities, strengthen networks, and permit direct contact with a legitimate arm of the state. For police forces, LGBTQ2S community-police alliances enable affirming interactions of awareness-raising and education with often difficult to reach target groups.

Police forces engage in second-order institutional allyship through four modalities: queer competency training, community event attendance, safety programmes, and inter-organizational relationship brokering, all of which illustrate the challenges of scaling up from first-order (typically superficial) to second-order (intentional actions). Queerburban police institutional allyship is, however, a place-remaking compromise. Despite a history of oppression by police, many queerburbanites make do with the support of a punitive arm of the state that offers occasional coalitional moments but primarily extends allyship via rainbowized inclusion to ameliorate its image of criminalization and securitization.

Churches

Strategic compromise remains part of queerburban place-making with religious institutions, notably Protestant Christian churches. Whether in its liberal or more orthodox and conservative forms, LGBTQ2S people have contradictory relationships with organized religion (Fetner & Kush, 2008), nowhere more so than in the suburbs. Whyte's (1956) sociological analysis of American middle-class suburbia depicts a "hollowing out" of religion (later described as "wholesale abandonment" (Dwyer et al., 2013)) that speaks to attendance decline but also a pulling away from ritual, hierarchical constraint, and liturgical rules (Gilbert et al., 2018). For Whyte (1956), the "church of suburbia" is a "socially useful church", meaning that the church provides a social focus in a world of transience. Churches also have "social issues" missions, with contemporary Protestant outreach ministries targeting "what we can do together" to create "safe and peaceful neighbourhoods" for families (Minister, 27 June 2017). While Protestant "megachurch" extrovert evangelism is a public imaginary associated with North American suburbia (Warf & Winsberg, 2010), in Canadian queerburbia it is the small, older Protestant churches of previous eras that implement LGBTQ2S programming to renew their relevance.

Religious contexts can, however, be hostile and traumatic places of queer damage and vulnerability. Christianity has been imbued with homotransphobia, macro- (excommunication and shunning) and micro-aggressions (appeals to "pray the gay away" that necessitate "walking a tight line of who to come out to") (28 November 2019)), and repeated negative messaging about deviancy and the hellish consequences of the wages of sin (Seitz, 2017). The Free Methodist Community Church in New Westminster, for example, is described by an Indigenous trans woman as an institution that "bashes" people like her. Even driving by this church reminds her "that there are people who hate me, and the world is not … safe and welcoming" (28 June 2019). In addition to propagating hate and being gateways to conversion therapy (Drescher et al., 2002), some churches are places of reparative care work, affording mental health protection against minority stress (e.g., depression, anxiety, and suicidal ideation), healing and support, and a deeper meaning or purpose (Lomash et al., 2019).

Due to the predominance of Christianity among Canadian settlers, parish churches were central to post-war suburbanization, becoming focal points for heterocisnormative associational life. While Catholic (predominant in Quebec), Evangelical, and Orthodox churches continue to interpret homosexuality as a sin, mainline Protestant churches (e.g., Anglican, Lutheran,

Methodist, Presbyterian, and United) respond variably to LGBTQ2S congregants, ordination, and same-sex marriage (Dickey Young, 2008). Since the 1970s, with rising secularism[21] and the declining community role of churches, congregations have shrunk. Many suburban mainline churches are now facing financial unviability, consolidation of land holdings, and congregational "diversification". These suburban relics (frequented by ever-smaller, aging, and commuting congregations) sustain their vitality through "adaptive dexterity" (Watson, 2009, 319). They integrate multiple uses and publics, share spaces with ethno-cultural congregations, forward Indigenous reconciliation, and embrace new family configurations (Rayside & Wilcox, 2011). To overcome heterocisnormativity (Dickey Young, 2008), queerburban church place-unmaking and remaking involves accepting LGBTQ2S ordination, performing same-sex marriages, and welcoming LGBTQ2S congregants.

Churches are more than merely "silent" or "tolerant" spaces (Acosta, 2020). Mainline Protestant clergy have facilitated for "homosexuals" (Cadge & Wildeman, 2008, 587),[22] advocating for sexual minority rights, challenging ordination and marriage doctrine, and providing political and social meeting spaces (Schroeder, 2012). The ordination of openly gay clergy has transformed church leadership, sensitized congregations through inside-activist embodiment, and linked suburban churches to regional interfaith and national affirmation movements. In Canada, such ordinations have transformed United (and some Anglican) churches, as ministers cycle through suburban appointments.[23] At the pulpit, ministers perform institutional allyship, giving sermons about "how to embrace difference, keep an open mind, and learn about, engage, and accept people's identities" (Minister, 27 June 2017). They guide congregations towards affirmation because it is "important to get people ready before a church indicates with symbols that they are a safe space" (Minister, 27 June 2017). In contrast with other institutional rainbow-washing, LGBTQ2S affirmation is a sustained process of collective reflection.

The Richmond Hill United Church on Yonge Street in Toronto's York Region exemplifies critical institutional allyship through LGBTQ2S leadership. It became the first affirming church in the York Regional Municipality in 2013 and, in 2016, hired its first openly gay married minister. A self-described "boring house husband", he "continues to be in your face about it", refusing to "apologize for who we are" as a couple or a church. Rarely called upon to perform same-sex weddings, his congregation's allyship focuses on "advocacy and being a community that would speak up for sexual and gender minorities rather than being a source of celebration" (Minister, 18 June 2018). Given that Richmond Hill is a "religiously and ethnically diverse community where being out isn't a gift" (Minister, 18 June 2018), he strategically uses rainbowization – a Pride sticker on the church doors, and rainbow communion candles – to navigate local antagonisms (Figure 5.6). He hosts a table and provides a service at the York Region Pride event "to engage with folks for whom church is a negative experience". Interfaith leadership activities (e.g., Mayor's breakfast) offer another opportunity to "raise the profile of LGBTQ issues with Imams

Figure 5.6 The adaptive dexterity of the affirming Richmond Hill United Church, 2018 (*Source:* Alison Bain).

and Rabbis who are often silent" (Minister, 18 June 2018). Within this suburban context, LGBTQ2S place-unmaking and remaking occurs through a constant embodied presence at the pulpit and out in the community.

Burnaby's Metrotown Centre in the Vancouver city-region is home to another (not officially affirming) United Church, Jubilee United, where a gay minister juggles LGBTQ2S inclusion alongside competing responsibilities for congregational mergers and land redevelopment. An insider activist, he remakes the church as LGBTQ2S-positive by: giving an annual Pride sermon; mounting a display on the struggle for LGBTQ2S liberation; offering BC Coastal Health Trans 101 training for new congregants; ensuring that church tenants are not homotransphobic; and presiding himself over all funerals and weddings. He (27 June 2017) maintains that "No matter who you are, no matter where you come from, regardless of your gender, your sexual orientation, your colour, your economic status, whatever you think might keep you from being close to God, you're welcome here". Such sustained critical institutional allyship is mobilized during the United Church's affirming process.

While Anglican, Methodist, and Presbyterian churches continue to debate LGBTQ2S inclusion and same-sex marriages, the United Church of Canada, in response to internal pressures regarding ordination and the legalization of same-sex marriage, has prioritized LGBTQ2S affirmation. Affirm (which became Affirm United) was created in 1982 to advocate for lesbian and gay members, but the national church became officially affirming in 1988. Local congregations could then choose to undertake the multi-year Affirming Ministries Program, a self-reflection exercise involving LGBTQ2S congregant consultations, discussions about theological discrimination, and meetings with other affirming ministries (Affirm United, 2019, 2021a, 2021b). The process concludes with a congregational vote of public commitment (with a vision statement, a plan of action, and an equal marriage policy) and a celebratory joining of Affirm United. Affirmation is a significant form of critical institutional allyship that leads to public, explicit, and intentional second-order change that is more transformative and reliable than mere rainbowization. Nevertheless, remarkably few Canadian congregations have undertaken this process,[24] the majority of which are located in Ontario and in central-city areas.

Across the three city regions, the greatest density of suburban-affirming churches is in Montreal's West Island municipalities. Of the 12 United Churches in the West Island Region of the Nakonha:ka Regional Council, six have openly gay or lesbian ministers and five are officially affirming.[25] Pointe-Claire's Cedar Park United became affirming in 2009, but it was "an after the fact catch-up" for a church with a liberal congregation (Minister, 6 August 2019). The church has a "passively explicit" rainbow iconography of a flag, a windsock, an Affirm lozenge-shaped logo that is used on all communications, and sanctuary praise banners. During Pride, it hosts a sermon for all West Island United Church congregations (Figure 5.7), scheduled prior to Montreal's celebrations to ensure high levels of attendance. Cedar Park's approach to affirmation embraces critical institutional allyship, treating the church as a place "where our questions are valued, our diversity is celebrated and where a certain humility about our convictions leaves ample room for apology, forgiveness, and reconciliation" (Minister, 6 August 2019).

In British Columbia, New Westminster's Shiloh 5th Avenue United Church has an ally minister, who has helped remake this place as "*the* LGBTQ2S church". It became "affirming" in 2004, without much debate because the congregation was "really open minded, very mission focused" and the "church was dying around their ears" (Minister, 28 June 2017). The congregation was made up of predominantly older women and middle-aged people who wanted to be part of a "mission-oriented" congregation committed to "radical inclusion" and LGBTQ2S affirmation. Affirmation gave this church its LGBTQ2S reputation. It attracted new members following print-media stories about hate mail as they concluded that this church is "standing up for the right thing" (Minister, 28 June 2017). Financial struggles meant the church gave up its historic building and rented shared space at nearby St. Barnabas Anglican Church headed by a lesbian-identified minister. Consolidation has increased LGBTQ2S resources through inter-organizational synergies with Narcotics Anonymous, Pride in the Spirit, a food bank, and anti-renoviction activism. The affirmation process, then, has helped unmake elements of suburbia's heterocisnormative

Figure 5.7 Cedar Park United Church decorated for the West Island Pride sermon, 2019 (*Source:* Julie Podmore).

religious landscape by forming well-resourced (although underutilized) institutional nodes of allyship.

Community engagement – to neighbouring single-parents, lonely seniors, low-income households, refugee newcomers, and palliative patients – through food and clothing provision, music, gifts, and companionship has long been part of church caring mandates, offering a precedent for LGBTQ2S outreach. For some queerburbanites, churches offer fellowship, music, "balance", and "quiet" (14 June 2019). In combination with Bible studies, coffee hours, and Sunday school, youth groups are also staple church socializing components that adapt to support increasing numbers of trans and non-binary young people. Some churches donate their halls as LGBTQ2S-welcoming meeting spaces. Others target youth and seniors for outreach or to redress queer poverty.

In response to the social isolation of her transgender nephew, the minister at St. James the Apostle Anglican church in the York region established We Are a Safe Place. With PFLAG York Region allyship training and Diocese of Toronto funding, they sought LGBTQ2S mentors and bought print and multimedia resources for young people: "they get it with each other; they go off and read, run around, scream 'I'm gay,' and play videos" (Youth group facilitator, 22 June 2016). Together they attend and march in the York Region Pride and a small spin-off parental support group has formed. While not officially an affirming church, the minister preaches about inclusion and the "congregation is on board" with the youth programme. Personal relationships are leveraged as coalitional change-making moments. During one service, for example, the minister's sister explained the journey of her transgender son, requesting that "if you can't use his new name and pronouns, please say nothing" – for which she received a standing ovation. These sisters have become LGBTQ2S allies, inviting their congregation to join them. As the minister explains, "I simply support LGBTQ people in my life and community … without asserting my agenda. I back off and let gay people have a voice without putting my words in their mouth" (Youth group facilitator, 22 June 2016).

The prioritization of LGBTQ2S outreach, especially trans ministry, was most visible at New Westminster's non-denominational Lighthouse of Hope (LH) church that later went exclusively online (broadcasting from Victoria, BC to 52 different countries where it is often not possible to be openly LGBTQ2S or Christian or both) in 2022. A satellite evolution of the downtown Vancouver Rainbow Community Church, LH grew out of a home-based West End gay and lesbian Bible study group established in 1985. Gay, HIV, trans, and anti-poverty activism have always informed its mission. It maintained a tradition of AIDS Hospice caroling and distributing toiletry "bags of joy" on wards for "people who don't get love and don't get gifts" to show that "there's somebody thinking about you, and you do matter" (Congregant, 16 August 2017). Of the once 50 congregants, many identified as trans and held leadership positions. Allies were attracted to the congregation's humility, its service, and its mandate to "make a difference in the world" beyond being a "holy huddle" that "meets on Sunday for an hour and goes home and checks it off the to-do list" (Congregant, 16 August 2017). Without a dress code, congregants come as they are: "If you're sad that day, that's part of the human experience and we've all had sad days, and maybe we can encourage you. And if you're struggling with addictions, that's okay. And if you're on welfare, that's okay". Through its outreach, leaders and congregants of New Westminster's Lighthouse of Hope aspire to "put our faith into action" and that "we can be people of faith and be who we are" (Congregant, 16 August 2017).

The amassing of LGBTQ2S service delivery through the church as a critical institutional ally is particularly apparent at the West Island's Beaconsfield United Church. Its long-time (now-retired) gay minister (10 December 2019) focused on "poking holes" in the West Island "bubble" and looking for "substantive change in congregational life". While he experienced significant homophobic hostility in 2000, a 3-year affirmation journey concluded in 2011 with the establishment of the West Island LGBTQ2+ Centre, the only one of its kind in Canada. This sustained political commitment to affirmation brought a continued willingness "to do" and "to live out your affirming status" (Minister, 10 December 2019). The Centre occupies the ground floor offices of the former parlour of the "ladies' auxiliary" (United Church Women) in Centennial Hall. The two-storey church hall (shared with a daycare, a dog training company, and a commercial gym) has a gymnasium, kitchens, and meeting rooms.

Inter-organizational networks animate the building and its LGBTQ2S place-remaking. “All the groups are interconnected in one way or another” (Minister, 10 December 2019): a women’s auxiliary bakes cookies for the queer prom and the drag show while people with intellectual disabilities prepare food for a seniors’ breakfast (before their line-dancing in the gym). The minister (10 December 2019) prioritized “gathering people into community”, recognizing that “buildings have their own ministries, if you let them do it”.

Institutional survival necessitates that many Canadian suburban United Churches become LGBTQ2S allies by changing their leadership, becoming affirming, conducting outreach, and offering care. Such adaptive dexterity is demographic, structural, and spatial, remaking places of “vexed attachment” (Seitz, 2017, 4) into nodes of allyship and resource landscape augmentation. Suburban churches, their halls and their service-oriented contemporary congregations, remain “socially useful” in ways Whyte (1956) never imagined, with their spaces remade for LGBTQ2S youth, parents, newcomers and refugees, and seniors’ peer-support and sociability. Their established and extended social service mandates to alleviate exclusion through poverty and harm reduction by caring for the sick, lonely, displaced, and unhoused have created “good enough” churches (Seitz, 2017) that sporadically ally with queerburban populations.

Conclusion

In queerburbia, a select handful of para-public institutions – public schools, colleges and universities, public libraries, police forces, and churches – compensate for limited political opportunity structures and non-medicalized resource landscapes of suburban municipalities by becoming LGBTQ2S second-order institutional allies. They create micro-spaces of LGBTQ2S recognition and “coalitional moments” that release – albeit temporarily – suburban places from their predominant heterocisnormative configuration. Together, suburban queer student clubs, GSAs and Pride proms, LGBTQ2S library reading lists, displays and book clubs, Police training, rainbowization, safety programmes, and flag raisings, and church Pride sermons and affirmations augment and cohere queerburban resource landscapes, often leading to the inter-organizational exchanges required for second-order change.

Suburban para-public programming, however, prioritizes LGBTQ2S youth to the neglect of sexual and gender minority adults and seniors with little regard for intersections with differences in racialization, immigration status, ability, health, religion, mother tongue, and housing access. For secondary and post-secondary educational institutions, youth are their clientele. Public libraries share this educational mandate, yet they serve the whole population, with collections disaggregated by age (children, teens, and adults) and dedicated youth spaces, resources, and programming within their neighbourhood catchment areas. Police forces simultaneously pathologize marginalized youth – including LGBTQ2S-identified youth – as “delinquent” while working with community para-public partners on trust-building outreach initiatives to foster positive working relationships with future generations. Finally, churches, like other religious institutions, are sites of heterocisnormative familial association, operating Sunday schools and youth groups to socialize young people into Christianity's morals and principles. Through the lens of these select para-public institutions, therefore, it is often young people with their malleable "innocence" and desexualized "purity" who embody suburban futures. They are its temporal directive and social purpose and their reproduction is written into suburbia's morphology and institutional fabric.

The para-public release of suburbs from heterocisnormativity, however, is highly institutionalized and dependent upon insider activists. As the chapter demonstrates, para-public allyship with queerburbanites is also geographically uneven, temporary, and seemingly haphazard, mirroring the significant variability in how and why institutions enact second-order change. Strategic extensions of the state, they are indirect sites of citizenship charged with taking "command of the social individual" (Moten & Harney, 2013, 17) to condition, discipline, and reproduce new citizens. Thus, para-public resistance cannot be radical; rather, they enact change by extending their preferential resource access and promoting inclusion representation by generating welcoming spaces. Para-public institutions are furthermore led by professionalized community leaders who embody order, authority, and guidance and are synonymous with suburban public life. In contrast with grassroots LGBTQ2S activists and advocates, these teachers, professors, ministers, librarians, and police officers (along with doctors, public health nurses, social workers, and psychologists) are invested with trust, competence, and responsibility, giving their institutions influence to build consensus around LGBTQ2S institutional allyship.

Notes

1 For example, institutions that provide health services (i.e., hospitals, community health centres, sexual health drop-in clinics, or social and family service agencies and their self-help and support groups) were omitted due to the complexity of comparing their place-making role across city-regions. While these institutions and social service agencies are part of networks and partnerships, providing support for LGBTQ2S individuals and their families, they are not the allyship linchpins of queerburbia.

2 In 1997, the British Columbia Teachers' Federation (BCTF, 2023) was the first Canadian teachers' union to pass a motion establishing a program to combat homophobia and heterosexism in schools. They continue to produce LGBTQ2S curricular resources, lead professional development workshops, and support LGBTQ2S staff and students through conferences, programming, and GSA guidelines.

3 Equity, diversity, and inclusion officers consider sexuality within a wider social inclusion mandate of religious, accessibility, and accommodation needs. They may offer training via workshops, events, displays, and resources, function as advocates and supporters for marginalized social groups, provide policy advice, and assist with discrimination and harassment complaint resolution aligned with provincial human rights codes.

4 The sexual and gender diversity offices of some colleges and universities prioritize the needs of LGBTQ2S students. At the University of Toronto, for example, the

Sexual and Gender Diversity Office was established in 1998 to serve the downtown campus, but it now extends to Scarborough and Mississauga and finances the university's positive space campaign.

5 LGBTQ2S-positive-space designation was later integrated into community-based health and social service agencies to indicate welcome in environments historically marked by barriers to access due to homo-, bi-, and transphobia (MacDonnell & Daley, 2015).

6 The question of gender identity protocols in Canadian schools is as contentious as in other Anglosphere countries. Anti-LGBTQ2S protests and counter-protests over LGBTQ school policies erupted across the country in the fall of 2023 ("Arrests, heated exchanges", 2023).

7 New York City high students founded the first school-based gay group in 1972 (Johnson, 2007), but it was not until the late-1980s that a Los Angeles pilot project to protect students from bullying influenced the founding of the first three Gay-Straight Alliances in Boston (Fetner & Kush, 2008). GSAs became more widely adopted in American schools in the 2000s.

8 Nearly two decades later, full acceptance of LGBTQ2S students remains elusive in the DCDSB, where it took until 2015 to approve Durham Pride Prom posters, and 2021 to adopt and fly the Pride flag.

9 The Change Room Project was initiated by the former Olympian and principal of the University of Toronto Scarborough campus, Bruce Kidd, in collaboration with a university-community Pan American Games Pride leadership group (University of Toronto, 2015). It critically engaged LGBTQ2S inclusion in athletics by foregrounding student experiences of campus locker rooms. Quotations from a student survey became vinyl wall displays in athletic change rooms and common areas across the university's three campuses during the Pan American Games.

10 Guildford staff who hosted the Surrey library system's first DQSH in 2017 received hateful e-mails but continue to host rainbow story times to celebrate Pride month. In 2023, protesters in neighbouring Coquitlam argued that DQSHs were "sexualizing" and "psychologically harm[ing] children" (Junos & Bowman, 2023, para. 3), resulting in counter-protests that required Royal Canadian Mounted Police (RCMP) presence. Montreal's West Island Dorval library faced backlash in 2022 when it announced that it was hosting a DQSH featuring the well-known entertainer and educator Barbada de Barbades. Her DQSH again made headlines in Ville Saint-Laurent when it was cancelled by the Mayor and City Council.

11 Yet records of repressive interactions between police and LGBTQ2S populations provide some of the only archival evidence of pre-Stonewall era queer lives (Houlbrook, 2005).

12 In the Toronto city-region, there are separate forces for the four regional municipalities of Durham, Halton, Peel, and York. In the Montreal city-region, on-island suburbs (such as the West Island) are served by the Service de la Police de la Ville de Montréal (SPVM) while those off-island have regional or municipal forces.

13 Founded in 1920, the RCMP is responsible for the enforcement of federal law throughout Canada (border patrol, customs regulation, dismantling organized crime networks, and counter-terrorism). Members are also local police and peace officers in every province and territory except Ontario and Quebec which are served by provincial forces. In all other provinces, any jurisdiction lacking a police force is served by the RCMP (e.g., peripheral municipalities, small towns, and Indigenous reserves).

14 The SPVM geographically limits rainbow flag raisings to its gay village detachment, Poste 22.

15 Mississauga Fire and Emergency Services featured rainbow colours on a fire truck for the month of June, 2023, touring it around the city and taking it to Mississauga's Pride flag raising and the Toronto Pride Parade.

16 Originally, the force had worn rainbow epaulettes, but in 2017, LGBTQ2S community members requested greater visibility on the upper arm because they had been invisible beneath traffic vests.
17 Serving on the RCMP's inaugural LGBTQ National Advisory Committee in 2016, Elenore received the Canada Pride Citation in 2020 in the name of her gay great-uncle Sergeant Van Norman, who was a victim of the Canada's "LGBT Purge" in the 1960s. To restore his honour while reconciling her family's past and her own role as an openly lesbian police officer, she published his Mountie service journal. She also served on the LBGT Purge Fund National Monument Visioning Committee and the Curation Advisory Counsel for the Canadian Museum for Human Rights LGBT Purge Exhibit. A former Air Force officer and member of the Mounties' Musical Ride, Sturko embodied national institutional reconciliation and inclusion of sexual minorities. She has spoken publicly on behalf of both the LGBTQ2S community and the police agency, often from a homonationalist position that stressed her heroism and queer liberal subjectivity.
18 Culture Guard was founded in 2011 when heteroactivists mobilized in opposition to the implementation of Sexual Orientation and Gender Identity (SOGI) policies by local school boards.
19 The Seattle program arose out of a desire to rectify past police harassment and neglect, and increasing levels of LGBTQ+ hate-crime.
20 Serving with Pride (2022) is an Ontario-based non-profit organization that supports the creation of safe working environments for 2SLGBTQ+ employees in policing, corrections, and criminal justice professions via education and awareness raising.
21 In 2021, 35 percent of Canadians self-declared that they are non-religious up from 17 percent in 2001 (Statistics Canada 2022).
22 Since the 1960s, North American progressive clergy have advocated for homophile and gay and lesbian rights in the face of heteronormative violence and oppression as part of their social justice role of defending marginalized groups and advocating for their rights (White, 2015).
23 The United Church of Canada voted in 1988 not to discriminate against ordination candidates on the basis of their sexual orientation. In 1992, Tim Stevenson became the first openly gay person to be ordained by the church's British Columbia conference. His ministry at St. Paul's United Church in Burnaby began in 1993 and from 2002 to 2018 he served as a Vancouver city councillor.
24 Of the approximately 2800 congregations serving 2 million people nationally, only 10 percent are affirming (Affirm United, 2021a, para. 2).
25 Beaconsfield United (Beaconsfield), Cedar Park United (Pointe-Claire), Roxboro United (Roxboro), Summerlea United (Lachine), and Union United/Merging Waters (Ste-Anne-de-Bellevue).

6 Organizing

Introduction

Beyond municipalities and the para-public sector, queerburbia's "place identity" emerges from the "improvisation, reinvention, and innovation" of the grassroots organizing of activists and advocates (Gilbert et al., 2018, 31; Dwyer et al., 2015). It is the vernacular creativity of these activists and advocates, their suburban scavenging and networking for resources to support LGBTQ2S populations, that remakes queerburban places from the ground up. For as Seamon (2018, 91) theorizes, place identity "relates to how people living in a place take up that place as their world; how they unself- and self-consciously accept and recognize that place as part of their personal and communal identity". Queerburbia's place-remaking occurs in the micro-spaces forged through relational interconnectivities or the complex connections, interactions, and entanglements of individuals and organizations wrought via socializing, volunteering, and fund- and friend-raising (Putnam, 2000; Retzloff, 2015).

This chapter focuses on the vernacular creative efforts of queerburban activist organizing to remake suburbs as queer and trans places (Bain & Podmore, 2021c). Vernacular creativity refers to the hidden and ignored everyday inventions that unmake and remake the socio-cultural life of places through the routine spatial practices and mundane activities of ordinary people who seldom occupy positions of power (Ettlinger, 2010). In Chapters 4 and 5, queerburban vernacular creativity emerged from the underground of grassroots activism, percolating to the institutional surface to gain middleground organizational support and sometimes upperground civic visibility. Institutions are the formal "upperground" of creative cities, while organizations form the intermediary "middleground" and individuals compose the informal "underground" (Cohendet et al., 2010, 92). The middleground is the repository of "innovative micro-ideas", a location where lived underground individual experiences are legibly codified for macro-level place-making institutions and market forces (Cohendet et al., 2010, 97). Its turbulent energy drives queerburbia's vernacular creativity, composed of "different actors with different agendas and goals; it is a place of contestation and negotiation" between the upperground and the underground (Charrieras et al., 2018, 135).

DOI: 10.4324/9781032632544-6

Imperceptible, fledgling, and scarce, LGBTQ2S-specific socio-cultural, health, service, and advocacy spaces are the organizational by-products of strategic collaborations with the queerburban underground and its intermediary middleground, producing LGBTQ2S-positive sites of information and tacit knowledge exchange, interpersonal relationship-building, self-expression, and resource access. Traditional urban social movement scholarship distinguishes between cultural or service provision, institutional advocacy, and social protest (Levitsky, 2007). *Cultural organizations* run activities and events (e.g., festivals, fundraisers, shows, dances, sports leagues, or tournaments) while *service organizations* provide (in)tangible resources and/or services (e.g., health care, counselling, legal aid, crisis telephone lines, support groups, or consciousness raising). *Advocacy organizations* lobby and litigate to politically influence institutions (e.g., media, religious, or political-electoral), while *protest organizations* use disruptive tactics (e.g., demonstrations, marches, or sit-ins) to shape politics, public officials, and public opinion. In queerburbia, however, large aggregate populations seldom discursively occupy public spaces and community services need disaggregation as "the heteronormative world" often "conveniently amalgamate[s]" them "into a composite" (Hess & Bitterman, 2023, 239). Using six non-exclusive categories, the Hess-Bitterman Taxonomy of North American Community Service Organizations (CSOs) offers an LGBTQ-specific organizational typology, but these too do not directly translate to queerburbia because there is less legal organizing, fewer protest movements and LGBTQ2S-owned businesses, and no research-based cultural institutions.

Adapting these typologies to queerburban activism (Brown, 2007), this chapter prioritizes how suburbia's heterocisnormativities are unmade by organizing queerburban places to socialize, receive care, express identities, and politically organize. It accentuates the relational dynamics of the underground, middleground, and upperground in place-unmaking and -remaking. The empirical discussion opens with Pride as queerburbia's highest profile LGBTQ2S organizing event. The eclectic array of institutional, para-public, and community partners who coalesce around adjacent Pride booths is metaphorically parallelled in the chapter's structured examination of the everyday socio-cultural infrastructure such solidarities create. Following Levitsky (2007), these include LGBTQ2S-specific socio-cultural, service, and advocacy organizations.

Queerburban Prides

As demonstrations of LGBTQ+ opposition to heterocisnormativity, Pride festivals and parades have diffused globally, becoming common grassroots, but also municipal, place-making tools (Bruce, 2016). Usually held at June's close to commemorate New York City's 1969 Stonewall Riot and modelled on the subsequent Christopher Street Gay Liberation Day March (Carter, 2004), Pride is a singular or multi-day event-based cultural tactic of public visibility, support, and celebration that ends with cleaning crews erasing all physical traces. Its volume, colour, and joy confront LGBTQ+ invisibility and offer a reprieve from heterocisnormative intolerance, leveraging play to validate queer existence (Turesky & Crisman, 2023). As the Pride movement has globalized, many municipalities have joined international networks encouraging inclusive place promotion (see Chapter 4), leading to localized civic interpretations of LGBTQ+ visibility.

Canada's largest cities launched Pride events in the late 1970s and early 1980s[1] with oppositional demonstrations that have become large-scale, corporately sponsored, mega-events attracting hundreds of thousands of spectators, participants, and tourists who follow a summer Pride circuit (Podmore, 2015, 2016). In Toronto and Vancouver, suburban Prides emerged from metropolitan fragmentation in the 2010s, with growing opposition to the commodification and apolitical spectacle of central-city Prides[2] and suburban desires to host independent events that often omit marches in favour of rainbowized family-friendly festivals (Bain & Podmore, 2022; Podmore & Bain, 2023). These suburban Prides are visible civic place-making initiatives but also articulations of queerburban vernacular creativity that temporarily unmake and remake suburbia through place-based and regional community solidarities (Bain & Podmore, 2021d). While they are celebratory moments of LGBTQ2S inclusion into the neoliberal upperground, they are realized through year-round inter-organizational middleground solidarities, the vernacular creativity of activists, allies, and communities on display in the booths and bodies that come together in these temporary event spaces. This "throwntogetherness" (Massey, 2005) creates an illusion of queerburban density in civic public spaces where Pride attendees express their queerness, transness, and/or allyship in response to the celebration's rhythms, music, and performances (Smith, 2020). Queerburban Prides, however, operate within spatially dictated limits, where the prioritization of family-friendliness and ethno-cultural and heterosexual citizen inclusion submerge overt expressions of sexual alterity (e.g., public nudity, and fetishes) and substance use.

The Vancouver city-region's three case study municipalities host distinct queerburban Pride events, each sharing a standard iconography but differing in organizational capacity, resources, and audience. Surrey Pride is one of Canada's oldest queerburban events, emerging in 1999 following the "Ban the Bigots" fundraising dance to oppose the Surrey book-ban. The Out in Surrey Rainbow Cultural Society hosted a small family-oriented event until it expanded with a 2010 move to Holland Park and renamed itself the Surrey Pride Society (SPS). This shift brought Surrey Pride closer to upperground political power and remade it as a day-long festival with booths uniting para-public institutions (e.g., the library, school board, regional health body, the Surrey BIA, the RCMP, and BC Foster Families), labour unions (e.g., teachers, nurses, postal workers, and public sector workers), religious organizations, recovery groups, political parties, and Indigenous and new immigrant groups. Surrey Pride remains distinctive for its activist history displays and occasional Indigenous leadership.[3] After years of struggle, its public realm claims have changed, first relocating to Surrey's main commercial area with sponsorship from the Central City shopping mall in 2019[4] and then to city hall in 2023 with the new mayor's updated events roster. It took 34 years of underground and middleground organizing to finally achieve municipal sanctioning and access to central civic space.[5]

Where Surrey Pride is derived from the tenacious vernacular creativity and interpersonal networks of local activists, New Westminster garnered swift upperground support for a citizen request to declare the 7 August 2010 as "New Westminster Pride Day". The Royal City Pride Society was founded by a "mature group" of white, cis-gender gays and lesbians who "moved from downtown because 'we've done it'" (New West Pride organizer, 25 June 2017). They wanted to organize a more "smalltown" Pride revolving around the local workers' pubs and quickly discarded the suggestion of a park picnic: "I'm sorry, I don't do little picnics in the park. If we're going to do it, let's do it. Let's create a society and make it official" (Café owner, 22 June 2018). The event was launched by the "Hills n' Heels" fundraiser based on a journalist quip about the topographical challenge of drag queens marching from the bars of Columbia Street uphill to city hall (Café owner, 22 June 2018). Since its integration into the city's festival roster in 2014, New West Pride has become the city-region's second largest Pride event. Now a week-long celebration, it begins with a city hall flag raising and concludes with a day-long, BIA-sponsored street festival. The festival is launched by civic officials and LGBTQ2S leaders gathering for a traditional Two-Spirit Down Ceremony (Figure 6.1). A revitalizing Columbia Street fills with over 30,000 people passing by multiple performance stages and the booths of local cultural, service, and advocacy organizations – "everything in one place" (Community organizer, 15 August 2018). In contrast to Surrey, Pride activists in New Westminster were encouraged to bring the LGBTQ2S underground directly into city hall's upperground, resulting in a civically well-resourced event with funding, permits, security, and access to infrastructure.

Burnaby is a newer addition to Vancouver's peripheral Pride roster, initiated in 2018 through a partnership between the middleground and upperground, with civic leaders capitalizing on the labour, knowledge, and energy of local para-public ally and insider activists. A para-public initiative, Burnaby Pride secured widespread upperground civic staff support who collaborated with local charities and service agencies to form a 25-member Pride Community Advisory Network (PCAN). It is a recontextualized event that tailors Pride iconography, performances, and rituals to civic concerns with immigrant integration and youth support. The city's social planner (9 July 2019) describes how Burnaby's youth and newcomer focus creates an event with "a different feel than New Westminster and Vancouver, which are party-oriented; Burnaby had tweens with their parents out to find information and support and a diversity of people who asked questions across cultures". These upperground social inclusion priorities enacted through para-public agencies are evident to LGBTQ2S residents: "I remember thinking that it was weird that most of the

Figure 6.1 New West Pride after the Down Ceremony, 2019
(*Source:* Julie Podmore).

booths were organizations that serve queer people rather than queer organizations ... it felt like they were here to get in touch with queers and recruit them ... rather than it's an actual community thing" (FG4 New Westminster, 6 June 2019). Without the input of the queerburban underground, Burnaby Pride is the by-product of para-public networked improvisation, bringing middleground pressure to bear on upperground reluctance towards LGBTQ2S-inclusive civic reinvention.

The Vancouver city-region examples provide three articulations of the vernacular creativity involved in Pride organizing and its underground, middleground, and upperground inter-organizational relations. The youth orientation of queerburban Prides initiated by upperground interests is paralleled in the Toronto city-region. The Regional Municipality of York, with the theme "Show your PRIDE, fly the flag!", was the first Toronto suburb to host its own Pride event in 2001. With a well-documented history (York Pride, 2024), it has become one of the GTA's largest peripheral Prides. In 2013, this event featured a parade down Yonge Street in Richmond Hill before being relocated to Newmarket's Main Street in 2017, which brought more streetscape visibility, storefront synergies, and upperground support. For queerburban youth who may not go downtown to Toronto's Pride, York Pride (like Burnaby Pride), is described as "a small dose of what they are looking for in a positive and

Figure 6.2 "Pride in the Square" in Brampton's Garden Square, 2018
(*Source:* Alison Bain).

family-friendly environment" (Minister, 22 June 2018). Young people "love participating" and the opportunity to "speak their truth, loud and proud" to the applause and tears of onlookers (Minister, 22 June 2018).

Racialized politics cross-cuts queerburban Pride organizing in the Toronto city-region as illustrated by the Peel and Durham regions. The Peel Pride Committee was founded in 2002 by a network of volunteers who, reflecting civic power relations, were predominantly white, middle-class, and cis-gender. Their first Pride event attracted approximately 100 people was a family picnic in Brampton's Heart Lake Conservation Area, with subsequent annual picnics drawing up to 500 attendees and funded by LGBTQ2S dances. A stronghold of South Asian culture, Peel Pride was in Brampton. In a gradual move from the white underground to the upperground, Brampton's civic leaders offered Garden Square and covered the cost of its 15th anniversary celebration "Pride in the Square" in 2017 (Figure 6.2). When the Peel Pride Committee disbanded in 2019, the municipality took over the organization of Brampton Pride and made Garden Square its home. In reaction to the lack of spaces and representation for queer and trans Black, Indigenous, and people of colour in Peel, activists formed QTBIPOC Sauga in 2016. As a QTBIPOC Sauga Instagram post asserted regarding Peel Pride (cited in Costello, 2020):

> organizations that don't run programs or offer services to QTBIPOC folks rush to declare the month with rainbows. This is one of the few times of the year when QTBIPOC folks get multiple gigs ... But there aren't policies to support QTBIPOC life. No representation in decision-making. No resource allocation. No significant change.

In an effort to counter "the idea that all queer and trans folk in Peel are white" (QTBIPOC Sauga, 2024), Moyo Health and Community Services partnered with the Black Coalition for AIDS Prevention in 2020 to host Ebony Pride, the first Black Pride counter-event in Brampton.

Some of the activists who formed QTBIPOC Sauga had organized the 2008 Pride alternative called Queer it up!, "a sweet, community event" (Activist, 10 July 2018) held in the parking lot of the East Mississauga Community Health Centre (an important queer and trans medical service hub). In partnership with para-public organizations, it aspired to demonstrate social and health service provider commitment to queer and trans people of colour. By the fourth annual Queer it up! in 2011, the event had moved to the more transit-accessible Port Credit waterfront location where it held a Pride march along its commercial, multi-lane Lakeshore Avenue and a community fair with 20 local organizations and performances highlighting the diversity of Mississauga's LGBTQ2S community. Archived in YouTube videos (Queer it up!, 2011) and photographs, youth volunteers described how the event offered empowerment, visibility, and community: "for the small people like myself in the suburbs to just have a voice, like that whole community vibe we're just missing in this end of Mississauga". This activist coalition (10 July 2018) carried organizing momentum into regional activities such as the Pride Week in Peel march (2011) and the Pride Week in Peel festival (2013). As in Brampton, the disbanding of the Peel Pride Committee in 2019 shifted activist leadership and para-public and municipal involvement. Rainbow Sauga continues to organize an annual Pride picnic while the municipality hosts a month of civic Pride events in Celebration Square. From the vernacular creativity of resourceful local activists, Pride in the Peel region is more proximate to the upperground, but its underground is fractured by socio-racialized fissures within LGBTQ2S coalitions.

Similar struggles for QTBIPOC recognition are part of Durham Region's Pride where the regional Pride association organized its first event in a Whitby park in 2004. By 2010, Whitby Pride was a week-long Pride Durham event, culminating in the region's first parade, a local vendor display, and an after-party at Club 717. As in Peel, Durham Pride is challenged by the suburban homonormativity of its leadership and para-public institutional allies. Originally, its mostly cis, gay, white "conservative-leaning members wanted to have a good time, party, make friends, and didn't want to do the hard political work because they didn't believe it was relevant to our context" (Organizer, 24 November 2019). Durham Pride was moved to the Town of Ajax in 2018,[6] which was perceived as an upperground "ally because of their extensive positive space training" for staff (PFLAG President, 15 August 2015). With a ten-float procession from the Ajax Community Centre along the main commercial artery to the Town Hall that included uniformed Durham Police, transit officials, and civic leaders, para-public allyship is a source of conflict. Although there have been internal debates about the impacts of uniformed police on community participation, the organizational committee "wouldn't let it go" and "everybody voted to let the police do what they wanted" (24 November 2019). With "staff, volunteers, and boards that are all white, don't reflect the community, and are disconnected from racialized communities" (Diversity and Community Engagement Coordinator, 4 July 2018), middleground LGBTQ2S organizations do not represent Ajax's QTBIPOC population. A gap exists between LGBTQ2S local communities and the allied Pride para-public leadership that gives activists access to the middleground and upperground.

Together, these seven queerburban Pride events held in places nowhere near the top of the world roster of "gay" cities are articulations of vernacular creativity. They are relationally embedded in transnational Pride histories that are rearticulated locally and traced on website timelines. They centrifugally extend Pride geographically within city-regions challenging both suburbia's heterocisnormativities and the homocisnormative exclusions of gay-village Prides. Into suburbia's voids of socio-cultural infrastructure, queerburban Pride events remake parking lots, parks, community and recreation centres, shopping mall plazas, and civic squares as queer places. Resourcefully materialized through scavenged individual and para-public networks that gather momentum in the organizational middleground, queerburban Prides become institutionalized by securing upperground civic support as funding, permits, public realm access, and publicity. As rainbowized celebrations with grassroots origins, Prides are increasingly enfolded into municipal place-making opportunities that dovetail with civic social inclusion stances and diversity mandates. Nevertheless, they remain important nodes of year-long LGBTQ2S activism. The work of organizing, planning, and solidarity-building for Pride depends on institutional upperground support and local networks of socio-cultural organizations that form queerburbia's backbone.

LGBTQ2S-specific socio-cultural organizations

At queerburban Pride events, the breadth of local LGBTQ2S-specific socio-cultural organizations is seen in sponsorship lists and the eclectic adjacencies of booths. These organizations offer sociability and entertainment through performance and ritual as key elements of community-building and place-remaking. Through their vernacular creativity, they remake suburbs as more hospitable for queers by generating a range of micro-spaces and visibility outcomes (Kelaita, 2024). As "third places" (Oldenburg, 1999) – those more communal sites outside of home and work that foster social infrastructure for encounters and "interaction scenes" (Fine, 2012, 22) – queerburban socio-cultural organizations reconfigure the parameters of suburban sociability. Their events form "interaction ritual chains", regular yet unscripted moments of commonality, attention, and fun that lead to shared emotions and alliances (Collins, 2004b). The ongoing and identarian character of their activities temporarily emplaces queer culture into the social fissures of suburbia via drag-animated nightlife, meetups, and ethno-cultural organizing.

In the early 2000s, queerburban activists in Surrey leveraged personal networks to hack "dying businesses" often "in the rough end of town" to "create a safe place to play" (18 June 2017). The "shabby bars" of Whalley's declining King George Boulevard were the "only spaces they could get" (Bartender, 25 June 2017). Bringing their own sound systems, taking over the cash register, tending the bar, and hanging Pride flags, they hosted karaoke nights, dances, and/or drag shows. One iteration of these "transitional" queer spaces (Vallerand, 2020) was the "Pride Lounge" at a former strip club in the Flamingo Hotel, facetiously described as "Surrey's first LGBTQ+ licensed club" (DiverseCity, 2018). Pride Lounge was hosted by a Two-Spirit youth support worker and his husband, but it eventually "fell apart" as the regular biker clientele "were not so receptive and it didn't feel safe" after they aggressively intervened (EPAMA President, 18 June 2017). As the bartender (25 June 2017) recalls, "the gay flags on the wall caused tension" leading to "gay bashing" when the regulars ripped down the Pride flags, taunted that "the faggot is here to host karaoke", and said "how dare you impose your crap on us" (Community organizer, 18 June 2017). Displaced, the search continued for queerburban social infrastructure until the local Days Inn Hotel offered commercial allyship prior to its demolition.

In the eastern suburbs of Vancouver, drag performances lie at the heart of queerburban organizing, networking, and leadership, and ultimately place-remaking. In Surrey, drag artist Amanda Luv hosted monthly shows at Surrey's former Fire-Side Café in 2006. A combination of serendipity, assertiveness, and attentiveness to possibilities hatched "the first gay-friendly place in Surrey" (Mz. Adrien, 18 June 2018). The café's interior was transformed with curtains, tablecloths, candles, disco-lighting, a sound system, and a temporary bar licence. The Luv Show, and later "Diva Girls", moved to the backroom of the former Heritage Grill restaurant-theatre in New Westminster. This secondary storage space was appropriated as a drag queen dressing room, while the windowed front stage, hand-painted with their silhouettes (Figure 6.3), became their showcase until 2021 when it burned down. Drag queens – their bodies, costumes, performances, and audiences – have moved between neighbouring peripheral municipalities, expanding in more recent years to the large casinos and bingo halls of Langley and Chilliwack.

Drag in Surrey is also anchored in the local chapter of the International Imperial Court System (IICS), the Empire of the Peace Arch Monarchist Association (EPAMA). A queer play on the fraternal society, the Court System integrates drag into its camp monarchal hierarchies (Druschel, 2016). Founded by José Sarria in San Francisco in 1965, the IICS unites sovereign courts in

Figure 6.3 Heritage Grill restaurant drag queen window mural in New Westminster, 2018 (*Source:* Julie Podmore).

70 Canadian, American, and Mexican cities that advocate for LGBTQ2S equality and fundraise for charitable causes (Gorman, 2014). Analogous to ballroom (Rio, 2020), it recreates queer chosen "family" relationships and centres the performative rituals of gender non-conforming competitive pageantry (Bailey, 2014; Kelaita, 2024). A third place for social life, the court is the largest philanthropic LGBT organization in the world, providing funds for members and LGBTQ+ charitable programmes (e.g., HIV/AIDS care or LGBT equal rights campaigns) and civically to local communities (e.g., breast cancer treatment and anti-poverty programs, or holiday food baskets) (Druschel, 2016). Annual Coronation Balls are the largest fundraising events, held to inaugurate new "royal courts" (Emperor, Empress, Princess, and Prince), imperial teams that make a year-long commitment to community service, travel to royal charity balls and coronations, and host drag ball fundraisers for their own communities.

Surrey got its own court when an activist who unsuccessfully ran twice for Emperor in Vancouver moved here, creating "the opportunity for a gay community to exist, which led to Pride, which led to the founding of a court out here" (FG3 Surrey, 8 June 2019). When his application to the Vancouver "mother court" was denied and its boundaries were redrawn to align with the city limits, the Empire of the Peace Arch Monarchist Association (EPAMA) was created, thus named because support came from the American Bellingham and Seattle chapters. EPAMA then "claim[ed] all the territory that was around … and open and available" from Richmond to Hope (FG3 Surrey, 8 June 2019) and Emperor I was crowned by his friend, the late José Sarria, achieving his "highest honour" (FG3 Surrey, 8 June 2019). Embedding this queer pageantry into the declining cultural infrastructure of this suburban region, EPAMA's annual coronations are well-attended events at New Westminster's Columbia Theatre. Throughout the year, it organizes social activities in local bars, pubs, and bowling alleys, carving out queer third spaces and raising funds to support the local community (e.g., foodbanks for people living with HIV and Hepatitis-C, LGBTQ2S youth leadership scholarships, or for individual members experiencing financial precarity). While court drag provides the opportunity "play royalty, dress up with pageantry" and "go out in a tux or a uniform with a crown" (EPAMA President, 18 June 2017), its routinized organizational structure and regular appropriations of ordinary suburban spaces also build networks and identifications with a collective process of queerburban place-remaking.

Internally, the spatial activism of EPAMA is not without socio-racialized and gender tensions while externally it also innovates across unconventional solidarities. Within EPAMA there is "back-stabbing and bullying", "bitching about … make-up and eyebrows" or being "too manly, not having enough foundation, and being too dark", and colonialist conflicts around fundraising that pushed a Two-Spirit couple from its leadership (Bartender, 25 June 2017). Meanwhile, commercial allyship has been built with the Filipino owners of Surrey's Turf Hotel who are "trying to keep this a safe space for us" (FG2 Surrey, 8 June 2019) even when their regular biker clientele derogatorily question "why gays and fags are here" (Erica St. Asia Divine, 18 June 2017). The Turf Hotel is temporarily remade as queerburban "community space" (FG2 Surrey, 8 June 2019) for EPAMA Royal Court events, Pride fundraisers, drag shows, and karaoke competitions. As drag queen Mz. Adrien clarifies (18 June 2018), "we needed a community. We needed a sense of gathering". Along with the Radical Faeries, Bears in the Burbs, and the Sisters of Perpetual Indulgence, EPAMA provides a grassroots framework for queerburban organizing. Undergirded by long-term friendships formed in downtown nightclubs, EPAMA offers group sociability, leadership, and community structure to its predominantly white, middle-aged, and cis-male members exemplifying people as suburban infrastructure.

The resourceful spatial activism of drag performers and EPAMA is mirrored in the appropriative and networked practices of New Westminster's Rainbow Event Network (REN). Established in 2016 as an umbrella organization curating an online schedule of weekly queerburban activities, it began as a monthly events email newsletter written by a Japanese-Canadian lesbian (21 June 2017) to cultivate community locally without "going downtown". When the organizer of Rainbow Karaoke night (21 June 2017) moved to the suburbs and launched Dining Under the Rainbow, a monthly event in local restaurants, options to meet other LGBTQ2S community members expanded, later adding beer bashes, pub trivia nights, and riverboat dinner cruises. Such underground yet conventional organizing and informational resource sharing is key to unmaking and remaking queerburbia, a way for individuals to stay connected to dispersed communities where "leathers, bears, and drag" (Activist, 18 June 2017) may occasionally coalesce in one place (McGlynn, 2021). Organized by older, white primarily cis-gay men, REN reproduces a generational understanding of LGBTQ2S cultural practices and rainbowized rituals and politics but its spatial activism also creates possibilities. A younger African American lesbian mother (27 June 2019) attends REN events with her child, as "a way of doing things together" and "show[ing] them the many different types that queer can be". Its events are also welcomed by Columbia Street's local business owners as opportunities to expand their market and transform the street's place-image from one of "drugs, prostitution, homelessness" (Bar owner, 30 June 2017). As the karaoke night organizer (21 June 2017) explains, it also raises local LGBTQ2S awareness:

> I take the rainbow flags with me and stick them on the table so that people know that this is where we are. This one guy came over and he says, 'You know, we're trying to figure out what country that flag is.' So, one of the group says, 'You know what, this is the Pride flag. It's in every country around the world'.

Queerburban activists practise spatial appropriation in local casinos, pubs, and restaurants, improvising relational connectedness while also raising awareness of sexual and gender diversity.

REN's commercial activism involves the routine appropriation of bars and restaurants, but in the Toronto city-region, Ajax provides the singular example of a dedicated LGBTQ2S nightclub. Hidden in a light industrial area, Club 717 is organized by the non-profit Durham Alliance Association Social Club to provide support and "anonymity … for people who did not want to be outed" (President, 24 August 2018). It remains one of Canada's only LGBTQ+ private members' clubs,[7] providing "privacy in public" and community control to facilitate LGBTQ2S social interactions in third spaces (Bassi, 2006). While the membership base diversified over its 35 years, by 2018 only 40 people paid membership fees. Even pre-COVID-19, the President (24 August 2018) "struggle[d] to keep the doors open" and to "activate energy and decrease apathy". Far from flashy, there is no air conditioning, food, or patio. Beyond the monthly rent (CDN$1400) and annual insurance (CDN$5000) fees, there are few savings for equipment repair. To stave off closure and generate revenue, the space is rented for weddings and fundraisers and has been repositioned as a "safe space" beyond the LGBTQ2S community for kink and BDSM communities. Club membership restricts access while permitting the adaptive dexterity of other revenue streams to maintain place-connections for Durham queerburbanites (Fine & Cortes, 2017).

Besides Ajax's Club 717, the spatial appropriation of the meetup model is more common among LGBTQ2S-specific GTA socio-cultural organizations as exemplified by QTBIPOC Mississauga Meetup (QTBIPOC Sauga). In 2016, two community activists (10 July 2018) were feeling like "lonely queer[s] in Mississauga" and wanted to combat "extreme isolation", so they announced a coffee shop meetup via emails to friends, acquaintances, and local listservs.

Explicitly organized outside of the non-profit sector (Beam, 2018), QTBIPOC Sauga's monthly meetup is intended as an antidote to societal "racism, homophobia, transphobia, and ageism" in a suburban context "where reaching across the expanse of the city, and of the region, is so hard" (Verma, 2017, 226 & 228). In contrast with REN, QTBIPOC Sauga employs a "collaborative commitment" to building "collective intimacy" through restricted sociability (Fine & Cortes, 2017). They meet in publicly undisclosed locations (often members' homes) to protect individual privacy and safety and allow "people to move in and out" of the group without "spending more time on public transit than together" (Activist, 29 January 2020). QTBIPOC Sauga[8] evidences the "long, complicated history of queer and trans organizing in the burbs" that not only "lives in the minds and bodies of individual people" but is also "a living archive" of writing and talking that constructs "a new future for QTBIPOC communities, outside of the oppressive structures of non-profit organizations" (Verma, 2017, 228).

QTBIPOC Sauga and others practise a different form of queerburban spatial activism from the older gay male communities forming around drag performance in New Westminster and Surrey or the Club 717 in Oshawa. They focus on creating safety and social reproductive care networks that block negative externalities in support of their community's resilience. The socio-cultural venues supporting drag shows, the court system, or Club 717, are all ludically focused on play and fun (Stevens, 2007), but they also reconfigure the parameters of sociability and make daring visibilities less dangerous. These queerburban social moments offer degrees of familiarity, intimacy, protection, and endurance, but they remain peripheral within the city-region. As Surrey drag artist Erica St. Asia Divine (18 July 2017) quipped, downtown LGBTQ2S communities "rarely know the other groups that are out here". Queerburbanites instead rely on their own local social and cultural organizations as "a support system and family, a safe space to come out" and a means to "see and communicate with older generations" (14 June 2019). On Vancouver's eastern peripheries, queerburban ethnocultural specific organizations, like Sher and Pinoy Pride, are especially important to its LGBTQ2S South and Southeast Asian residents.

Sher Vancouver is a non-profit group established in Surrey in 2009 by gay social worker, activist, and first Sikh Grand Marshall of the Vancouver Pride Parade. Witness to his gay clients' sense of alienation, depression, and isolation, he saw a need for a youth-led peer support group for LGBTQ2S South Asians (ages 16 to 25) and their friends, families, and allies:

> I didn't feel accepted in my community where it is all about marriage. In my culture, you stay at home until you're dead and the grandparents raise the kids. It's a different world. It's hard to be flamboyant and gay.
>
> (16 June 2017)

Initiated through Facebook with small municipal grants, Sher then became a website (Sher Vancouver, 2023) and finally an in-person meetup – "a place for young people who feel marginalized in youth groups in the Lower Mainland to go where they feel safe to be gay and brown" (Organizer, 16 June 2017). Like QTBIPOC Sauga, a "safe space" for its 90 South Asian LGBTQ2S members, depends on a restrictive communal intimacy with the addition of privacy, emphasizing "not being outed by walking through the door" to their meeting in the recreational room of the organizer's condominium building. In this "(in) visibility schema" (Cisneros & Bracho, 2019, 721), "people want privacy and confidentiality – they don't want a sign saying 'gay' because people aren't out in the brown community" (16 June 2017). Sher's spatial activism operationalizes "strategic outness" – the "continual, contextual, and social identity management" process (Orne, 2011, 685) – upon which its ethnoburb members rely as a "survival mechanism and protective strategy" (Cisneros & Bracho, 2019, 721). Sher's founder, however, is a visible public advocate for LGBTQ2S people in the eastern suburbs, appearing at civic ceremonies and in inclusion documentation, participating in Surrey's Sikh Vaisakhi parade, and making films to combat homotransphobia in suburban communities.[9]

Also based in Vancouver's Eastern suburbs, Pinoy Pride offers Filipino-Canadians ethno-culturally specific LGBTQ+ support and advocacy. It was founded in 2010 for the Vancouver Pride parade, initiated by lesbian Filipina MLA Mable Elmore "for the purpose of being visible". Pinoy Pride began promoting events (e.g., dance parties, bowling tournaments, drag shows, and beauty pageants), providing social support, and outreaching to the Filipino diaspora community. Working with the Filipino radio program, Jeepney Café on Vancouver's Co-op Radio (100.5FM), the group launched a Pinoy Rainbow segment featuring their members' coming out narratives (Co-Director, 14 August 2017). For the Co-Director (14 August 2017), becoming more visible within the "conservative", Catholic Filipino diaspora community provoked a sense of "a second coming out", feelings of "marginalization", and "internalized oppression". For queerburban Filipinos, downtown is the "comfort zone" whereas suburban areas with high concentrations of the Filipino diaspora community such as Surrey's Whalley-Guildford area[10] require being *nagtatago* or "concealing oneself", deploying a spatial bracketing of LGBTQ+ and ethno-cultural identities (Co-Director, 14 August 2017). Their events centre on inter-community relations for pockets of the larger Filipino community. What Sher and Pinoy Pride share is the "for us, by us" approach, creating "culturally affirming spaces" for ethno-cultural communities to promote "mutual recognition and relevant support" rather than the "one-size-fits-all" model of generalist organizations (Kojima et al., 2017, 75).

From Pride events and the Imperial Court to members-only nightclubs, and ethno-cultural gatherings, LGBTQ2S-specific socio-cultural organizations hold place for queerburbanites. They temporarily unmake suburbia's hetero-cisnormativies by remaking place through queer interpersonal networks, friendship, and community. They create space for sociability, creativity, and fun, adding tensile strength to what might otherwise be loose knit queerburban connections. Such instances of collective public pleasure shape group engagements, cement allegiances, and offer shared reference points. Queerburban cultural activities create "embedded histories", stories of the connections, spaces, and events that confirm "the gathering's groupness" (Fine & Corte, 2017, 71 & 78).

Service organizations

With upperground government funding and targeted populations, service organizations provide queerburbia with an institutional support framework, underpinning and multiplying its resources. Targeting a combination of community social, psychological, and health needs, they are places to receive care and peer-support, access to health services, and gender-affirming gear, but also to mobilize politically, create innovative support networks, and develop partnerships. They include service organizations targeting LGBTQ2S populations, universal services where insider activists identify unmet needs, and those that reconfigure their clientele in response to queerburbia's emerging populations. Adaptable and resourceful, it is primarily community-based AIDS Service Organizations (ASOs) that have transformed themselves along with youth groups and seniors' centres.

ASOs originated to address the institutional neglect, societal stigma, and inequities of the 1980s HIV/AIDS pandemic (Flicker et al., 2009). Their activism catalyzed LGBTQ+ organizing, foregrounding social justice, advocacy, human rights, and equitable access, redefining sexual citizenship, and opening new political spaces (Brown, 1997) including ethno-specific ASOs with their own "place-making" and "belonging" practices (Catungal, 2013, 253; Catungal et al. 2021). By the early 2000s, the shift from the antiretroviral AZT (Azidothymidine) to the HAART (Highly Active Antiretroviral Therapy) triple cocktail led to a reorientation from drug advocacy and hospice accompaniment to long-term support for people living with HIV, testing and prevention, and sexual health promotion (Di Feliciantonio, 2020). These "self-help" organizations, primarily staffed by gay male and/or HIV-affected volunteers, provide non-discriminatory social support services and public health information (Poindexter, 2007). With limited funding to counter political inertia, ASOs serve multiply-marginalized communities (Kay & Musgrove, 2020) facing addiction and recovery, housing and food insecurities, unemployment, and mental health challenges. While suburban ASOs are historically autonomous regional centres serving multiple populations, they have recently integrated more LGBTQ2S social programming.

In Toronto's York and Peel regions, for example, ASOs now provide closed-group discussions for regional LGBTQ2S social support. Moyo Health and Community Services (formerly the Peel HIV/AIDS Network (PHAN)) was initiated in 1991 by three public health nurses in Brampton who identified a regional gap in HIV/AIDS counselling services (Figure 6.4). It has a history of focusing "on the most marginalized" asking "how can we create a space for people to desire and flourish with intersecting identities and complex

Figure 6.4 LGBTQ2S-positive entrance to PHAN, 2018
(*Source:* Alison Bain).

racializations" (Staff, 10 July 2018). Rebranded Moyo (a Swahili word meaning heart, life, and spirit) in 2019 to counter mistaken assumptions about HIV/AIDS status that "create[d] a barrier for folks who come through the door" (Staff, 19 September 2019), it reoriented towards the promotion of health, education, and support for people living with or at risk of contracting HIV. Its new Queer Minds social group was part of the reorientation towards becoming "a comprehensive community service provider to LGBTQ2S communities" (Peel HIV/AIDS Network, 2017, 3). A similar social mission is shared by CAYR (pronounced "care") in the York Region (rebranded from its 1993 founding name of the AIDS Committee of the York Region), which also experienced pressure to "operate beyond a disease model" (Executive Director, 22 June 2018). Staff received frequent requests from their queerburban clients for social space because "there are no gay bars to hang out in to meet people" (Executive Director, 22 June 2018), and low densities prohibit a community-centre model. In 2011, CAYR decided to take "its programs to where the people are" with "floating rainbow rooms" rather than "fixed space" (Executive Director, 22 June 2018). These low-barrier, peer-led rainbow rooms in public libraries, recreation centres, and cinemas are intentional spaces of friendship and discussion to address queerburban social isolation (especially for youth and seniors), while also offering a degree of anonymity.

ASOs on the peripheries of the Montreal city-region (Saint-Jérome's Centre sida amitié, Terrebonne's Le néo, and Laval's SIDA-Vie) have also been transformed into LGBTQ2S support centres now served by social workers and sexologists. SIDA-Vie (meaning "AIDS-Life") Laval began in 1995 to support "endemic" populations such as MSMs, sex workers, and intravenous drug users living with HIV-AIDS in Laval's Pont Viau district. In 2018, its name changed to "Sphère: Santé sexuelle globale" to reflect its larger "sphere" of influence on the sexual health of Laval's population, especially its youth (Gohier-Drolet, 2018). With public consultations indicating a need for LGBTQ2S youth and seniors' services, it has become the peripheral municipality's leading "transgenerational" LGBTQ2S organization (Director, 29 June 2022). It now runs support groups for people living with STBBIs and for trans, non-binary, and questioning young adults and their "entourage" (friends, supportive parents, siblings, other family members, or guardians) (Sphère, 2020) as well as weekly drop-ins for sex workers, and for LGBTQIA+ youth. Sphère is Laval's resource for accessing safe sex, injection and inhaling paraphernalia (condoms, gloves and dams, syringes, Pyrex pipes, and Naloxone kits), and gender-affirming gear (binders, gaff, TransTape bodytape, STP – Stand to Pee, breast forms, and packers). It also runs two LGBTQ2S youth support groups, *Espaces 2SLGBTQIA+* (a drop-in, social space for ages 12 to 25) and *Projet plur-iels* (a facilitated support group for trans, non-binary, questioning, and gender-creative youth). With this reorientation, Sphère has become the leading LGBTQ2S organization in Laval celebrated in 2023 by the first annual cabaret night featuring youth talent for IDAHOBIT that transformed a municipal hall in a suburban park into a QTBIPOC space.

A second model of queerburban health service provision involves satelliting central-city organizations. One example is Vancouver's Health Initiative for Men (HIM), which provides Sexually Transmitted Infection (STI) testing, vaccines, pre-exposure prophylaxis (PrEP) treatment for HIV prevention, and mental health counselling for male-identifying people over age 19. Originating in the gay village, it expanded eastward to Commercial Drive and outwards to Surrey, Abbotsford, and New Westminster with satellite health clinics in nondescript buildings and pop-up testing centres in hotels, medical buildings, and hospitals. Satelliting involves ethno-cultural outreach to suburban diaspora communities (notably Persian, Lebanese, Taiwanese, and Afro-Caribbean) and assurances of privacy since "guys in the 'burbs' want different things"; while they may "feel alone and segregated, especially if they are of a certain ethnicity", they "don't want to be seen to be out at a gay space" (Manager, 29 June 2017). With diverse bodies and ethnicities, their poster, packaging, and social media campaigns (Figure 6.5) aim to be "relatable" to their clients along with discreet pocket-size resource guides in Chinese, Farsi, Punjabi, and Spanish (Manager, 29 June 2017). To "bring community to spaces that don't have one" (Manager, 29 June 2017), HIM began offering community-building activities beyond "hooking-up" (e.g., condom packing, movie screenings and discussions, yoga, board games, bowling, life-drawing, and workshops on financial

Figure 6.5 HIM Condom + Lube packages, 2017
(*Source:* Alison Bain).

health). In New Westminster, for example, a cohort of "out" older, white men incorporate two hours of packing condoms at Judge Begbie's Tavern into a queerburban "Sunday marathon of events" that includes Queers and Beers, and a drag show. Yet each suburban locale differs in its clientele and outreach strategies. Surrey clients are often "focused and nervous and don't want to talk", oftentimes "worried" that "if they see someone of the same ethnicity, they will say something" (Manager, 29 June 2017). Here, HIM hosts "Life in Colour",[11] a peer-led program and safer space for queer men of colour with food and conversation to build relationships between clients. In the words of one South Asian staff member (9 July 2019):

> It's a group that I wish I had had when I was younger where South Asian men of colour can relate to each other as not just extras to a cis-gay, white world … Gay white guys don't understand that men of colour are going through different issues, we're fetishized and stereotyped.

HIM's satellite organizing breaks the social disconnection between queerburban older white gay men and heteroflexible men of colour: "People don't come out to the suburbs. I became isolated. I discovered HIM and put myself out there … I fit in by doing things for people" (Volunteer, 4 July 2019).

A third model of queerburban service provision targets two underserved groups, youth and seniors, by launching special projects, forming internal committees, and offering tailored discussion and social spaces. On Montreal's South Shore, Le JAG (Jeunes adultes gaies), for example, is a gay youth organization created in the regional municipality of Saint-Hyacinthe in 1995. It opened a new service centre in Longueuil 25 years later when it was mandated by a Quebec government grant to serve the Montérégie region of 1.5 million people and include "all of the LGBTQ+ acronym, the young and less young, and also all the allies" (Director General, 25 July 2022). Struggling to connect with youth who are "more multicultural" and community group ecosystems that are more politicized by class and "ethno-cultural grievances", Le Jag is still in search of "the right language to connect with Longueuil" (Director General, 25 July 2022). The West Island LGBTQ+ Centre operating out of Beaconsfield United Church has had more success reaching multiple populations. It supports youth aged 14 to 21, but in 2013, it established the West Island Rainbow Seniors (50+) program and the Friends and Family Support Group. Since then, they have formed a young adults social drop-in program (2016), trans youth and parents' support groups (2019), and the LGBTQIA+ Migrant and Refugee Support Program (2021). These meetings provide peer support and meal-based socializing, with participants directed to the region's limited social and psychological services. Now operating as an independent charity with its own board, the Centre provides the main regional English-language training workshops (for high schools, businesses, and community services) and partners with schools to facilitate their GSAs and host two annual events – the Royal Flush drag show fundraiser and the Queers in Space queer prom – as well as pub quiz and movie nights (Figure 6.6). These middleground place-remaking practices have mobilized and expanded inter-organizational networks of LGBTQ2S allyship.

Other LGBTQ2S youth programming is offered through community partnerships with healthcare providers, police, and schools. Durham's weekly Colours Youth Group for youth ages 13 to 20, for example, was initiated in 2013 by a Carea Community Health Centre (CCHC) social worker who realized that outside of the local PFLAG chapter and the AIDS Committee of Durham Region, there were no drop-in groups or resources for LGBTQ2S youth. Service groups were "working in silos" and the lack of public transportation isolated LGBTQ2S people who as a community were "invisible", so they "began to work together to bring people out of the shadows and get them involved in programs" (Social worker, 12 July 2018). In 2008, they held a one-day youth conference, "SAY WHAT?! Exploring Sexual and Gender Diversity among Youth in Durham Region", at the University of Ontario Institute of Technology (UOIT) which ignited LGBTQ2S youth regional organizing (Durham Pride Prom, PFLAG's Camp Rainbow Phoenix (2012), and Youth Pride Durham (2018)). The Durham Regional Police Service (DRPS) launched the Colours Youth Group, an evening of activities (e.g., meditation, professional make-up, bowling, and concerts) and workshops (e.g., perceptions of safety, social media bullying, sexual assault and abuse, and tactical and canine unit operations) coordinated by a uniformed ally officer in the Equity and Inclusion Unit

Figure 6.6 The West Island LGBTQ2 Centre's annual Royal Flush drag show fundraiser, 2022

(*Source:* Julie Podmore).

to "break down barriers of what the uniform represents" (Police Sergeant, 28 August 2018). In Burnaby, this kind of after-school drop-in social space for LGBTQ2S young people operates in a storefront partnership with the school district and the Lower Mainland Purpose Society. Open to all marginalized youth as an LGBTQ2S-supportive "hang out" space of casual social conversation, the Burnaby YOUth hub is a place where "nobody is putting you down. I don't experience hate. I don't fear that people won't accept me or isolate me" (Trans youth client, 12 July 2019). From Durham to Burnaby and beyond, LGBTQ2S youth are a middleground and upperground rallying point. It is only in Surrey, however, that LGBTQ2S youth are explicitly mentored to become the next generation of activist leaders at Youth 4 A Change (Y4AC, 2024).

Established in 2011 by lesbian parents, Y4AC is a place where predominantly white, queer feminine and/or trans masculine youth members (aged 13 to 21) socialize and organize – "a place to be … a place to grow, and a place to achieve success" (Co-founder, 16 June 2017). Located in a strip mall on a bus route, Y4AC holds weekly two-hour gatherings in a storefront youth drop-in centre with access to a kitchen and meeting rooms. While this fixed physical space supports the establishment of reliable social routines and friendship networks, it is Y4AC's emphasis on activism and advocacy that nurtures a sense of belonging and empowerment for its LGBTQ2S and ally youth members.

Their meetings involve preparing political interventions, speaking engagements, and creative activities that temporarily remake queerburban educational, commercial, and public spaces. Y4AC participants variously: offer high school GSA workshops on homotransphobia; research and discuss LGBTQ2S history; monitor school board social inclusion policies; lobby city councillors and school board trustees about LGBTQ2S youth issues; participate in local campaigns (e.g., for gender-neutral restrooms and changing facilities, and for city council to fly the Pride flag at city hall); collaborate intergenerationally with elders; train community youth workers; consult on public safety; attend Pride; do LGBTQ2S art bombs; and organize community events. In contrast with stereotypes of bullied suburban LGBTQ2S youth, Y4AC participants are active agents in remaking queerburban places.

Queerburban seniors meanwhile strive to find new sites for the community. One place where they have had success is New Westminster's Century House, the only seniors' centre in Canada with its own Seniors' Gay Straight Alliance (SGSA). Located in Moody Park's Uptown neighbourhood, surrounded by high-rises populated by low-income seniors, Century House is a municipal daytime activity centre for residents over age 50. Its SGSA was inspired by a mandatory, city-wide "Safe Harbour" diversity initiative and Century House's own inclusion enhancement project. A recreation programmer (15 August 2017) explains that "We always felt that we're pretty inclusive here, but that we could always do better", particularly with respect to ethno-cultural and sexual diversity. Organizers presented *Gen Silent* (Maddux, 2011), a documentary film profiling six Boston LGBTQ+ seniors in the long-term-care system, and then leveraged well-attended in-house film screenings to form the SGSA activity group. Its monthly meetings offer members (50 percent of whom are allies) support, friendship, and learning with guest talks about LGBTQ2S care in seniors' centres, caring for partners with dementia, and adapting facilities for gender-variant people. Their code of conduct stresses discretion ("anything that happened in the space stays in the space"): no one is required to "come out" and the presence of allies provides a cover for seniors not wanting to disclose (Co-Chair, 15 August 2017). It provides peer-support and enhances social opportunities beyond the seniors' centre with members attending REN's activities together. In 2019, an SGSA sub-committee bridged the upperground with a municipal report on New Westminster seniors' care facilities documenting the lack of inclusion, queer and trans competency training, and outreach initiatives.

Queerburban service is anchored in the adaptive dexterity of ASO middleground organizing that has pivoted to operate beyond a health model, offering bridges to the underground and upperground. As "educators and catalysts" in queerburbia, ASOs continue to provide responsive community-based leadership and service delivery to meet LGBTQ2S service gaps despite organizational challenges and internal conflicts (Poindexter, 2007). Service agencies prioritize engaging LGBTQ2S youth and seniors through pilot projects, mandate revision, satellite service delivery, or GSA replication outside of schools and for a different demographic.

Advocacy organizations

Important bridges to the upperground, advocacy and awareness-raising organizations drive queerburban change, particularly for vulnerable community members facing health challenges, risking violence, or lacking financial security. Advocacy organizations confront discrimination by mobilizing their respective communities to remake suburbia:

> advocacy gave me a lot of strength. I lived and survived the HIV virus, so I thought it was my duty to make the world a better place for the LGBTQ community. I've committed my whole life to that … I'm an absolute passionate advocate. I have stood up to all of the negative powers in this city to make it a safe place (FG3 Surrey, 8 June 2019).

From such individuals to more formalized organizations and service providers, advocates witness and respond to queerburban living conditions. As this section demonstrates with reference to PFLAG, Centres des femmes, and immigrant and refugee programming, advocacy organizations publicly articulate LGBTQ2S legal and service needs and stress intersectional understandings of sexual and gender identities (as youth, women, immigrants, and refugees).

PFLAG (Parents, Families, and Friends of Lesbians and Gays) is the largest non-profit national support, education, and advocacy social movement group in North America (with over 70 local chapters in Canada) (Broad, 2002; Field & Mattson, 2016). Founded in the United States in 1973 to provide information and support to parents of lesbian and gay children, its mandate now includes siblings, children, and friends of LGBTQ+ people and offers resources to schools (Shi, 2021). The contemporary PFLAG organizational symbol in Canada (three rainbow inset circles accompanied by the slogan "family for all") anchors its website, newsletters, and pamphlets foregrounding Two-Spirit people in the umbrella acronym 2SLGBTQ+ (PFLAG Canada, 2024). While PFLAG chapters have different operational practices, they prioritize inclusive "family values" and undertake meaning-making work with their parental ally participants (Broad et al., 2004).

PFLAG is the key advocacy organization in Toronto's suburbs, particularly in the Durham and York regional municipalities. Founded in 2002, PFLAG Durham has become "the biggest resource in Durham" and the "centre of the hub" (Community Engagement Manager, 17 August 2018). It was catalyzed by the 2008 physical and verbal assault of two lesbian mothers in Oshawa outside their children's elementary school by the father of another student, leading media scrums, and providing legal support. PFLAG Durham partners with the Durham Regional Police, Durham District School Board, and Ajax Public Library. It hosts monthly coffee sharing nights for over 100 people at the University of Ontario Institute of Technology (UOIT) and runs LGBTQ and Trans 101 training workshops. It also runs a peer-to-peer mentoring program where volunteers with lived experience of the LGBTQ2S community offer support and resources to new members. In the absence of regional resource lists, "PFLAG consistently links people" to the community support and information they need (Community Engagement Manager, 17 August 2018). Organizationally, however, it grapples with being unable "to draw in folks who aren't white and privileged LGBTQ", recognizing that "until you see yourself in that organization you won't be drawn to [it]" (President, 15 August 2018).

To respond to the ethno-cultural and age diversity of Ajax, PFLAG Durham's board has sought funding for LGBTQ2S resource deficiencies (e.g., programming for young adults who have aged out of youth-oriented activities, older adults who need a voice and support, and opportunities to bring both groups together). In 2019, it began organizing all-ages QTBIPOC drop-ins with a small community grant at the Ajax Public Library. A volunteer transgender co-organizer (2 November 2019) explains:

> Our goal is to create an anti-oppressive space for people to come and just be … it's not a 'come and throw your trauma at each other kind of thing', but it's more, 'can we have a space for queer and trans people of color that isn't centred around drinking culture or capitalism'?

This drop-in is attended by a handful of people who appreciate PFLAG Durham's diversified programming, community-building moments, and support (e.g., the development of a QTBIPOC Relief Fund during COVID-19). Trans adults and older adults are also identified as having few local mental health and post-operative supports, necessitating trips downtown for experienced specialist medical care. This service gap became clear when an 89-year-old trans woman (who came-out at the age of 85) "called me because her wife just died … and there is nothing for her and nowhere to go" (Community Engagement Manager, 17 August 2018). Suburban MSMs also need support to access "full healthcare" because they often "hide [their sexuality] by going to drop-in clinics" (Community Engagement Manager, 17 August 2018).

In collaboration with the AIDS Committee of the Durham Region (ACDR) and space from UOIT for mentoring volunteers, in 2010 PFLAG Durham established Camp Rainbow Phoenix for youth ages 12 to 17 who attend this free overnight camp. Whether Black, South Asian, or Indigenous, "this generation knows it has a home" (PFLAG President, 15 August 2018) where they have "five days to be safe and have conversations" (PFLAG Manager, 17 August 2018). Expanding from seven in its first year to over 60 in its eighth, Camp Rainbow Phoenix draws from within and beyond the Durham catchment area, offering a place where LGBTQ2S young people can "be their authentic selves and can present in any way that they want" (PFLAG Manager, 17 August 2018). Young adults lament hearing "oh, you're too old" and wish there was "a queer young adults camp" (FG1 and 2 Ajax, 2 November 2019). In Durham, the camp is an organizational rallying point that mobilizes upperground partnerships with school boards and invites PFLAG chapters to fundraise and send youth.

PFLAG York Region is another active GTA chapter and hub. Without a permanent York region LGBTQ2S space, PFLAG collaborates with CAYR, using its living room space for monthly meetings. Its board leverages upperground relations with municipal officials to provide policy feedback (as with respect to gender inclusive signage in the City of Markham), offer LGBTQ+ training workshops, fundraise for Pride and galas, coordinate flag raisings, and attend York Region District School Board social justice fairs. As the former president (3 July 2018) explains, "we do everything because we don't have a 519[12]... we advocate, consult, educate, provide resources ... and enact conversations in different spheres". Like other PFLAG chapters, the six-member board grapples with its institutional whiteness in an ethno-culturally superdiverse suburban region. To "break through cultural barriers" (Former President, 3 July 2018), it sought funding to connect with different ethno-cultural and English-as-a-Second-Language communities by broadcasting translated advertisements on local radio stations (e.g., in Tamil and Chinese). Despite "curate[d] images to make sure that people of colour are represented", the majority of new service users remain white, with high proportions of parents of transgender children and trans children themselves who are unable to access downtown resources (Former President, 3 July 2018).

Queerburban PFLAG chapters are particularly active on Toronto's peripheries relative to Montreal and Vancouver. Many of these chapters have boards and a membership that is composed, at least partially, of queerburbanites, but PFLAG advocacy is primarily parental allies undertaking community caregiving out of concern for their children's well-being. PFLAG thus "blurs the boundaries between family and community as well as caretaking and politics" (Broad et al., 2008, 501). In the Montreal city-region, however, beyond reoriented ASO services it is the neighbourhood-based *Centres des femmes* that advocate for its lesbian, bisexual, and trans women constituencies. Launched by Quebec feminists in the 1970s, the province's Centres des femmes are part of L'R, a pan-regional umbrella organization of over 80 centres. As collective spaces of empowerment, their goal is to "know and understand" the "needs, problems and demands" of predominantly cis-gender women in their respective areas (L'R des Centres des femmes du Québec, 2016, para. 13) and form working groups with targeted themes for mobilization. Some have historically been spaces for lesbian organizing, including the Centres in both Longueuil and Laval. Since Bill C-16 was passed in 2016, all have ontologically grappled with the legal imperative to "open the doors" to trans women (Co-Coordinator, Centre des femmes de Laval, 22 August 2019), reshaping their social inclusion policies, requiring training and education to effectively and safely welcome and resource them. There is, however, significant regional variation in LGBTQ2S inclusion and group formation related to regional histories and the average age of the membership (particularly the proportion of second-wave feminists). The Centres des femmes in both Laval and Longueuil have histories of lesbian organizing and attempts at trans inclusion that render them significant queerburban sites.

Located in a small commercial building at the heart of Laval's Pont Viau district, the Centre des femmes de Laval (CF-Laval) is known for its diversity and innovation, tackling issues of domestic violence, housing inequality, and the conditions of incarcerated women, with an intersectional approach to inclusion. CF-Laval was the first in the province to adopt an inclusion policy for trans women due to the recognition "that there are different systems of oppression that shape your life" and a desire not to reproduce these among themselves (Co-Coordinator, 22 August 2019). They "worked hard to inform ourselves, to train our community workers, to educate our participants" because they did not "want to welcome anyone into an unsafe space" (Co-Coordinator, 22 August 2019). In 2019, they launched an LGBT pilot project bringing together older cis-gender lesbians and trans-identified women in a daytime activity group. Due to older participant discomfort with attending "a group that was just for LGBT people" and the absence of younger participants, their activities were rescheduled to evenings and allies were invited (Co-Coordinator, 22 August 2019). The group's actions included distributing pamphlets at Montreal Pride, organizing workshops on LGBTQ2S identities, and participating in a pocket cinema workshop to make their own films of Pride. This pilot project taught organizers a valuable outreach lesson noted by other queerburban advocates: "just because there are no knocks on the door, doesn't mean there is no demand" (Co-Coordinator, 22 August 2019).

The Centre des femmes de Longueuil (CF-Longueuil) provides support for women experiencing housing and relationship transitions, homelessness and poverty, and domestic abuse. In 2012, insider activists announced the founding of a lesbian support group, "Les gaies luronnes" (cheerful gays), in the local press. A coffee klatch to "hang out" and talk about their lives, it lasted for five years with regular attendance from 15 participants, most of whom were over 60 years old and experiencing isolation, and a potential return-to-the-closet in seniors' housing and assisted living facilities. To counter internal lesbophobia, they organized sensitivity training, invited LGBTQ2S guest speakers, and celebrated IDAHOBIT. In 2017, ageism and bi- and trans-phobia led to the group's dissolution with the promise of further critical reflection about how to diversify the category of "woman" to better include younger, transgender, and immigrant women (Centre des femmes de Longueuil, 2023). Different activities now contribute to the Centre's LGBTQ2S inclusion mandate, including an annual cocktail party for the *Journée internationale de la visibilité lesbienne* (Lesbian Visibility Day), the marking of IDAHOBIT with a decorative flag on the Centre's door, a monthly discussion group on "Realities and challenges of LGBTQ+ people", and the integration of LGBTQ2S themes into their newsletter content, enhanced by a new partnership with Le JAG.

Where PFLAG and the Centres des femmes play key advocacy roles on the peripheries of Toronto and Montreal respectively, in Vancouver it is immigrant settlement organizations. MOSAIC's I Belong program in Burnaby provides the only comprehensive program for LGBTQ+ newcomers and refugees of all of the case study peripheral municipalities. It began as a pilot project in

2014 when settlement workers realized that their LGBTQ2S clientele had additional barriers to employment, housing, and safety. To increase suburban understandings of the LGBTQ+ newcomer and refugee populations, MOSAIC organized a stakeholder community dialogue and co-created educational video portraits about queer immigration with their clients. Local public library partnerships and Pride event attendance further strengthened their middleground presence as has multi-scalar connections to Rainbow Refugees, and the Vancouver Association for the Survivors of Torture (VAST). Officially, I Belong runs three monthly activities (peer-support, trans and non-binary, and counselling) but "it's a pretty free-flowing group. We just provide a space for LGBTQ newcomers to get together, to exchange resources, to get to know each other, to network, to basically to create a community" (LGBTQ Facilitator, 17 October 2017). Since younger refugees often live with family, I Belong activities attract people ages 30 to 50 who identify as gay, lesbian, or transgender women, many of whom come from African and Middle Eastern countries where same-sex acts are prohibited by law. As fear of being out and the lack of a common language are major group barriers, MOSAIC staff use arts-based strategies and movement to help participants express their "anguish and mourning" (LGBTQ Facilitator, 17 October 2017). Participation in I Belong can also be used to demonstrate to the Refugee Board "that they are part of queer community" (LGBTQ Facilitator, 17 October 2017). But mostly, as a lesbian newcomer from Columbia recalls, it is the sense of safety and hope that I Belong offers that is the draw. It is "the first place where I came out to a lot of people at once" and made a "friendly acquaintance" with a gay compatriot (FG3 Burnaby, 8 June 2019). It is an opportunity to connect with other queerburbanites striving to live freely in an unfamiliar place.

Across the peripheries of Canada's largest cities, local PFLAG chapters, Centres des femmes, and settlement service agencies spatially support queerburbia. In their different capacities and in dialogue with their respective upperground and middleground, they increase access to resources, support legal challenges, lobby governments for recognition, and demystify compounding client stigmas. By serving LGBTQ2S families, women, and immigrants, they advocate for specific LGBTQ2S populations, creatively resourcing the group activities and activism that remake queerburban places.

Conclusion

Vernacular creativity – with its scavenging and improvisatory dynamics – drives queerburban organizing across dispersed, temporary, and re-invented spaces. Through its lens, the uniqueness of queerburban place identities is revealed in all of its solidarities and organizational adjacencies. In queerburbia, the typology of organizing is different. While the lower-density, limited visibility, and lack of social infrastructure constrain cultural organization and social protest movement formation, the selective presence of responsive service providers and advocacy institutions contributes to the remaking of suburbia as queer.

In contrast with the Hess and Bitterman (2023) Taxonomy of North American CSOs, there are fewer categories, the coverage is more uneven, and there is a greater reliance on middleground and upperground mainstream institutional allyship.

Queerburbia's organizational mix relies upon Pride as an activist anchor with the greatest potential for upperground civic legitimacy as an annual ritual. Transnational and generational networks carried such rituals to suburbia as part of queerburban place-unmaking and -remaking becoming institutionalized as rainbowization. Many cultural organizations also draw upon Pride rituals and the surplus visibility of rainbows and drag performances to mark peripheral space as queerburban. Alongside these instances of more overt rainbow visibility, queerburban cultural community activism is distinguishable from that of the central-city by a requirement for privacy, selective visibility, and boundary work within suburbia's ethno-racialized social tectonics. QTBIPOC and ethno-cultural organizing is especially dynamic where diaspora communities are concentrated, using monthly meetups, gatherings, and events to build "collective intimacy" through restricted sociability (Fine & Cortes, 2017).

With some variation, queerburban service organizations largely emerged from ASOs, pivoting and rebranding to meet the more-than-HIV/AIDS needs of their clientele, most especially youth. While one peripheral municipality has actively taken up the question of queerburban seniors in the institutional upperground, it is the youth demographic that has garnered the greatest queerburban service attention. Striving to improve the everyday for suburban LGBTQ2S youth, women, immigrants, and refugees, advocacy organizations have assumed the middleground organizing mantle to augment well-being and belonging. Queerburban advocacy stems from ally institutions who appreciate the paucity of social infrastructure and confront negative framings of LGBTQ2S identities, desires, and behaviours, and their "perpetual discomforts" (i.e., fear, anxiety, shame, rejection, insecurity, and exclusion) (Ahmed, 2004, 147). Highly fragmented and idiosyncratic, these rare moments of strategic advocacy emerging as intersectional pilot projects to include LGBTQ2S youth, women, and immigrants rarely have longevity. Taken together, queerburban organizing is motivated by a desire to remake suburbia through microspaces of informational and tacit knowledge exchange, interpersonal relationship building, self-expression, resource access, and ultimately fun.

Notes

1 Canada's first Prides were chronologically Vancouver in 1978, Montreal in 1979, and Toronto in 1981.
2 Critical opposition has generated inner-city neighbourhood alternatives such as Vancouver's East Side Pride and Montreal's Pervers/Cité.
3 Surrey Pride's opening ceremonies showcase rotating leadership. In 2017, it featured local drag queen performers while in 2018, organizer Holly Bear II collaborated with the Greater Vancouver Native Cultural Society to feature Two-Spirit performances.

4 Surrey Pride went virtual in 2020–2021 during COVID-19 lockdowns.

5 Surrey's community activists offered Pride as a civic place-branding opportunity, yet political leaders remained concerned about hetero-conservative local morality politics and refused access to municipal space, march permits, insurance, or funding.

6 In this same year, the Durham Children's Aid Society ("who haven't until now been involved with LGBTQ") collaborated with the local PFLAG chapter and the City of Oshawa to organize the first Youth Pride Durham event in its parking lot (Health promoter, 12 July 2018).

7 Club status permits the membership to restrict entry and defend against the licensing restrictions and police surveillance that in the mid-twentieth century resulted in raids and public outing (Jennings, 2006). Such strategies have also been crucial to the protection of LGBTQ2S communities in smaller Canadian cities (Korinek, 2004, 2018).

8 For a brief period, QTBIPOC Bramptonians also had "a place to just hang out, talk, and chill" when, inspired by the QTBIPOC Mississauga Meetup, another was organized at the Brampton Library (Organizer, 19 October 2019). This Meetup folded, however, when it lost its Planned Parenthood funding in 2019.

9 In 2018, Sher produced the documentary short *My name was January* to commemorate the life of activist January Marie Lapuz, a trans woman of colour and sex worker living in New Westminster who was murdered in her apartment in 2012. In 2021, Sher released the feature-length documentary *Emergence: Out of the shadows*, which traces the journeys of three LGBTQ2S-identified South Asians from Surrey's conservative Punjabi Sikh cultures.

10 Since Surrey's Guildford neighbourhood is one of four major Filipino enclaves in the region, the Co-Director of Pinoy Pride was selected to appear alongside Sher's founder in the "We are Surrey" diversity campaign for a poster with the slogan "We are Proud" exhibited on bus shelters across the municipality in 2017.

11 Originally named "Queer Life in Colour", the queer was dropped because many of the men identify as "heteroflexible" rather than as bisexual or gay.

12 The 519 is Toronto's LGBTQ2S community centre located in the heart of its gay village.

7 Living

Introduction

Like all places, suburbs are spatio-temporal loci of human experiences, activities, and meanings that constitute the lifeworlds of their LGBTQ2S residents. Lifeworlds[1] are the taken-for-granted physical environments and social spheres shaped by the "common-sense reality" of people's daily lives, interactions, and life course trajectories (Schütz, 1932 as cited in Harrington, 2006; Seamon, 1979). Queer geographers have attended to such lived emplacement as "liveability" (Browne et al., 2021, Banerjea & Browne, 2023), a conceptual framing that skirts the homo/metronormative hierarchies that deem "peripheral" settings as queer impossibilities. Liveability focuses on the multiple embedded ways that LGBTQ2S people deliberately seek a "life that is a life" in different places through complex dynamics of (in)visibility and disclosure (Browne et al., 2021), ensuring that tacit or unspoken sexualities are not necessarily invisibilized or victimized (Coleman, 2013).

As this chapter illustrates, queerburban life "takes place" (Seamon, 2018) relationally through "place interactions" made possible by infrastructure's social affordances. Social infrastructure sustains public life (e.g., in pools, libraries, parks, markets, cafés, and sidewalks) by offering moments of encounter and affordances for connection (Klinenberg, 2018; Latham & Layton, 2019). Infrastructural affordances are the functional and relational capacities for agentic human action enabled by objects (Hutchby, 2001) which, in place-based terms, explain how people may appropriate and inhabit social infrastructure. Urban and suburban spaces are not, however, equally interconnected through their affordances, nor do they foster the same kinds of public life for everyone. Suburbs are structured by manifold exclusions and spatial mismatches, especially for new immigrant populations, elders, people with disabilities, and youth (Lo et al., 2015). Their "sparse" social infrastructure reinforces spatial segregation "by design, not just [as] a function of lower density or a random outcome of unintentional land use policies" (Hartt et al., 2023, 41). Queerburbanites not only live with and contest the unjust affordances of suburban social infrastructure; they "glitch" suburbia's social system (Stevens & Dovey, 2023) creating unintended byproducts, "much-needed erratum", and

DOI: 10.4324/9781032632544-7

positive departures (Russell, 2012, para. 9; 2020). Glitches are the exposed faults of suburbia's heterocisnormative social infrastructure that offer alternate affordances for queerburban place-making (Elwood, 2021; Alvarez Leon, 2022).

This chapter demonstrates how queerburbanites hack the suburban social infrastructure of homes, parks, social venues, and transit through their embodiment and the malleability of everyday objects to reproduce care for people and emotions, ultimately remaking suburban places as queer. It foregrounds the visualization of everyday practices that sustain queerburban life in the absence of explicit and proximate "queer infrastructure" of commercial night venues and regular LGBTQ2S-positive programming (Campkin, 2023). It argues that queer failure to properly use suburbia's social infrastructure subverts its heterocisnormative logics (Halberstam, 2011) resulting in "quiet" place-unmaking and -remaking through the "regular, everyday comings and goings" of queerburbanites "carrying out their own ordinary needs, obligations, and activities" (Seamon 2018, 103). As documented by queerburban photo-elicitation interviews,[2] valued for their capacity to surface feelings and memories that constitute place (Bain et al., 2015), it showcases queerburban lifeworlds that extend from domestic to public realms.

Homes

Canadian suburbs are inscribed with settler-colonial histories of heterocispatriarchal relations of homeownership and homemaking. While extended and blended families, singles and roommates, childless couples, and empty nesters generate demand for suburban housing density and variety, detached homes with multi-car garages still predominate in queerburbia. Behind this illusion of uniformity, however, are many households aspiring to make homes with different visions of domesticity and inhabitation (Vallerand, 2013). Irrespective of household composition, durable, sanitary, and non-violent homes provide the physical, psychological, and emotional security so elemental to human survival and social reproduction (Blunt & Dowling, 2022). Sustaining privacy, relationships, well-being, identity, and memories, homes are "inward looking" places that also have "outward-looking" social properties (e.g., refuge, freedom, ownership, status, and security) (Porteous & Smith, 2001). These properties render homes politically contested identity sites imbued with conflicting emotions ranging from attachment and belonging to alienation and fear (Blunt & Dowling, 2022). As Johnston and Valentine (1995, 111) observed three decades ago with respect to lesbians, home is "where our identities are performed and come under surveillance and where we struggle to reconcile conflicting and contradictory performances of the self". Across variable LGBTQ2S identities and household formations, queerburbanites glitch the materialities of suburbia's heterocisnormative domesticity, embodying an otherness that exposes its contradictions, but also generates alternate byproducts – affordances that remake suburban domesticity as queer (Moon et al., 1994).

For many LGBTQ2S young people, heterocisnormative suburban parental and extended-family homes can be spaces of discipline, limitation, and sometimes outright rejection where they cannot fully express their needs or desires nor comfortably invite friends or partners (Schroeder, 2015). These homes may function as "closets" – spatial expressions of the social metaphor of LGBTQ2S "denial, concealment, erasure, and ignorance" (Brown, 2000, 1), but so too are they places where (in)visibility is creatively reworked (Boussalem, 2021). "Going home", then, may not just be about "hiding that rainbow … putting it back in the backpack and taking it off" so as to "fit back into the box neatly" and "appear normal instead of creating drama" (28 November 2019). It also involves homotransphobic interactions with parents and siblings, necessitating resilience to resist (micro)aggressions, violence, and denials of subjectivity that accrete over time, potentially contributing to depression, loneliness, substance abuse, or suicidal ideation. Take, for example, a pan-sexual teenager in Surrey (19 June 2019), for whom the living room couch in her adopted family home is a contradictory symbol of closeting and coming out. It "belongs" to her Baptist parents. It is where their heterosexual parental coupledom is routinely reproduced as they "watch their shows". This couch is also symbolic of the queer glitch she created when she articulated her sexuality and they responded by saying "it was a phase, that I would get over it. They told me they wanted grandchildren, and I couldn't deprive them" (19 June 2019). Furthermore, it represents public re-closeting: "My parents haven't told anyone. I think they're ashamed" (19 June 2019). Such domestic erasures and expectations in concert with other personal challenges contributed to a period of being unhoused and entanglements with self-harm, drug use, and addiction. Remaining resilient, she asserts that in a "dog eat dog world, I'm going to pick me" while questioning the necessity of "coming out": "I'm going to stop telling other people my business because my life isn't an open book" (19 June 2019).

Another Surrey young adult in her 20s (25 June 2019) from a traditional multi-generational Hindu home is locally celebrated in the press as a "social justice activist" and LGBTQ2S ally while domestically grappling with the "duality of life" as a queer and spiritual person (Figure 7.1). At home, there are pressures to "present and be a certain way" and "expectations about what it is to be successful" that necessitate that she "live life strategically, which can be exhausting" and "suffocating"– always "on guard" (25 June 2019). Her father has specified that her gay friends "can do what they want in their life, but none of that comes into our house because I just want grandchildren". He later apologized for this procreative pressure, but verbally reinforced "don't do that" and "just don't be that". The duality of her life is carefully navigated with Pride stickers hidden in her bedroom, lesbian books buried in bookshelves, and relationships photographically framed as platonic. Where, when, and with whom activities are undertaken is planned to ensure that she is not outed within the Hindu diaspora community. She has long-distance relationships and goes for dates where she will not be seen. While excited to participate in Surrey Pride festivities, she is "always concerned and afraid", particularly when the media is

Figure 7.1 Intersections of queerness, Buddhism, and Hindu spirituality (*Source:* Participant with permission).

present. She uses clothing – rainbow t-shirts and running shoes – to be seen by other queer people. From a place of self-harm and suicidal ideation in high school, participation in Y4AC taught her to be a queerburban youth leader and mentor – "a living reference book for others" (25 June 2019).

In Brampton and Ajax, two gender-non-normative participants described their childhood homes as "hell" (26 November 2019) and "really shitty" because "it never felt like home" and "we were never comfortable" (24 January 2020). They describe domestic psychological violence and verbal aggression as prohibiting them from "fully own[ing] the non-binary part of myself … and saying, 'they is my pronoun'" (26 November 2019). Resisting parental denials and physical, verbal, or sexual abuse necessitates finding housing alternatives. As one young lesbian who ran away from the abuse in her mother's Surrey home to find shelter in the homes of friends of friends explains, "I couldn't move forward in my life … I didn't want to keep going home" (4 July 2019). Other queerburban young people who share bedrooms with siblings or whose parents prohibit locked doors experience limited domestic privacy and use spatio-temporal routines to hack familial homes so as to avoid domestic closeting. A second-generation Chinese-Canadian Brampton lesbian (29 November 2019) in her twenties explains:

> I live in a very conservative household and my parents frankly don't like LGBTQ people. I tend to find community and find connections outside, and I tend to hang out in those spaces until I come back home very late … I like to stay outside as much as possible … away from that household where I have to be confined into my parents' expectations … of what I should be.

Hacking her closeted home involves amplifying her work and social routines to build a life of "elsewhere"; at home, she retreats into privacy behind a closed bedroom door.

The micro-geographies of homes, bedrooms, and bathrooms afford visual and sonic privacy for queerburban youth self-care and self-expression – "a safe space to be me" (24 June 2019). As demonstrated by feminist cultural studies scholarship on the "bedroom culture" of adolescent young women, bedrooms are among the only places to make "a place of their own" (McRobbie & Garber, 1976, 213). Many queerburban youth also find respite here: "My bedroom is somewhere I feel one hundred percent safe and comfortable. It is somewhere I go for peace and to calm down" (24 November 2019). The possessive pronoun "my" – "it's my bed, my bedroom, my safe space" (21 January 2020) – underscores the sense of personal ownership. For queerburban youth who are neither estranged from their families nor unhoused, bedrooms are a key social infrastructure of emotional care. A non-binary youth in Baie D'Urfé uses a bedroom bulletin board to display their sketchbook drawings created during downtown bus commutes. Each image represents an ideal, "what I wish I looked like, what I wish my relationships looked like … an amalgamation of everybody I've ever taken inspiration from and every version of me I've ever wanted to be" (8 November 2023). The bedroom (Figure 7.2) is where they protect their aspirational images of a queer self-in-formation because it is "the only queer safe space in my home" (8 November 2023).

Bedroom social infrastructure (walls, beds, shelves, doors, desks, and windows) is hackable (visually, aurally, and fragrantly), offering an expressive pluriverse of affirmative images and objects that validate queer and trans lives – photographs of friends and lovers at Pride and other events, LGBTQ2S identity flags, flyers, posters, stickers, artwork, books, and games. As material cultural artefacts, these assemblages of queer objects "are elements of wider networks of time, space and sociability" (Brickell & Collard, 2019, 19). They engender different emotions, variously evoking "curiosity, familiarity, reassurance, safety, or pride" (Brickell & Collard, 2019, 18). Such mementos combined with drawers and cupboards of clothing, shoes, make-up, hair and skin products, and scents, trace individual queer and trans trajectories, and create a personal atmosphere. As an adult gay Sri Lankan newcomer and activist (21 November 2019) who continues to share his mother's Brampton home reminds, the bedroom is where "I can love or dislike whatever I want because you haven't journeyed my path". Behind bedroom doors, different looks and identities are experimented with. The surfaces of floors, desks, and beds offer places to

Figure 7.2 Inspirational and aspirational bedroom notebook sketches

(*Source:* Participant with permission).

sleep, dream, and enact sexual desires. Sound waves from music, their lyrics, and melodies create immersive sonic bubbles of care and retreat. Smells bring experiences to life with the scents of laundry detergent, soap, flowers, candles, incense, cologne, and perfume offering comfort and reminders of people and places.

Cell phones, tablets, and computers also extend bedroom spaces to other social media worlds and communities with a degree of "personal territory, intimacy, and individual control" (Hodkinson, 2017, 274). The bedroom cultures of contemporary middle-class queerburbanites blend the material, imaginative, and virtual realms as social media blurs distinctions of public and private, social and intimate. From the privacy of their bedrooms, queerburbanites can virtually "exhibit, play out, and reflexively map identities and transitions", becoming "base points for sociality and identity" that undergird their multispatial lives (Hodkinson, 2017, 286). Many queerburbanites, particularly those at the intersections of marginalized communities (e.g., asexual and aromantic people of colour, and neurodivergent youth), live online, treating their physical location as a material necessity, but secondary to their queer and trans existence. When financially possible, smartphones (with their dating apps and geolocation features) and computers become electronic safety nets, portals to self-curated yet surveilled worlds through gaming, social media platforms, and private online groups. A means to overcome the friction of distance, virtual worlds provide a queer hack for suburban isolation and disaggregation.

For instance, a queer African American parent (27 June 2019) who relocated from Vancouver to Surrey treats her "big screen computer" as "where the magic happens". Online applications and the Internet accessibly permit self-care and support (e.g., the affirmational texts and care packages of the Dear Black Women project based out of Michigan). For an older trans woman (28 June 2019) with a child and little money, "my community is Facebook" while for a gay Portuguese-Canadian man (28 January 2020), Grindr,[3] Tinder, and Scruff make virtual cruising possible where there is "so much potential in the grey area between sex and friends". For a lesbian Filipino newcomer, Instagram (with validation from responses and likes to image posts) helps maintain transnational kinship linkages. For a gay Sri Lankan-Canadian (21 November 2019), Instagram "images carry an intimate story of who I am" and offer a way to counter hate speech, take a stance, spark stories, and possibly shift his followers' thinking. Online lives connect friends and build communities of interest around shared stories, pictures, and resources, worlds of people "who make me feel seen" (24 January 2020).

As social infrastructure, homes are where domestic intimacies are produced through embodied performances and the decorative deployment of material cultural objects. For a white bisexual living with her mother and finding social support through participation in Y4AC, her bedroom is "everything to me" a "place to express yourself" (26 June 2019). Her bed, covered in her favourite sheets and childhood stuffed animals, is surrounded by Pride and Bisexual flags from Surrey and Vancouver Prides. For one Brampton, first-generation Chinese-Canadian participant (20 December 2019), an asexual Pride flag on her bedroom desk works with a Pride sticker that reads "queer" to indicate her "quiet" opposition to her homophobic parents' heterocisnormativity. An Indo-Caribbean drag artist (27 December 2019) describes how he decorated his suburban youth bedroom as a visual "archive … of all the queens that I had met and I looked up to". Refusing complicity with the cultural logics of his heteronuclear family, he used queer flamboyance to hack domestic "disidentification" (Muñoz, 1999). He confined his domestic exploration of drag to lockable rooms while posting the results as self-portraits on social media:

> I'd have to go into the bathroom and do my face for three hours, take a picture, wipe it all off, go back upstairs. These are some of the first wigs I bought, but I had to hide them in my room, because my Dad is quite homophobic. There's been moments where he's found certain things and it's been a huge battle (27 December 2019).

In this example, hacking involves performative negotiations of domestic (in)visibility. Inside the physical home, drag performances and the objects that afford them are invisibilized, hidden in bedrooms and bathrooms while online spaces afford an interface with the wider world where queerburban life can be documented and shared.

Moving out of parental homes affords some queerburbanites avenues of domestic self-expression to create their own "oasis" behind closed doors with chosen keepsakes (23 June 2019). Quite simply, domestic independence means "now I can hang what I want" (24 June 2019). The Indo-Caribbean drag artist (27 December 2019) can now leave out drag paraphernalia as "curated decorative objects" and meaningful "visual cues" to sexual and gender identities (Andrucki & Kaplan, 2018, 783). Bedrooms can be converted to drag dressing rooms, displaying boxes of wigs, racks of gowns, and promotional posters, such as Amanda Luv's archives of suburban drag personas and performances in her Surrey apartment. Yet "finding a place that feels like home" (24 January 2020) in suburbia is challenging for many sexual and gender minorities. For a lesbian, pan, polyamorous, ace couple living in their first two-bedroom apartment together in Pointe-Claire, a living room couch symbolizes queerburban homemaking: "it's ours" and "it's where you can go to feel at home and to feel safe" (6 December 2023). This couch is "the hub of our home" where they watch television and engage in the "end of day rituals", such as "winding down after work" and "all that other shit that relationships have, conversations, fights and make-ups" (6 December 2023). Being in common in queerburban domestic space also requires creating distinct spaces for themselves: one occupies an extra bedroom as a study space while the other inhabits a "little hidey hole" behind the couch surrounded by journals, sketchbooks, and seasonal decorations (6 December 2023).

Adult queerburban domestic spaces offer opportunities for queer privacy, self-actualization, self-expression, and being-in-common. For queerburban young adults transiting to financial independence through employment, Canadian suburbs offer more affordable housing options, increasing domestic autonomy and liberating the expression of queer identities. Domestic autonomy provides opportunities to comfortably "just veg" with partner(s) and friends (24 January 2020) but can also be milestones of self-expression and acceptance. Looking back at Surrey from the window of his Burnaby apartment, a South Asian gay man (9 July 2019) evaluates "how far I've come from what I hid when I was in Surrey – to be in my own space. I made it". A white cisgender gay man (13 January 2020) and his partner bought a condominium that overlooks "downtown" Mississauga after renting in a subdivision that was "wall to wall with conservative families" and feeling "intensely visible at all times". Queerburbanites, he explains, are "all just living these home lives, and we want to invite … our friends to be part of it" in a "conventional, social way" while still feeling "connected to my actual sexuality … but … very much with other people" (13 January 2020). For his queer friendship circle, the routine of watching RuPaul's Drag Race together on television over food and drinks in this "safe space" brings queerburbia to life (Figure 7.3). The kitchens and

Figure 7.3 Queering a suburban condominium living room by watching RuPaul's Drag Race together

(*Source:* Participant with permission).

living rooms of non-normative households are relatively affordable social spaces of friendship where "it is affirming to be obnoxiously loud and gay in each other's company" (24 November 2019).

Such "home-making practices" (Gorman-Murray, 2006) and private gatherings for potlucks (Kirkey & Forsyth, 2001), "dinner-party circuits" (Brown, 2000, 81), and movie and boardgame nights are among the distinct social practices identified in the scholarly literature on queerburbia (Dines, 2005, 2009). As a queer-identified married young woman who rents a suburban townhouse with her partner and two friends echoes: "my community is all over. If we come together, it is in someone's home" to talk, eat, and play games (24 November 2019). A polyamorous lesbian on a disability pension in New Westminster (14 July 2019) uses her micro-bachelor apartment as a "safe" and "secure space", with a door that "is always open, people can call me day or night [and ask], can I come over?" (14 July 2019). But there are geographical limits to outer suburban home-based socializing, as an older, white gay male (4 July 2019) learned after buying a condominium in central Surrey:

> People don't cross the Fraser River. There is a straight wall that queers won't cross. I wanted to have friends over and then I discovered that people won't come out here. So, I live in New Westminster and sleep in Surrey … I just happen to live out here.

Queerburban social relations have some geographical stretch until distance limits in-person community formation. Unaccustomed to socializing in private homes, older generations of gay queerburbanites who forged their social networks in the gayborhood commercial environments of bars and bathhouses find the move to outer suburbia challenging (Ghaziani, 2014b).

Queerburban "care-full" (Williams, 2017) sociability spills out from domestic interiors onto residential balconies, terraces, driveways, and yards creating "safe spaces to talk", "decompress", and "be seen through a queer lens" (26 November 2019). Outside of homes, however, there are no guarantees of security as interactions with landlords and neighbours can produce anxiety about the presence of partners, accurate pronoun use, and even harassment.[4] Neighbours are uninvited audience members who may reinforce the heterocisnormativity enveloping private and intimate homes through interactions that involve "second-guessing" displays of affection and questioning exterior decorations (13 January 2020). Gardens and driveways are, however, sites where homo-hetero social cohesion is negotiated through everyday interactions while doing property and vehicle maintenance (Gorman-Murray, 2013). Impromptu or more considered rainbow iconography (e.g., flags, stickers, DIY posters, and chalk drawing) can signal queerburban belonging and function as micro-political rejoinders to heterocisnormative suburban imaginaries. In one Markham cul-de-sac, a rainbow chalked message read "welcome to the big happy house" in the public celebration of the support, acceptance, and shelter that one trans masculine youth found within his boyfriend's lesbian-led parental home. Participating in his classmate's queer family afforded him the opportunity to "feel part of something bigger than just himself" (20 November 2019). The making of queerburban family homes is an overt remaking of suburbia's reproductive heterocisnormativities. Beyond mere dreaming, it is the living enactment of different futures for the next generation of queerburbanites.

In the absence of formal queerburban social infrastructure, homes are places that reproduce LGBTQ2S social networks whether virtual or in-person. In the hands of glitch subjects, homes may afford multiple hacking registers to heterocisnormative exclusions: strategic outing and temporal management enable LGBTQ2S youth to sustain themselves in non-supportive homes; the privacy of bedrooms and bathrooms affords opportunities for alternate embodiments and performances; queer objects hidden or displayed in bedrooms provide entry into other worlds; and autonomous homes are sites of self-expression and sociality that compensate for the absence of commercial and community infrastructure. As queerness spills out onto balconies and into the yards, driveways, and the public realm, parks become another key hackable social infrastructure of place-unmaking and -remaking.

Parks

Central to the reproduction of heterosexual families, public parks are sites of symbiotic ecological, social, and economic importance (Pincetl & Gearin, 2005). The form, management, and design of parks often communicate a community's heterocisnormative middle-class reproductive and moral values. Embedded in residential neighbourhoods centred on the collective social infrastructure of schools, libraries, and community centres, parks are civically ordered spaces of public "respectability" within which to recreate, see and be seen, and encounter nature (Catungal & McCann, 2010). Concomitantly, heterocisnormative spatial imaginaries can be hacked through cruising or public sex solicitations (Berlant & Warner, 1998), teenage escape, or even LGBTQ2S activism. For example, Y4AC activists annually art bomb "The Grove", a cluster of cedars in Surrey behind the Newton Recreation Centre known for property crime, drug use, and sex work. A community reclaiming of this site began with a "Friends of the Grove" installation commemorating the life of Newton mother Julie Paskell who was randomly murdered in the parking lot in 2013 while awaiting her son after hockey practice. Contesting community violence, Y4AC decorates it with rainbow streamers, flags, wool art, poetry, and, more specifically, a collage for the anniversary of the Stonewall Riots, an "inclusion" gate, and a space dedicated to trans art. This "artivism" is a way for "other youth to see the possibilities of positivity in living a queer life" and nurture community belonging (Marchbank & Muller-Myrdahl, 2019, 47).

In contrast with the prescriptions of domestic spaces, suburban public parks are often the public social infrastructure of queerburbia. Whether they are the scenes of annual events such as Pride or part of everyday routines (e.g., dog walking, running, socializing, or cruising), suburbia's plethora of green and open spaces provide a social infrastructure that blurs publicity and privacy. Parks offer all suburbanites access to recreational opportunities, the grounding interactions with "nature" that stimulate reflection. Queerburbanites often interpret parks as spaces of emotional safety, sites of "sanctuary" and "joy" (19 June 2019) where they can feel connected to their full selves. As a non-binary POC in Ajax (2 November 2019) explains, "I think a lot of queer people are told that our bodies are not valid or like normal, but I find personally whenever I'm in nature I feel most grounded". While there may be "nothing inherently queer about the [park] space", for one non-binary white Bramptonian (26 November 2019) it is a place to freely "disappear … for a couple of hours" and not "have to show up with an offering or a performance of anything". Through seemingly banal routines of walking, running, reading, daydreaming, meditating, smoking marijuana, or drinking, queerburbanites hack public green spaces, queerly unmaking and remaking them through their embodied presence at different times of the day and night.

In parks where there is limited surveillance, the "unassimilated otherness" (Young, 1990, 251) of public space offers queerburban possibilities. A white gay Mississauga condominium dweller (13 January 2020) clarifies:

> you see all these other different groups of people who are using this space … where everyone feels like they're here for their own reasons and everyone's getting various things out of it, but no one's really interfering … [or] policing how you're using it.

The co-mingling of park publics affords some queerburbanites anonymity and autonomy to assert agency in their lives through the oscillating dynamics of movement and rest, interaction, and retreat that come with activity. As one non-binary queerburbanite specifies (26 November 2019), running in open green spaces "physicalizes and gives me back my agency and sense of control over my body". Whether it is walking a forest path or balancing atop riverbank stones (Figure 7.4), nature affords ecstatic moments of queer and trans embodiment away from the scrutinizing heterocisnormative gaze where "you can let your guard down" (13 February 2020). The ordinary routine of dog walking in Beaconsfield's Angell Woods old-growth forest affords a South Asian trans-identified resident (13 October 2023) with "a place where I can just go and exist as a trans person". It is a place to be public with limited social liability: "It's not like I'm walking through there with a Pride flag or anything, but you run into people and say 'hi'. Then it's like 'Oh, you just interacted with a trans person, and nothing happened'!" (13 October 2023). A similar sentiment is

Figure 7.4 The creek: "a spot where we connected with each other and nature and documented it for ourselves"

(*Source:* Participant with permission).

articulated by a trans-identified resident of nearby Pointe-Claire who runs on a local bike path: "I exist there as a trans person and, therefore, it is a queer space no matter what anybody else who walks on it can think" (6 December 2023). Nature's lack of judgement and the spatial expanse of green spaces affords freedom for temporary place-remaking.

When not in homes, many queerburbanites are outside alone or in the company of friends, partners, and lovers, transforming suburban green spaces with their own embodiment and the objects they bring. To commemorate their first in-person meeting after a long-distance relationship, a lesbian, pan, polyamorous, ace couple left a small figurine in a knot hole on a wooden bridge in a West Island park. This personal artefact was "a teeny, tiny treasure to find in the woods" that "transformed the park a little" and "left their mark" as well as a commemorative trace on Instagram (6 December 2023). A queer Ajaxian describes appropriating park space with their friends for queer expression:

> We are making a space that isn't anyone's but ours. We bring things to eat and drink and sprawl and stick our stuff everywhere with explosions of things and taking pictures … showing ourselves in spaces. Showing where we are and sending them to each other. These spaces exist because we photograph them (24 November 2019).

Figure 7.4 is one such moment of queer place-remaking that was captured in a photograph, posted online, garnered comments and likes, and inspired other friends to drive out from the city to visit. They too admired the rocks, listened to the rushing water where the salmon spawn, and documented their queerly communal yet "private feeling" in nature. When suburban "nature" is unprogrammed and under-landscaped, when it is large and remote enough to afford alone time or communal socializing, it is a place of queerburban regeneration.

Interactions with nature – from brief exposure through to more intentional and sustained activities (Tillmann et al., 2018) – accentuate the full-body physiological reactions and potential positive mental health outcomes (e.g., managing stress, anxiety, and depression, enhancing self-esteem, and fostering resilience) that parks as social infrastructure afford queerburbanites as they navigate the daily "ins" and "outs" of the closet. For example, the presence of water within parks – fountains, ponds, creeks, rivers, lakes, and oceans – releases queer emotions. Feeling the ocean's calmness while playing frisbee with high school friends, one young Surrey woman found the courage to yell out "I'm pan-sexual" and to experience their acceptance (19 June 2019). For a queerburban Bramptonian (20 December 2019), a small pond behind her subdivision conveys "a sense of calm" walking to the bus stop "after leaving such a restrictive conservative household" and makes her hopeful that she too "will go into the outside world and have fun". The fluidity of water inspires self-reflection, shaping places to "think about who I am and what I like to do, to get my morals and values back" (26 November 2019). The edges of water bodies – sandy, rocky, muddy – offer places to run, skid, jump, and balance, places of freedom where "nobody is around me … judging me" (19 June 2019) and "no rules apply, and anything goes" (28 June 2019).

Nature's positive mental health outcomes are, however, conditioned by expectations of behavioural conformity to heterocisnormative imaginations of parks. Public displays of same-sex intimacy or the physicality of queer desire invite scrutiny and social judgments of "sexual deviance" (Catungal & McCann, 2010, 87), often necessitating temporal tactics to garner greater freedoms. A Sri Lankan-born Bramptonian (21 November 2019) cautions that the local park is "a family environment … for social units of a traditional sense". The park is "not necessarily queer un-friendly" but its dominant coding as a place for "families" means that "if you show your queerness by doing things like holding hands, people perceive it as 'putting it in their face'". Public intimacy expresses social reproductive care and enhances emotional well-being, but queerburbanites may have to disguise, hide, or downplay their sexuality to blend in due to the potential for heterocisnormative judgement or even verbal and physical harassment. Public displays of queer affection require self-confidence, a willingness, as one white bisexual Mississauga resident (4 December 2019) describes, to "stand out … draw an eye from people … divert from the norm, [where] people will notice it very quickly". Going public in the park is something a Filipina newcomer aspires to ("I hope to one day") since her lesbian relationship remains "an open secret that nobody talks about" (30 June 2019).

Queerburban imaginations of public parks as cruising locations of same-sex public eroticism counter worries about daytime queer publicity. In suburbia, park cruising grounds have been a consistent social infrastructure for gay men and MSMs to create queerburban worlds. Burnaby's Central Park once attracted a hundred men a night to the "sex scene – not a social scene or a community" (5 July 2019), but night-time parking lot gating has reduced that number to a handful, who often "come" earlier in the evening before sunset. An older, white gay man recalls how he accessed this evening homosocial scene: "I had good night vision … I knew my way around every tree. I never tripped over a log. I was safe in the dark. There was no smoke. There was no loud music … I could talk to anybody" (5 July 2019). Across the three city-regions, cruising for public sex was a core place-remaking practice of gay queerburbanites, whether in Brampton's Heart Lake, Mississauga's Streetsville Park, or Longueuil's Parc Marie-Victorin. In Surrey, both the more central Bear Creek Park and Blackie Spit Park in Crescent Beach are "secret place[s] inside of the city where you can do anything and be with anyone" because "nobody cares in the forest" (26 June 2019). For a low-income, white bisexual sex worker, these are places of "exploration" and "freedom" where privacy is publicly found by repurposing picnic tables, benches, trees, and trails for sex. Through the voyeuristic gaze permeating public sex grounds, she observes queer, trans, and non-binary young people using parks for sexual exploration – "they're not hiding, but they're not exposed, so it's a nice balance … [and] I get to watch the younger generation" (26 June 2019).

Despite accounts of dissonance within suburban parks, especially with respect to public displays of affection, they provide a social infrastructure for individual publicity and the experience of nature as well as group gatherings and "homosociality" (Qian, 2017). Individuals dare to be themselves while walking dogs or running in mundane green spaces while groups use parks for informal meetups, public sex, dates, social gatherings, and programmed Pride events. Such "identity driven congregation" within parks underscores their adaptive re-use (Brown-Saracino, 2018, 101) as meeting places, while individual park exploration supports queerburban bodies and minds by calming emotions, enhancing well-being, and providing an expansive and less proscriptive place to be oneself.

Social settings

Queerburbia's LGBTQ2S-supportive commercial and institutional infrastructure is minimal, necessitating improvisation within a range of public and private spaces. Queerburbanites find moments of sociability and personal enrichment in suburbia's "thinned-out places", those with "perforated surfaces" that lack "densely enmeshed infrastructures" (Casey, 2001, 684). Open to continuous "reshaping and reconnecting", such places are thickened – if only temporarily and on occasion – through social and affective relations that shape their mood, degrees of privacy, and activity possibilities. As Duff (2010, 893) asserts, "[t]o feel connected to [a] place *is to experience a sense of belonging in place* that itself generates resources of immense value in the promotion of health and well-being". From streets to coffee shops to restaurants to gyms, as well as the municipal infrastructure of recreation centres, libraries, theatres, and art galleries, queerburbanites are attuned to their welcome and possibilities for appropriation. This place attentiveness is based on landscape cues such as aesthetic coding and material cultural artefacts as well as from the "ambient" sense of community generated by co-present publics.

Suburbia's car-dominated streets, drive-in coffee shops and restaurants, and strip and shopping malls encircled by parking lots are sites of anomie and haphazard interaction offering occasional relational moments of recognition. The micro-geographies of suburbia's liminal margins offer both acceptance and indifference with opportunities to temporarily appropriate accommodating spaces and create meaningful yet fleeting senses of communality. Whether it is the temporary presence of other queer bodies, a poster, or a soundtrack, queerburbanites use subtle strategies of communication and interpretation to become visible to one another. So scarce in suburbia, queerburbanites report "hanging on to little threads wherever I can find them" (21 November 2019). As one asexual white Mississauga young person points out (21 November 2019), "people give off a vibe, whether a haircut or an androgynous look" while others strive to "do gay stuff as visibly as possible to meet queers in the suburbs" (24 November 2019). Shopping malls and parking lots can even be affirming spaces when queers see each other. An older Goan-Canadian gay man described how his "gaydar" went off in Laval when "cross[ing] paths with other LGBT people in the car park at the Maxi or the Super C" (22 October 2023). Although they "were obviously trying to pick me up in a fun way and I do not do that", the recognition this interaction signalled was memorable. In the anomie of the parking lot, being visible to one another nurtures recognition in the shared transgression of heterocisnormativity (Podmore, 2001).

Without the commercial bar, club, and restaurant "central place" (23 June 2019) equivalency of downtown gayborhoods, queerburban life – especially for seniors and young people – has little to "orient yourself towards" (Marchbank & Muller-Myrdahl, 2019, 40). A retired white gay man in Burnaby (5 July 2019) observes that "none of my friends go to gay bars"; instead, "I go for coffee every day, read by myself, or talk to others. It's easy to meet people". He relies upon Rainbow Event Network posts and attendance at biweekly events as "a good way of re-igniting community and maintaining it". Joining social networks and creating informal spaces for queer socializing remains challenging because of "personalities" and "cliques" that make it "hard to break into groups and relate to the conversations" (4 July 2019). When a member of the Vancouver Radical Faeries (4 July 2019) wanted to organize a Surrey evening coffee meetup, he "couldn't compete" with the city centre's magnetism nor could he "scout" a suitable venue. For queerburban young people under 18 who may also lack the finances to experience adult gay culture, meet-ups in suburban coffee shops offer affordable places to seemingly "talk about anything" (11 November 2019). These semi-public settings, however, do not ensure adequate privacy for participants who are overheard by other customers and where "as a queer person you are on high alert" for "any dangerous vibes" (21 November 2019).

Inside and outside of commercial venues, streets can be completely benign or pose the threat of homotransphobic violence. For a transwoman living in New Westminster, walking down 6th Street at 3rd Avenue always reminds her of encountering a pedestrian wearing a "straight pride" t-shirt. Even as New Westminster codes itself as a place of queerburban belonging, heteroactivism's hatred surfaces. Likewise, a residential street in Burnaby where an abusive neighbour jumped out of the bushes, kicking her dog while calling her "an abomination" is a powerful reminder of the homotransphobia of the gated community where this Persian lesbian lives with her partner's family (29 June 2019). In Baie D'Urfé, a truck driver hurled homophobic insults at a young non-binary person walking along an empty street which now represents the displaced loneliness of their queerburban childhood where, ironically, "I felt like I was home every place, but home" (8 November 2023).

For a trans-identified couple in New Westminster, homeliness is found at a local hamburger restaurant on Columbia Street. Although there is nothing countercultural about this space of comfortable sociability, the increasing embodied presence of local queers, from inside and outside of their friendship circle, makes it feel like a queerburban meeting place. The street contains one of Vancouver's only 24/7 suburban commercial gay male spaces (the Steam 1 bathhouse),[5] and its businesses attract queerburbanites from neighbouring peripheral municipalities for whom it "feels like a city, it's more sophisticated" and offers "more things that I like to do" (23 June 2019). Whether it is finding a bar to write in – "a place where people are in their own little world" – or a gym to workout in, queer-friendly public spaces for social reproduction are found through a myriad of hacks. Queerburban anonymity is found by walking "the Surrey Strip" where the survival tactics of the unhoused produce "a place where no one cares what you are doing" (26 June 2019). Queerburbanites thicken mainstream and discarded public spaces and chain commercial venues, creating "space in which to belong to each other when there is little else to belong to" (Marchbank & Muller-Myrdahl, 2019, 39).

When queerburban belonging cannot be seen and felt locally, many queerburbanites hack it virtually or through selective planned commutes downtown to access services, programming, and events. This at-a-distance living involves time and expense and a willingness to negotiate hyper-sexualized, racialized, and alcohol- and drug-infused gay village spaces. Nevertheless, gayborhoods, their institutions and streetscapes, are symbolic places that meaningfully resonate for some queerburbanites. For example, along Vancouver's Davie Street, Pride flags offer permanent reminders that "We're here, we're out in the open" (4 July 2019). Queer businesses such as Little Sisters' Book and Art Emporium – part sex shop, book shop, resource centre, and event space – and the drag bar The Junction signal queer and trans safety and a break from heterocisnormativity. Along with Church Street in Toronto, the businesses and queer institutions that compose its streetscape afford queerburbanites reminders "that I can exist amongst everything" (4 July 2019). Such queer resonance is also found in queer-friendly neighbourhoods such as Vancouver's Main Street and Commercial Drive. These streets are where you "see a lot of people who are like you", where LGBTQ2S people have "a higher visibility" and it is a more "neutral" and "alternative space" that is not necessarily heterocisnormative (29 June 2019). The LGBTQ2S community centres located in gayborhoods are also centripetal resources drawing queerburbanites to their programming, and services. The 519 in Toronto and Qmmunity in Vancouver variously function as intersectional, affordable, and life-affirming social hubs for queer and transgender newcomers and refugees, older adults, parents, youth, and children. In Montreal, the Centre communautaire LGBTQ+ welcomes participants in community groups while youth also congregate around Maison Lambda downtown, P10 in Ahuntsic, or the West Island's LGBTQ+ Centre.

The unequal city-regional distribution of explicitly queer social infrastructure is, however, acknowledged by residents and social service providers alike. A QTBIPOC community activist (21 November 2019) in Brampton opines:

> it's a known fact that Peel does not have safer or inclusive spaces, so folks often migrate to Toronto to access those services … Those are spaces where folks want to go and need to go: where they feel comfortable enough to be truly who they are.

Central city social infrastructure is described by another Mississauga queerburbanite (21 November 2019) as "the mothership downtown". Such social infrastructure is not in competition with the ephemeral and scattered queerburban social venues, but rather it is tethered to mobility infrastructure to create queer city-regional constellations (see Gieseking, 2020a, 2020b). Significant geographical distance needs to be overcome through journeys on buses, trains, and cars to access LBGTQ2S-supportive social programming in downtown Toronto, Vancouver, and Montreal that offers moments of learning, growth, and networking. These opportunities enable some individuals "to come back as one of your teeny-tiny Starfleet … and create a little bit of change" (21 November 2019), however fleeting, in queerburbia.

While significant queer and trans regenerative possibility exists in Canadian central cities, gay villages are still crosscut by manifold disidentifying queerburban exclusions (Sandoval, 2018). A Punjabi community service provider (9 July 2019) dismissively characterizes Vancouver's gay scene as "flakey, pretentious … and image-based", a place where South Asian men are "fetishized or stereotyped" as little more than "extras to a cis-gay, white world". Toronto's gay village is also "oversaturated with white-cis-gay men" who an Indo-Caribbean Mississauga resident (27 December 2019) sees as "often very inappropriate, very racist, very misogynistic, and very transphobic". Similarly, a Goan-Canadian gay senior from Montreal's St. Eustache suburb describes his sense of gay village alienation and disdain where the dynamic is "cliquish" and "sexually charged" (22 October 2023). Despite working at a local restaurant and attending Tuesday Karaoke at a bar, he maintains that "there's nobody whom you can connect with about anything very deep"; instead, "it's all commercially created" and nothing is "geared towards community-building" (22 October 2023). Exoticized by other men, he is repelled by the white homonormative pleasure-seeking of the Village's bar and sauna social infrastructure and laments the lack of places "to hang out for a few hours, have a coffee and exchange with people in a meaningful way".

Queer disidentifications with gayborhoods push queerburbanites to other inner-city areas, notably gentrifying queer-friendly districts (like Leslieville and Parkdale in Toronto, Commercial Drive and Renfrew Heights in Vancouver, and Rosemont-Petite-Patrie and Hochelaga-Maisonneuve in Montreal) where "ambient communities" (Brown-Saracino, 2018, 52) foster inclusive spaces for queer and QTBIPOC communities in local bars, clubs, and theatres. One Mississauga drag artist finds life-affirming sustenance with other Caribbean queer people adjacent to Toronto's gay village (Figure 7.5):

> There were a ton of people there. Most, if not all, [were] people of color, black and brown people. My friend who's also a performance artist—a drag king—produces a show for queer women and trans women and non-binary people … The theme for this show … was 'soca versus dancehall.' She's Jamaican, and she was like, 'I'm going to do dancehall and you're going to do soca.' It was really exciting, because this was the first time, actually, that a fellow Caribbean person has actually booked me for a Caribbean show.

Figure 7.5 "Favourite performance"
(*Source:* Participant with permission).

The hosting of QTBIPOC events, however temporary, is a labour of love. Such networked social venue hacking affords queer and trans and/or QTBIPOC people the exuberance of dance in the absence of the surveillant, eroticized, and racialized gaze of village gays. A key resource for queer social reproduction is friendship itself, which helps transform one-off social venues into places of intimacy and care (Andrucki, 2021).

The remaking of suburbs as queerburbia occurs through the social reproductive practice of adaptively occupying and reusing commercial social infrastructure (e.g., bars, coffee shops, gyms, and theatres). By hacking together, the affordances of these (privatized) public places, and convening one-off community events, queerburbanites challenge the social reproduction of heterocisnormativity. In these ways, queerburbanites take up space in ephemeral and embodied ways to hack together a sense of community. But queerburbia includes networks that extend to central-city gayborhoods and queer-friendly inner-city spaces requiring travel within and across city-regions.

Public and private transportation

Given the uneven city-regional distribution of LGBTQ2S social infrastructure, public transit is integral to sustaining queerburban lives because it connects private-oriented homes to more public-oriented education, employment, and social opportunities. In Toronto, Montreal, and Vancouver, transportation infrastructure is dominated by highways, parking lots, private automobiles, and disparate public bus, light rail, and commuter train routes that feed into central-city networks of elevated or subterranean trains, streetcars, and buses. In these metropolitan peripheries, queerburbanites calculate their commute times or assess their budgets to travel to the urban core. Connections between peripheral municipalities are also relatively underdeveloped, which creates additional mobility barriers for lower-income queerburbanites who cannot afford to own a private vehicle. Suburban public transit infrastructure does not afford the same capacities for city-regional mobility as automobile infrastructure.

Affordances of queerburban access and mobility vary considerably. Vancouver's elevated SkyTrain is described by a Punjabi-Canadian gay man (9 July 2019) as "an escape" from queerburbia, "a way of going downtown to be myself and to figure out myself", a transitional space to "mentally prepare to shift from closeted gay to open gay". For other queerburbanites, its driverless and unstaffed stations or the regional trains and peripheral bus networks that feed the Toronto Transit Commission's (TTC) subways, streetcars, and buses that are places of homotransphobic discrimination, verbal abuse, acts of harassment, and physical assault. Transportation infrastructure is especially unsafe for gender non-conforming people (Shakibaei & Vorobjovas-Pinta, 2024). Queerburbanites have mental lists of stations and bus routes where they experience "trouble" and "incidents" of being physically "accosted" and verbally "harassed" for how they look or for public displays of affection, especially at night, that puts them "on guard" (23 June 2019). As one Ajaxian (24 January 2020) with a QTBIPOC non-binary partner details:

> [We] have had a lot of encounters with shitty people when we're commuting, like in the subway, on the GO train, Union Station. Yeah.... being heckled at; if we're holding hands, we'll get yelled at; or, creepily approached—always by a man ... We've had people who've outwardly come at [my partner] for gender-based violence ... screaming 'what are you'.

Such overt aggression that publicly questions the very essence of who one is occurs in the presence of passers-by and fellow passengers yet rarely elicits a public response. A myriad of non-verbal cues and unfriendly looks in conjunction with misgendering amplify stress (Lubitow et al., 2017). In buses and trains, "you are trapped" (19 June 2019) with few escapes from "weird looks and mutterings" (24 June 2019) and interrogating conversations. The relentless scrutiny of the heterocisnormative gaze eventually forced an older trans woman off of public transit: "I stand too tall, and it makes me uncomfortable" (28 June 2019). Following her transition, she obtained a bank loan to buy an electric car to commute to her high school teaching job in Surrey. But younger and older suburbanites rarely have the luxury of transportation alternatives. Instead, they routinely endure personal vulnerabilities on public transit to access central-city education, work, and social infrastructure.

Queerburbanites spend significant portions of their day in the absence, as one Bramptonian remarks (21 November 2019), of "any markers of queer- or trans-inclusiveness on public transit". Like other suburbanites engaged in lengthy, expensive, and crowded commutes (upwards of 1.5 hours one-way), time is spent reading or digitally tethered to music, podcasts, or social media filter bubbles while also in one's own thoughts, sometimes "spinning and digging oneself into ruts thinking and processing and analyzing" (28 November 2019). Different modes of travel span fragmented public transit systems, producing logistically complex journeys of delays, interruptions, and ultimately exhaustion. By evening, the central-city night-time economy often seems inaccessible leading to wistful remarks that downtown events look "fun … adventurous, or sexually satisfying" (22 January 2020) but ultimately the commute means they "tend to stay away from those late-night events" (20 December 2019).

Where private automobile use is possible, cars enable social reproductive care for bodies through eating, talking, listening to music, and passenger sleeping. For a car-owning white gay man, private transportation opens up city-regional social infrastructure: even though Brampton is "a little bit out here", it is "quick and easy to drive downtown on a Friday or a Saturday night" and he has "made friends with Toronto people". With distinctions made between central-city and suburban friends, queerburban commuting patterns are centripetal with queerburbanites responsible for moving themselves closer to extroverted concentrations of queerness. Car ownership is also liberating for queerburbanites because it affords a sense of personal empowerment, autonomy, and sociality. As one pansexual youth (19 June 2019) details, "when I'm having an off day I go and drive. I park in a park, blast music, and cry. It's a true escape". This liberating sense of emotional release is confirmed by a non-binary Bramptonian (26 November 2019) who uses their car as a space in which to write in their journal and self-reflect (Figure 7.6):

> …it feels like freedom. For me, having that reconnected me to family, it reconnected me with people that I wouldn't see a whole lot … [and] to maintain an element of control. So, having the car meant that I could just go for a drive for two hours, and I could just disappear … In the back of my mind, I always thought: 'If I don't have a place to live, I have a car I can sleep in, at least' … it might not be much, but it's mine.

To maintain and insure a vehicle is to retain a sense of autonomy over who one travels with, where, and when. As a white lesbian youth in Surrey (26 June 2019) who received a second-hand car for her sixteenth birthday asserts, "I'm in charge, I can do whatever I want". Personalizing a vehicle's interior with sound and objects can affirm one's sexual and gender identity on the journey. As Bramptonians specify, not only does the car create "the freedom to travel where I want, when I want, how I want" (28 November 2019), but "my car is a marker of my queerness. Within its closed doors … I feel comfortable and safe" (21 November 2019). The privacy of a car's interior enhances personal

Figure 7.6 "Meep": the freedom, control, and privacy afforded by car ownership (*Source:* Participant with permission).

safety, extending possibilities for holding hands with lovers and partners, enhancing individual comfort, and protecting queerburbanites from the scrutinizing heterocisnormative gaze of public transit riders.

Public transit can, however, be hacked with micro-practices of personal decoration, travelling together, and sonic and social media escape. Inspired by a backpack he saw "with extreme colours that reflected LGBTQ allyship and identities" on Toronto's subway, a Bramptonian (21 November 2019) wears LGBTQ2S lapel pins and political buttons. Likewise, a white asexual young woman in Mississauga (21 November 2019) affixes ACE pins in the recognizable grey, purple, and black colours on her bags and jackets as articulations of "stealth Pride" – "it's there, but I'm not drawing attention to it". Sometimes it sparks conversations between people who recognize each other in the crowd as queer. To decorate clothing and bags with queer signs that articulate the wearer's affiliations symbolically expresses both difference and belonging (Brickell & Collard, 2019). The material culture of buttons, patches, stickers, and t-shirts carries messages of self-identification, social justice, and aspirations through heterocisnormative places. Whether it is overtly provocative t-shirts with statements like "news isn't fake, and you need to fix yourself" and "God doesn't hate gays, he hates hateful people" (26 June 2019) or a Black ACE ring on the middle finger of the right hand to signal asexuality, these gay material cultural objects encode deeper meanings for people in the know (Hermes, 2019). An ACE ring can be something to look for on other hands, a conversation starter, but also a talisman to touch when encountering homotransphobia, a personal reminder that "aceness is always there" (21 November 2019). When combined with queer body modifications – piercings, hairstyles and colouring, tattoos, and henna – these material objects help to hack "straighter" spaces. Immediate environments can be changed "to validate me" via that self-challenge of being "out in that space" (21 November 2019).

Queer objects and embodiments on transit may also expose individuals to violence. After repeated slurs and spitting, one trans masculine Burnaby youth (12 July 2019) describes feeling "overwhelmed" before "tackling" the journey home. Despite enjoying "being visible for others", for his own self-protection "I'm going to take my gay off" on public transit because it renders him vulnerable. Where he has opted for more defensive dress and greater vigilance, others hack the heterocisnormativity of public transit by travelling with queer friends and "chosen family", "people who I choose … and want to spend time with" (28 November 2019). Still others escape into the queerer private sonic, reading, and communication bubbles of smartphones. Blurring online and offline worlds, queerburbanites live a significant portion of their lives through smartphones, treating this artefact of modern technology as an extension of their consciousness (Clay, 2019). While destinations motivate travel patterns, the who of travel can sometimes supersede the where. Relationships, then, are the social infrastructure that reproduce the queer world for some people and not others, helping to retain its motion and practical connection to itself (Berlant, 2016).

Conclusion

Social infrastructure "allows us to think not that there will be other worlds later but that other environs are emerging now whose shapes are made by the living" (Berlant, 2022, 24). It offers conceptual and material durability to identity- or community-based analyses of social change. Although suburban infrastructure may glitch when it encounters queer and trans bodies, needs, and behaviours, it still affords modification capacity in support of multiple uses that may overlay the public and the private and their heterocisnormative "systems of familial organization" (Vallerand, 2013, 68). To echo both Berlant (2016) and Halberstam (2018), an infrastructural analysis is invaluable precisely because it reveals that injustice must not just be critiqued or challenged, but unmade and remade. Instances of queer domesticity, recreation, transportation, and sociability serve to unravel suburbia's heterocisnormative social infrastructure, offering unpredictable affordances that facilitate queerburban lives.

In contrast to gayborhood extroversions "with neon signs and full-length windows" (Rushbrook, 2002, 193), queerburbia lacks the geographical boundedness, public streetscape visibility, and temporal predictability of LGBTQ2S-supportive social infrastructure. Instead, queerburban life happens across public/private binaries in queer and non-queer homes, parks, adapted social venues and streetscapes, and in-transit journeys. While private sexual acts of desire may become public, queer lives and their kinship networks may also remain tactically hidden and largely unnoticed, strategically playing with the politics of (in)visibility. The facades of suburban built structures may not beckon with rainbow flags and neon lighting, but they are strategically hacked to provide anonymity and community remaking possibilities enabling some queerburbanites to live in plain sight. This queer and trans reclaiming of neglected and transitional suburban spaces may counter the need for fixity and underscore the importance of ephemerality while simultaneously dreaming of more permanence.

Notes

1 In Husserl's (1967 as cited in Harrington 2006, 341) original conceptualization, lifeworlds are the "tissue of intersubjective background understandings that first makes scientific objectifying knowledge meaningful".
2 Responding to a request to represent their LGBTQ2S city, photo-elicitation interviews became agentic queer worlding opportunities of envisioning suburban liveability – bringing to life those places where queerburbanites live, love, socialize, and organize.
3 Developed in 2009, the social networking smartphone app Grindr uses geolocation to offer its gay male users a "gamified" tiled screen of the torsos of the nearest 100 users based on customized demographic preferences, allowing people to variously "cruise" for sex, drugs, employment, and flatmates, and to receive compliments, criticisms, and messages (Clay, 2019). It may also result in rejection, ghosting, and blocking with a conversation left unfinished. Grindr's success inspired similar apps used by gay men like Growlr, Recon, and Scruff.
4 In 2014, an extreme example from Longueuil involved a lesbian couple who received an anonymous and threatening letter from a neighbour in response to the rainbow flag they hung on their apartment balcony.
5 Established in 1995, the Steam 1 bathhouse is a popular cruising site for MSMs and bi-curious men. After the original on Front Street burned down in 2014 in a gang turf war, it relocated to Columbia Street. Like many gay bathhouses, it has a stream area, small rooms with single beds, and lockers lining the corridors. Not all visitors like the layout: "the central hallway makes it hard to cruise because it's like being in a library with everyone watching" (5 July 2019). Bag checks (for needles, weapons, and alcohol), a staff nurse, and an ask-before-you-touch rule promote sexual and physical safety. Drag queens, considered "a big thing for straight guys", are encouraged to remain in one room. Co-ed events – all gender "Wild Wednesdays" and "SeXXXy Saturday" – are popular with trans and cis women alike (12 August 2019). Although located near New Westminster's rainbow crosswalk, Steam 1 does not sponsor or participate in LGBTQ2S events.

8 Dreaming

Introduction

Dreaming engages with an "ache for that which we cannot have, or that which is yet to happen, or has the potential to happen" (Sundén, 2018, 71). More than an individual desire to escape oppression or an aspirational process of imagining life otherwise (Sliwinski, 2017), "social dreaming" can reparatively build towards queer and trans futures. First explored in queer psychology, it is a practice of jointly, openly, and safely examining the personal and collective meanings of socio-political aspirations or "dreams" so as to provide the "shared imagery, shared vocabulary, shared stories, to help us meet the challenge[s] we face" (Slade, 2005, 33). For queerburbanites, struggles with spatial exclusions have made social dreaming an especially relevant place-remaking tool. In its queer futurity, social dreaming can imagine "a world where queer lives, politics, and possibilities are representable in their complexity" (Muñoz, 1999, 1). It can unmake suburban heterocisnormativities by revealing queerburban realities all the while pulling towards place-remaking by envisioning life differently. By mapping queer spaces of possibility, social dreaming challenges dominant systems through queerly "worlding from below" (Manalansan IV, 2015, 1), providing a vision of queerburbia that is imaginatively distinct from the visions of urban planners, urban designers, and policymakers. Queerburban collective dreaming underscores how place-making/unmaking/remaking is "grounded in 'belonging together'" (Seamon, 2018, 162) and inhabiting the "messy urban vernacular mappings" that disrupt suburbia's "normative arrangements of bodies, things, spaces, and institutions" (Gopinath, 2005 in Manalansan IV, 2015, 567–578).

Social dreaming about places implicates "imageability", or "the generalized mental picture of the exterior physical world that is held by an individual", which contributes to way-finding navigation within them (Lynch, 1960, 4). Images of place derive from senses and memories, guiding actions and establishing a relationship between the individual and the exterior world. The familiarity and distinctiveness of cognitive maps are sources of "emotional security" that deepen and intensify human experience, rendering them "the obverse of the fear that comes with disorientation" (Lynch, 1960, 5). As queerburbanites

DOI: 10.4324/9781032632544-8

move through heterocisnormative suburbia, they encounter contradictions emanating from places and their bodies that may manifest as existential crises, alienation, or "haunting" (Bell, 1997). The normative search for unified imageability by queerburbanites is one of constant disruption and disorientation that, when shared with others, offers reorientation, reconnection, and visibility.

Through its facilitation of social dreaming and queer worlding, this chapter proposes that collective queer mapping renders queerburbia imageable. Following Corner (2011), it treats mapping as a collective cultural intervention to diversify suburban worlds by denaturalizing "that which appears given" and exposing "the enormous work involved in making the world of appearances seem permanent and natural" (Katz, 2004, 257). Prioritizing the affectively charged queerburban evaluative elements and their lexicon of "queer visuality" (Kinkaid, 2018), collective counter-mapping activates differential perceptions and makes "new spaces and social relations possible" (Kinkaid, 2020, 13). Counter-mapping is an especially important tool for LGBTQ2S populations because it can undermine the chrononormativities (Freeman, 2010) that privilege certain spatialities, "genealogies, chronologies, narratives, and bodies" by rescripting the past in present voices (Sizemore-Barber, 2017, 123). By representing an accumulation of LGBTQ2S place associations, collective counter-mapping can empower queer and trans communities through the recuperation of LGBTQ2S places across space and time. Such mappings of queer infrastructure have a reparative place-remaking function, enabling people to situate themselves, learn from each other, and "strengthen common presence" (Seamon, 2018, 151).

The chapter investigates collective counter-maps produced for the yourL GBTQ2Scity project. By linking agency with storytelling and visualization, these maps contest the rhetoric of suburban heterocisnormativity. They are neither unified nor uniform, orderly nor predictable, comprehensive nor extensive and are doubly located in the past and the present. As such, the maps visualize socio-spatial knowledge, mixing the imaginary, virtual, and actual to collectively and temporarily remake suburban places and dream about their unknown queerer futures. It begins by describing the yourLGBTQ2Scity collective mapping project and then elaborates on the use of Lynch's (1960) typology of cognitive elements (paths, edges, districts, nodes, and landmarks) as the analytical framework for the chapter's reflection on queerburban dreaming.

yourLGBTQ2Scity: Remaking place through collective wayfinding

Participants were invited to draw their own "LGBTQ2S city", documenting the meaningful places they inhabit and move through in their daily lives. From individual participant maps, a group map was created from selectively added elements. Each addition was explained, and its placement, orientation, and symbolic encoding were negotiated with the group. By collectively aggregating the "whereness" of elements, a more extensive cartographic representation of queerburban environments emerged. In their reconnected juxtapositions and renewed adjacencies, the maps are necessarily incomplete and distorted; their discontinuous surfaces cognitively transform distance and direction, producing subjective deviations from a Euclidean view of the "real" world (Downs & Stea, 2011). The maps queerly re-centre the suburbs from the outside as representational by-products of organic, sometimes intergenerational, intercultural, and inter-identity conversations and information exchanges – "I could not have predicted how this discussion would go … I'm absorbing all of this, it is all new to me" (FG1 Burnaby, 8 June 2019). The resulting yourLGBTQ2Scity collective maps document plural, contested, and reconfigured suburbs that dream queer and trans collective lifeworlds into existence.

In cognitive mapping, a distinction is made between two complementary types of information: locations and attributes. Locational information combines distance and direction. Attributive information is "descriptive" ("affectively neutral") and "evaluative" ("affectively charged") (Downs & Stea, 2011). The evaluative elements in Lynch's (1960) classic understanding of the imageability of cities include paths, edges, districts, nodes, and landmarks. Paths are networked lines of movement that give a holistic sense of order by serving as the city-region's spatial skeleton. Edges demarcate physical barriers between areas and form lines of transition between insides and outsides. Districts are morphologically unique areas that form the building-blocks of cities. Nodes are physically distinct gathering places where paths meet, constituting conceptual anchor points and/or strategic foci. Finally, landmarks are singular yet contrasting components of the landscape that are used for self-navigation. To have orientational value, these elements need to be imageable for individuals and groups, meaning that they should be sufficient (distinguishable enough to be useful), clear (readable enough for interpretation), safe (consistent enough to be relied upon), open-ended (adaptable enough to change and be reorganizable), and communicable (distinct enough to be conveyed to others) (Lynch, 1960, 9).

One Lynch's (1960, 7) hallmark observation in *Image of the city* is that individual images exhibit strong similarities when aggregated by group characteristics (social class, age, gender, and ethnicity), resulting in "common mental pictures" for subcultural groups. Beyond individual cognitive maps, the act of co-creating in focus groups was a dynamic process of understanding and envisioning how queerness takes up space on a blank page but also in suburbia. It involved negotiation, discussion, and knowledge sharing. In its use of queer visuality, such co-creation deployed dreaming to imagine lifeworlds collectively and otherwise, encapsulating their place-remaking potentialities. The maps do not simply confirm what is already known; rather, their colourful simplicity and liminal empty spaces visually depict a desire for more community, visibility, and conviviality (Podmore, 2006). As "a collective aesthetic that spanned time and lives" (Futch & Fine, 2014, 43), the maps invited queerburbanites to dream in their varied representations of relationships to places, people, and time.

Cognitive mapping focus groups not only ask participants to locate, recall, and give meaning to locations, they invite storytelling about self- and collective-emplacement in ways that reconnect the disjuncture between queer infrastructure and locations (Gieseking, 2013, 2020a). Queerburbanites talked about how they experience places from their various LGBTQ2S identifications: degrees of comfort and welcome; visibility in content, resources, or representations; and "issues" with respect to personal safety and well-being. The places identified range from being "open" to anyone, "inclusive" along different axes of difference, but most especially those that host queer- and trans-inclusive events. These can be micro-places of momentary recognition but also spaces of conviviality and social fun. They may include places with positive-space iconography displayed "on the door, right when we first enter" (FG3 Mississauga, 20 October 2019), and/or places where pronouns are requested upon entry. A sense of community and connection can be built out of ephemeral conversations that may or may not extend into the future, much like the in-person connection among the focus group participants.

Socially dreaming queerburbia

For many sexual and gender minorities, queerburbia "feels like a lot of heteronormativity a lot of times" because queerburbanites rarely experience similitude: "you don't see a lot of your own kind represented back" (FG1 Mississauga, 20 October 2019). Suburbia's numerical lack of LGBTQ2S representation in its public realm amplifies feelings of alienation. Moving across the thresholds of institutions and commercial spaces, disorientation builds; derived, in part, from a lack of LGBTQ2S social inclusion training. Queerburbia's LGBTQ2S-inclusive institutional memory and allyship are unstable with the safety of places oscillating: "it becomes the safer space, and then it doesn't become a safer space, and then it becomes the safer space, and it doesn't" which necessitates that "you have to constantly vet locations and spaces" (FG6 Mississauga, 20 October 2019).

Such iterative assessments of places involve repeated evaluative calculations of "How much of myself can I be?" (FG1 Mississauga, 20 October 2019). Given their lack of "crowd" (Podmore, 2001), queerburban public spaces are less sites of liberatory alterity than places of exposure and vulnerability, requiring vigilance about who is present and who might be unexpectedly encountered. One Bramptonian lesbian (FG2, 19 October 2019) elaborates that her greatest "struggle" with everyday queerburban life is "if I'm outed somewhere or somebody sees me in an 'out' location … that could affect my employment". "Coming out", even to strangers in public spaces, is anything but "normal, routine, and unremarkable", requiring identity management, what Orne (2011, 685 & 688) describes as "strategic outness". Where ethno-racialized communities have suburban concentration, dense social networks pose a heightened risk of public witnessing and information transmission raising fears of familial rejection and the potential loss of emotional and economic support.

For queerburbanites who "have the privilege of passing", whose queerness and "transness is kind of invisible" (FG4 Brampton, 19 October 2019), suburbia is navigated in the relative absence of conflict, discrimination, and job or housing loss threats. Heterocisnormative passing, however, surfaces feelings of misrecognition and invisibility not to mention the guilty haunting of betrayed alliances and broken solidarities. As a white, non-binary resident of Brampton (FG facilitator, 19 October 2019) details: "I feel like my non-binaryness … haunts different situations", in that "I know it's there … [yet] it's something that nobody else can see a lot of the time because I pass as very cis". They reflect upon the places where "it doesn't feel like my gender is a specter that's hanging over", but rather it is "fully invited into the space" without having to be explained or people wanting or needing to know about it. They describe such places of freedom from the "tyranny of gender" (Doan, 2010) as "transcendent space[s]" where they may enjoy experiencing the resonance of belonging. These ghosts of personal recognition are also social responses that establish affective ties with the physical world, animating it with memories and feelings (Bell, 1997).

The ghosts, then, of queerburbia include those of present and future lives. Such ghosts shape how they "make claims about the territories of social life" (Bell, 1997, 823) – about who is welcome and who must remain on guard when navigating the spatial disorientations of heterocisnegativity. Interwoven within such highly subjective and compellingly specific personal journeys are the memories of queer places recounted in community stories: like the burned-down Heritage Grill in New Westminster; the straightened Pink Flamingo in Surrey; or Brampton's long-since gone weekly Pride night at the Wild Moss bar. Across its socially dreamed paths, edges, districts, nodes, and landmarks, queerburbia's counter-topographies remake its often invisibilizing and haunting disorientations.

Paths: Queerburban (im)mobilities

Peripherally situated, queerburbanites require paths of connection for mobility and connection with highways and major north-south and east-west arteries combined with public transit nodes acting as shared points of reference. Collective maps feature sketches of bus and car silhouettes on these major routes (Figure 8.1). They symbolize access to downtown LGBTQ2S resources and social infrastructure, and connections with geographically dispersed queer chosen family and friends. For queerburbanites, cars invoke dreams and enable the realization of mobility and escape (Ingemarsdotter, 2017). They signify queerburban autonomy and personal expression, represented with rainbow-clad drivers and multi-coloured exhaust, contrasting with the intense, "spot-lit" scrutiny of public transit and the stern visages of fellow passengers. The unsolicited attention of strangers (often accompanied by fears of aggression) is a feature of queerburbia's (im)mobilities that can be transformed into more liberating and transformative feelings of joy by queer co-presence and togetherness (Sheller, 2018).

For queerburbanites, the car remains both an anchor point and a means to navigate suburbia's disjointed complexities and "cis-het energy" (FG1

Figure 8.1 The paths of movement by bus and car in Ajax, 2019
(*Source:* Alison Bain).

Brampton, 20 October 2019). A shared second-hand car affords a non-binary young adult access to queerburbia and downtown Toronto: "My car is the only place I have ... the centre of where I feel comfortable ... and, if I feel uncomfortable, I can leave in my car because I have that level of privilege of accessibility" (FG1 Brampton, 20 October 2019). Automobility is a feature of the queerburban life cycle where youth growing up in middle-class families get their driver's licence at age 16, progressing to car reliance, and seniors become increasingly entrapped when they later shift to public transit. An instigator and perpetuator of lives lived in "'instantaneous' or 'timeless' time" (Sheller & Urry, 2006, 745), automobility generates the "pre-conditions for its own self-expansion" (Urry, 2004, 27), a continual solution to the fragmentation of suburban life. The quintessential object of excessive consumption, competition, and the exertion of power over others (Redshaw, 2007), the car's autonomous mobilities enable flexible queerburban travel. In the words of a Black lesbian parent, "when I drive, I fill my car, and those people who are the most targeted or the youngest, I drive" (FG4 Surrey, 9 June 2019). Her car is a mobile site of social care and community building, driving people anywhere and never knowing when she will return home after a night downtown. This car's mobile socialities, like that of a white lesbian couple with three children and a faded rainbow licence plate, make it "gay everywhere we go" (FG1 Markham, 3 November 2019).

These "car-enabled connections" (Ingemarsdotter, 2017, 57) permit queerburbanites to overcome significant geographic distances while fostering utilitarian cabin space intimacies. A car is a small room on wheels with windows and doors. Its forward-facing chairs oriented towards a windshield facilitate side-by-side conversation between driver and passenger(s). Stimulated by speed and seclusion, drivers and passengers "temporarily escape the 'normal' dimensions of life" (Ingemarsdotter, 2017, 59). Separate from LGBTQ2S homes and events, cars are bespoke and discrete spaces made possible through the politics of ownership and extended through communal sharing (Jukes, forthcoming). Public yet private, visible yet separate, the seclusion of car interiors "can induce different modes of being" for queerburbanites that public transit inhibits (Ingemarsdotter, 2017, 42). As one gay man explains: "when I'm a pedestrian or a public transit taker in Mississauga, I do not feel that level of freedom to be myself, to have conversations with the people that I'm with, to just be physically affectionate or expressive in the way that I would be in Toronto" (FG1 Mississauga, 20 October 2019).

Vancouver queerburbanites represent the region's SkyTrain as a tight space of scrutiny replete with incidents of aggression, especially at night and in the relative absence of security personnel. A New Westminster resident (FG3, 5 June 2019) characterizes it as "a really dangerous place" that is "just not friendly", where people "all look at each other" and "take off their Pride stuff to hide it" as they prepare to exit stations. As a white lesbian couple explains, "We are [only] safe if we're not being actively queer … demonstrating queerness that would draw extra attention to us" (FG1 New West, 6 June 2019). Other lesbo-queer women confirm that "transit can sometimes be really scary – I've gotten looks, sneers, people who looked angry" (FG1 Burnaby, 9 June 2019) while others face micro-aggressions like under-the-breath murmurings of, "they're holding hands" (FG1 Burnaby, 9 June 2019). Central Surrey where the SkyTrain meets the main bus loop is notorious for aggressive LGBTQ2S harassment and security negligence ("reporting stuff doesn't do shit") (FG4 Surrey, 9 June 2019). When a Persian lesbian was waiting with her partner for a bus here "people were laughing at us and saying stuff. I said, 'we're just chilling here'…. if you were a man and a woman, nobody would say anything'" (FG2 Burnaby, 8 June 2019). Due to their contrast with suburbia's compulsory heterosexuality and cisgender norms, companion-dependent public displays of affection result in a hypervisibility that is only mitigated by self-discipline and vigilance to curtail physical contact (Stasinska, 2022).

Suburbia's scrutinizing heterocisgender gaze falls on queerburbanites across all three city-regions. The sense of "people looking at me, staring at me" (FG4 Brampton, 19 October 2019) is intimidating; even more so during night's darkness when other transit riders make comments or passing cars hurl insults.

One white, non-binary Bramptonian describes having "garbage thrown at me from a car when I was holding my partner's hand at a bus stop in the evening" (19 October 2019). The spatial fixity of waiting for public transit is exposing, especially for genderqueer young adults in "booneyville", one of whom recalls having icy drinks and insults propelled at them from a passing convertible (FG3 Beaconsfield, 13 October 2023). For public transit-dependent queerburbanites, suburban immobility is reinforced by the risks of waiting for the journey (Weintrob et al., 2021). They describe "the busing, the system, and everything" as "a hassle" and "hard because I want to feel safe travelling, getting there, and being there" (FG3 Ajax, 2 November 2019). While destinations may offer safety and conviviality, the paths taken to get there are more nightmarish than dream-like, replete with discomfort, harassment, and even violence. These contours of safety and danger are articulated as edges on collective maps of queerburbia.

Edges: Bounding suburban safety

LGBTQ2S safety is associated with a sense of welcome and visibility, well-being and belonging, ontological security and validation, solidarity and relative freedom from dominant hegemonies (Cisneros & Bracho, 2019, 2020). It invites feelings of inclusion rather than exclusion, encouraging people to socialize and simply "be" using their preferred pronouns and as their full and "authentic" selves without fear of reprisal or discrimination. As a non-binary youth from Ajax details (FG3, 2 November 2019), safety is about feeling comfortable "with people … who know me" and can offer "acceptance just for me being me and taking the time to understand me". For younger generations, safety often involves a lack of judgment, a request for pronoun preferences, and/or a greeting that uses a chosen name. It entails seeing other LGBTQ2S people and being able to comfortably inquire about LGBTQ2S-friendly places, programmes, and resources while also accessing gender-neutral washrooms and change rooms.

Across the three city-regions, the colour green, much like the "go" of traffic lights, was used to indicate the welcome comfort of safety. The colour red, on the other hand, indicated feeling unsafe while yellow or orange suggested gradations of caution. In addition to colour codes, symbols like hearts, Xs, and stars appear in informal map legends to demarcate spatial (dis)comfort, signifying "caution zones" where it is important to remain "on guard" and to not draw unnecessary attention to oneself (FG2 Burnaby, 9 June 2019). For some queerburbanites, even queer-friendly, ally-owned local businesses with Pride flags affixed to windows and entrances remain places of caution because of "the looks, mumblings, and a lot of judgment" by heterocisgender long-time customers who may predominate much of the time, despite the hosting of regular, yet still occasional, LGBTQ2S events (FG2 Burnaby, 9 June 2019).

For queerburbanites of colour, safety trade-offs are constant. As one Black lesbian explains, she was advised by friends not to move from downtown

Vancouver to Surrey, but "if things are happening to me two or three times a day racially downtown, but I only get gaybashed once a week, then I'm going to go with gaybashing" (FG4 Surrey, 9 June 2019). As a parent of colour, she feels safer in Surrey, but as a queer person, she does not. A gay South Asian man living in Burnaby recounts how when he returns to Surrey alone or with gay friends, they are unlikely to be as "flamboyant and out-going", but rather "position ourselves so that we would not get noticed" by wearing a hoodie sweatshirt or another unremarkable item of clothing (FG2 Burnaby, 9 June 2019). Surrey is described by a white gay man as a place where "our community gets in the car and gets on the SkyTrain and goes to Vancouver because it is safe … the preconceived notion in Surrey is that you are going to get beat-up, if you go to a gay bar" (FG2 Surrey, 8 June 2019). While queerburbanites may embrace the local ethno-cultural and racialized diversity, they must confront "an overbearing heteronormativeness" that renders them not yet "comfortable enough to go to an openly queer space" locally (FG1 Ajax, 2 November 2019) in contexts that feel consistently "family-oriented … more conservative, and a little bit reserved" (FG2 Markham, 2 November 2019).

Suburban community centres, public pools, ice rinks, and medical centres can be particularly discomforting sites for queerburbanites. As one non-binary hockey coach (FG2 Ajax, 2 November 2019), who experiences anxiety and social dysphoria using gender segregated public washrooms and change rooms, remarks, "I'm scared for my safety, but just want to pee" (FG2 Ajax, 2 November 2019). "It's kind of like leading a double life … I don't tell people I'm non-binary because I fear for my job … and I don't have anywhere else to access … to play hockey … because … I'm kind of in the middle of trying to pass as a woman, but also not being a woman". When they do not have access to gender-neutral change rooms or washrooms or chosen names or preferred pronouns on medical records in healthcare facilities, it feels "as if we don't exist" (FG2 Ajax, 2 November 2019). Since medical facilities are particularly alienating for gender-nonconforming queerburbanites, many travel downtown to receive specialized HIV+ treatment and/or gender-affirming healthcare.

Across all of the collective maps, LGBTQ2S homes – their own, the homes of friends and family – mark the edges of safety and welcome. Homes are described as "my little temple" (FG1 Surrey, 8 June 2019), "our little nest" (FG4 Mississauga, 20 October 2019), a "space to just be you" (FG1 New West, 5 June 2019), and "a safe place for everyone in it" (FG1 Surrey, 8 June 2019). Home-based safety is derived, in part, from feeling comfortable and fully accepted and not "feel[ing] like I have to hide myself" (FG1 Ajax, 2 November 2019). Homes offer a manageable sense of stability – an opportunity "to be in my own place and to really make it my own and just to grow as a young person" (FG2 Ajax, 2 November 2019). As significant places of personal growth and friendship practices of hosting and care, homes are "a meeting place and a learning place", a place to "just be me without judgment where I can express my own special qualities" (FG1 Surrey, 8 June 2019). As domestic neighbourhood nodes, they become places to "change so many attitudes" as a "quiet activist", simply by being consistently queerly present – in the liminal spaces of front lawns gardening, backyards and balconies socializing, and driveways departing and arriving (FG1 Surrey, 8 June 2019). As this gay man specifies, "I walk out in drag and some people get out their cameras" or approach him to do their wedding make-up; he believes that "just by introducing yourself and being present you change the environment because they can't ignore you" and "I want to know who and why you are" (FG1 Surrey, 8 June 2019). For those queerburban activists who "live a public life" in the media, their homes are places of retreat "to hide and be safe" (FG3 Surrey, 8 June 2019) from intrusive curiosity and animosity.

In suburban Montreal, homes represent safety and are havens from which to redraw heterocisnormativity's boundaries (Figure 8.2). A small square on the map represents an apartment shared by three queer best friends who live as "a family unit" (FG3 Beaconsfield, 13 October 2023). Here they "create new traditions with each other as queer folks", co-offering a "sense of safety", and

Figure 8.2 West Island youth map of a nodal network of suburban homes and churches, 2023

(*Source:* Julie Podmore).

ensuring that they can "enjoy the way our queerness interacts with the world" (FG3 Beaconsfield, 13 October 2023). Other people's homes can be sites of safety, places where LGBTQ2S-positive parents provide havens for other young people still living at home or represent positive experiences of coming out within families. A neurodiverse young gay man in Montreal's West Island has included two parental homes, and the West Island LGBTQ+ Centre at Beaconsfield United Church in a circle of safety: "These are both places I felt really comfortable with because I came out to both of them … and I feel more accepted than anything" (FG2 Beaconsfield, 12 October 2023). A suburban townhouse of a best-friend's family represents what a queer adult describes as "the place that I would call my first group of people that were chosen family"; a place that is still "like a second home, or a third home, or another home, just a little space where it's my family" (FG3 Beaconsfield, 13 October 2023). Another box located downtown represents a friend's punk music-studio where queers scream their "little 'enby' hearts out", "feel understood, and not have to really talk to one another and not have to explain every reason why we want to be … heard" (FG Beaconsfield, 13 October 2023). When shared or opened to others, homes offer opportunities to explore common interests and passions and to find connectedness.

Such a sense of homes as domestic havens was not felt by all queerburbanites. A male-identified domestic violence survivor explains that "I don't feel safe at home … because I literally have nowhere to go", referring to the lack of resources or support both inside and outside of his home (FG3 Brampton, 19 October 2019). His attempts at police reporting have elicited ridicule: "I've been laughed at. I've been mocked. I've been told to fight back" (FG3 Brampton, 19 October 2019). Likewise, car break-ins on nearby residential streets invite caution from some queerburbanites who remove LGBTQ2S identity markers from clothes, bags, or vehicles. As a white queer parent specifies, "once I walk out my door there is nothing identifying on my car anymore … [or] on me" (FG2 Brampton, 19 October 2019). When she is alone with her child, she "has the privilege of straight-passing" but heterocisnormative negative attention is generated when she is joined by her ex-partner to form a family unit as "there was the automatic assumption of being a queer couple with a child" (FG2 Brampton, 19 October 2019). Yet bounded "home areas" are supposedly renowned for psycho-social functions of familiarity, identity, and belonging that extend edge-like affordances of safety beyond the domestic realm (Kearns & Parkinson, 2001).

Beyond homes, residential neighbourhoods are the socio-spatial building blocks of cities (Whitehead, 2003). They provide schools, libraries, and places of worship as well as leisure and social networking opportunities. For LGBTQ2S adults, suburbs are haunted by memories of homotransphobia experienced in schools, often represented on maps as unsafe spaces in red. A young gay man describes himself as "a survivor" of a West Island Catholic school that was "very white, Anglo-Saxon meets Italian-Catholic" and "mostly unsafe" (FG1 Beaconsfield, 12 March 2020). His ambivalence about coming out was reinforced by an incident of homonegative speech in the classroom which left him unable to trust his classmates. Others recall teachers and librarians who offered student support. For one West Island genderqueer adult, their high school was represented by a book signifying the safety of the library where an ally librarian made "special efforts to try to get books that she thought we would like, based off of what was going on in our lives" (FG3 Beaconsfield, 13 October 2023). Even if the queer students "weren't exactly out", this librarian would suggest books, increasing access to the school's queer library materials, and permitting youth to explore "a variety of queer experiences" (FG3 Beaconsfield, 13 October 2023). Nevertheless, through the lens of safety, high schools and middle schools, especially Catholic ones or ones lacking GSAs, are often identified by queerburban young adults as unsafe places where "a lot of people hold homophobic or transphobic mindsets", use "outdated terminology", and "make fun of non-binary people" (FG1 Brampton, 19 October 2019).

For other queerburbanites who benefited from access to GSAs, schools symbolize "a big turning point" in terms of recognizing one's sexual and gender identity and offering "a little safe space" (FG6 Brampton, 19 October 2019). Seen by some as a place of queer friendship and comfort, "I put my high school because that was the first GSA I ever went to where I actually found my gayness and discovered myself" (FG1 Ajax, 2 November 2019). Similarly, an LGBTQ2S teacher and her sponsored "Take Action" club at a Markham middle-school became a site of safety due to student advocacy to celebrate Pride (with a flag-raising and inclusive-promoting speech) and to better represent the LGBTQ2S community by "bump[ing] up the amount of queer books in the school" (FG1 Markham, 3 November 2019). Yet, for another Taiwanese-Canadian lesbian teacher in this peripheral municipality, her school was a site of student and parental scrutiny where she is cautious about revealing personal information. As she clarifies, "if I was white, and I came out to my students and my students' parents, they would completely understand, even though they are Asian. But the fact that I'm Asian, there's a double standard … LGBT is not part of Asian history or Asian culture" (FG2 Markham, 2 November 2019). Her observations resonate with the colleague who runs "Take Action", observing that in this predominantly Chinese community, they are "very traditional in their beliefs and especially in keeping your closeted information closeted no matter what it is" (FG1 Markham, 3 November 2019). Safety, then, is multi-directional. Sometimes privacy generates a stronger sense of personal security just as publicity in the LGBTQ2S-allyship of para-public institutions may also simultaneously afford safety in visibility, community, and solidarity.

Transitioning to a two-year regional college is especially important for Montreal queerburbanites where high school ends in Grade 11. On the West Island, John Abbott College represents both positive and negative experiences. College life provides students with opportunities to find an LGBTQ2S community beyond the boundedness of suburban neighbourhoods. One gay man recalls college as "a positive experience", a place to develop "an actual community of queer people" (FG2 Beaconsfield, 12 October 2023). A queer female student fondly recounts "hanging out in the Queer Hub club room" or in class "in the visual arts programme, [where] most of my classmates are also queer" (FG3 Beaconsfield, 13 October 2023). While her college rendering highlights institutional recognition of IDAHOBIT, it is tempered by the phrase "overhear homophobic transphobic things from other students sometimes" written in red. Students who commute downtown also specify finding queer community and safety within particular institutional micro-spaces, like the smoking corner, where "it's okay to be who you are" (FG3 Beaconsfield, 13 October 2023).

Suburban churches are also sites of welcome and caution, most especially for trans residents. A New Westminster church that hosted a speaker "who said trans people shouldn't exist" continues to evoke "an uncomfortable feeling" (FG4 New West, 5 June 2019). In response to this "mis-information … hate tour", another queerburbanite organized a counter-rally across the street with banners proclaiming "love is acceptance", "SOGI saves lives", and "love wins" (FG1 New West, 6 June 2019). Memories of coming out in suburban Montreal churches are also represented as boxes with cross steeples. A gender-queer suburbanite recalls the supportive role of a local lesbian minister: "I was really lucky growing up that I went to a very affirming church" and had "queer adults in my life that were religious" (FG3 Beaconsfield, 13 October 2023). Sites of unmaking heterocisnormativity, such churches work to create "an even better church in terms of acceptance" by putting pronouns on name tags every Sunday (FG3 Beaconsfield, 13 October 2023). Even with memorable grassroots activism and LGBTQ2S leadership in churches and schools, residual feelings of threat linger, defining the contours of safety for queerburbanites.

Districts: Distant gay villages

Despite their distance, homonormativity, and exclusionary necropolitical dynamics (Haritaworn, 2019; Rosenberg 2017, 2021a, 2021b), gay villages are emplaced off-centre on most collective maps. Toronto's gay village is relegated to map edges accessed via a metaphorical yellow brick road or a black line referencing the Highway 401 axis. Generic messages such as "welcome to Toronto" above a high-rise skyline filled with multicoloured rainbow streamers and a purple "the village" or rainbow "Queer Village" signpost convey a sense of celebratory arrival underpinned by the vivid extroversion of its place-identity (Figure 8.3). As a district, Toronto's resource-filled gayborhood stands out with references to the Sherbourne Health Centre, the 519 LGBTQ2S community centre, the Glad Day Bookshop, Crews & Tangos gay and lesbian drag bar, and the Church Street Garage restaurant and bar. Vancouver's gay village is indicated on the maps of the most proximate peripheral municipality, Burnaby, and in lists profiling Little Sister's Book and Art Emporium along with the nightclubs, bars, and restaurants of Davie Street and "East Van's" Commercial Drive.[1] In New Westminster and Surrey, queerburbanites mark

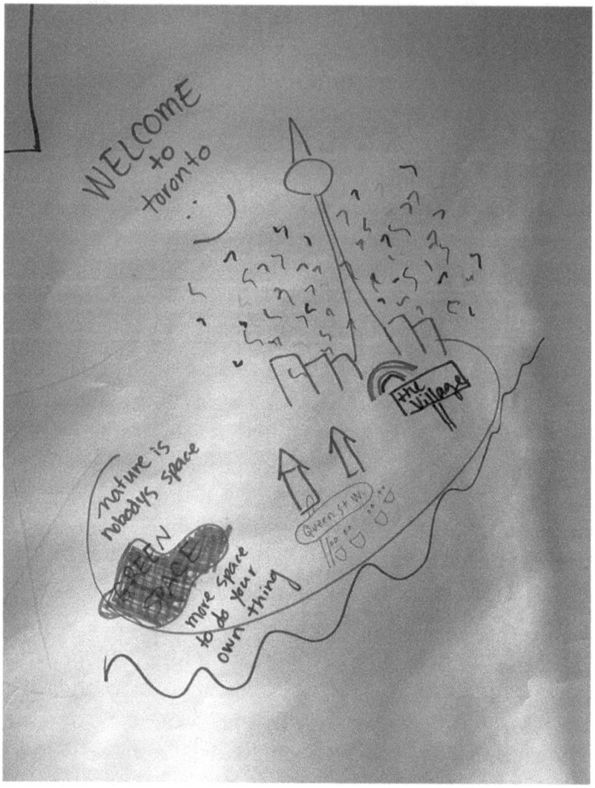

Figure 8.3 "Welcome to Toronto": Downtown Toronto landmarks, 2019 (*Source:* Alison Bain).

Vancouver's gay village with a large pink arrow leading to Qmmunity (the LGBTQ2S community centre), a small flag and smiling faces representing Vancouver Pride, and symbols of the SkyTrain stations and bridges that offer central-city connection. Canada's gay villages are mapped as LGBTQ2S-amenity agglomerations of gay entertainment venues, institutions, and organizations (Collins, 2004a; Lewis, 2013).

Branded with street banners and rainbowized municipal infrastructure, the queerness of Canada's gay villages has visible streetscape pre-eminence that queerburbanites describe as culturally "vibrant and obvious" (FG3 Mississauga, 20 October 2019). For one white, gay male, the gay village feels like "home away from home" because "you don't stand out so much" (FG4 Mississauga, 20 October 2019). An Asian-Canadian lesbian couple remarks that same-sex couples are so common in the Village "that people don't make a big deal out of seeing us together" (FG1 Markham, 2 November 2019). Therefore, they can be "more touchy on the street … because everyone … is kind of doing the same thing" (FG2 Markham, 2 November 2019).

Queerburbanites can be "comfortable in your skin" in the gay village, but so exceptional is this experience that it requires "switch[ing] those gears on and off" and numerous mental self-reminders to "calm down … chill a little bit" and remember that "you're a little bit safer right now" (FG6 Mississauga, 20 October 2019). Unlike queerburbia, gay villages are identified as "a safer place to be" because there is a density of queer events and services (FG2 Brampton, 19 October 2019). One queerburbanite remembers going "down to the Church and Wellesley area" when in high school to "pick up stuff from the AIDS Committee of Toronto and the 519 that was LGBT- and HIV-related" (FG3 Brampton, 19 October 2019). The 519 community centre is described as "a safe place where I was not judged" and could "have fun at different ages and … just be relaxed" (FG1 Ajax, 2 November 2019).

Much like LGBTQ2S community centres, the independent bookstores of the Toronto and Vancouver gay villages (Glad Day and Little Sister's), with their histories of combatting government censorship and seizures of LGBTQ2S books imported from the United States (Toronto Public Library, 2024), remain important informational resource sites for queerburbanites. Part bookstores, event spaces, and Pride merchandise emporiums, they variously ignite and represent LGBTQ2S identities and imaginations. As one young white lesbian recounts regarding Little Sister's: "I had an epiphany when I was there … it's an opening to a world that I didn't know really existed when I was in my younger high school days" (FG4 Burnaby, 8 June 2019). For another Iranian-Canadian young lesbian, the whole Davie Village remains "important because of the clubs and drag shows. It's a place of representation. I'd go with a group of friends from the community and that's where we'd go to hang out…" but notes that "it's mostly gay males" (FG2 Burnaby, 8 June 2019). Two non-binary young people also describe Toronto's gay village as "gay-male central" (FG1 Ajax, 2 November 2019) "which kind of sucks" (FG2 Ajax, 2 November 2019).

Montreal focus groups were ambivalent about the city's gay village. One focus group comprised only of gay men reluctantly included the Village, raising issues related to access ("It's too far," the bus service ends at 12:30 am) and location (it is in the "East End" which is "not an attractive space" unless it is "during festival time with the balloons and stuff") (FG1 Beaconsfield, 12 March 2020). It was described as inadequately social, unwelcoming, and too focused on sex: "There's just sex, bars, and songs" (FG2 Beaconsfield, 12 October 2023). For another gay man, a nightclub and a drag bar represent the Village, noting that he rarely goes there now but in his youth he needed to "revel in just having a queer space" when it was not possible to be "openly queer, openly gay, or hold your partner's hand, or do public displays of affection in suburban areas" (FG2 Beaconsfield 12 October 2023). A middle-aged, upper-middle-class, coupled gay white man describes it as "one of the few places I don't feel safe in Montreal" because "if there are any haters out there, that's where they're going to go to hate and all the straight people go there to prove that they're trendy and open-minded" (FG1 Beaconsfield, 12 March 2020). Although he and his partner "used to try and go once in a while to support the community," the Village offers them little because "we're snobby … and we like good restaurants and it just didn't offer anything other than the bars" (FG1 Beaconsfield, 12 March 2020).

One Montreal focus group did not even include the Village (FG3 Beaconsfield, 13 October 2023). Clarifying its absence, a queerburban young adult explains: "I've heard that the village historically is very important. For older generations, it's very important, but the couple of times that I have gone down there for community day or Pride nights it's not felt the safest". Another gender-queer youth adds, it is "very high energy", and everyone is "there to party and holler and be loud and foolish" which is "not my scene" and "it doesn't hold the majesty that it does for older generations" (FG3 Beaconsfield, 13 October 2023). These youth also reference poverty, homelessness, and other social issues as deterrents, determining that it is too expensive and "not worth it when I can do something nicer in another area and feel probably safer than I would in the Village" (FG3 Beaconsfield, 13 October 2023). Young queers feel alienated by the "mainstreaming" of popular Village queer clubs, increasing the potential for "homophobic slurs" (FG3 Beaconsfield, 13 October 2023). For Montreal's queerburban young adults, it is "a nice monument" whose expense and lack of commercial space mean that "there's no third place" within which to socially connect (FG3 Beaconsfield, 13 October 2023).

Outside of Montreal's gay village, there are queer-friendly downtown districts offering belonging, support, and self-actualization. For example, Turbo Haus in Montreal's Latin Quarter is one non-binary person's "favorite little bar, club, band space, stage, ever", where a punk community offers queer acceptance (FG3 Beaconsfield, 13 October 2023) just as the burlesque performances of the Plateau-Mont-Royal district's Wiggle Room feel "like home" to a genderqueer adult because "everyone's a bit queer" (FG3 Beaconsfield, 13 October 2023). For a non-binary South Asian young adult, it is the "very queer-affirming space" of a "nondescript" downtown gender clinic "perfectly at the intersection of queerness and medical medicine" where laser hair-removal treatments also offer reminders that "I can exist as a queer person" (FG3 Beaconsfield, 13 October 2023). These various central city sites and their districts are entangled with individual LGBTQ2S identity narratives, attesting to the productive porosity of gay district edges.

Despite place-brands bound up with whiteness and masculinity, their commercialism, and shifting relevance for younger generations, queerburbanites in Toronto still see the gay village as a counter-cultural reference point. Instead of "clutching at dribs and drabs of things" in queerburbia, this "orbit" for nightlife, entertainment, and Pride festivities offers proximity to "gay lifestyle, history, and culture" (FG1 Mississauga, 20 October 2019). The absence of such density in queerburbia is experienced by this gay male as "daunting" and "scary" because "there's no charted path here … no history to fall back on … no established narrative" (FG1 Mississauga, 20 October 2019). To combat this dearth, queerburban communities co-produce nodes of sociability grounded in safety and friendship.

Nodes: Co-presence, friendship, and embodied comfort

Within city-regions, the care and thriving of sexual and gender minoritized bodies does not just occur within gay districts. The nodal character of queerburban spatial practices emerges from the collective maps formed by the homes of family and friends as key locations of sociability, intimacy, and care between residents and their visitors. The "corporeal proximity" (Urry, 2002) of this domestic world is depicted through smiling, stick-figured bodies of friends, chosen family, and kin whose immediate, in-person presence represents support. During shared activities and reciprocal exchange, when having fun and conversation together, "bodily co-presence" (Bowlby, 2011, 612) becomes the materiality through which friendships are formed and maintained (Moreira, 2020). The intimacies and compassions generated by such voluntary relationships – from those that are emotionally intense and built over years through to those that are more fleeting (Bunnell et al., 2012) – are made in many social contexts and carefully sustained by queerburbanites in their homes, neighbourhoods, and places of work.

The queerness of queerburbia depends upon a "relational relay" (Kinkaid, 2020, 8), an intricate web of embodied relationships among co-present LGBTQ2S residents and a search for like-minded others. But in the absence of such relational possibilities, there is also misrecognition and isolation. As a white lesbian mother in Markham observes, "we're the only ones" (FG1 Markham, 3 November 2019). When her partner takes their children to the public library, community centre, and pool, "nobody else is queer" and "I don't know if people actually recognize that we're a queer couple" (FG2 Markham, 3 November 2019). While they lament a lack of queer people in their social circle to bring into "the big gay house" (represented as a rainbow rectangle on a cul-de-sac outlined by little black boxes), their home has become a site of respite for local queerburban high schoolers (brought home by one of their children) starting their coming out journeys. It has been harder to find queerburban adult sociability, leading them to conclude that it is "just what has happened in becoming a part of the suburban community" (FG1 Markham, 3 November 2019). Neither one has ever experienced living in a queer community; they have grown accustomed to being the queerburban minority.

A sense of individuated isolation and disconnection from the fun-filled sociability of queerburban community-building is a shared experience, especially across the divisions of the LGBTQ2S acronym. As a trans woman in New Westminster bluntly states (FG5, 5 June 2019), "There is no community here that I am part of … there is no connectedness … We keep in our own little silos. The trans community, in my experience, is much more private than other parts of the LGBTQ2S community". Likewise in this most sociable of queerburban peripheral municipalities, a bisexual resident underscores her apartness, "I really don't know where the bi people are … There's no bi- dinners or bi- dances" (FG1 New West, 5 June 2019). Yet for her, as for other queerburbanites with LGBTQ-friendly employers, work is a relational node of co-presence, where she can show up in all of her "queer magicalness" (FG1 New West, 6 June 2019) to support others in the community. Queerburbanites value a workplace where one does not have "to explain my queerness" or have straight work colleagues remark, "'oh my gosh, you're gay, wow, you're an icon'" or "'you're so brave'" (FG1 Ajax, 2 November 2019).

In keeping with the suburban ethos of individualism and a seeming "lack of the mutuality, the permanent but intangible 'community spirit' that is characteristic of the urban village" (Knox & Pinch, 2014, 191), these nodes of connecting with colleagues and friends are mapped as "isolated places"; queerburbanites tend "to draw things on little islands" (FG2 Mississauga, 20 October 2019). On one map, a solitary black stick figure with a suitcase labelled "safe spaces" passes floating clouds containing roofed structures while walking a narrow middling path that culminates in Toronto's gay village where two people are talking (Figure 8.4). The suitcase conveys "the way that we carry our spaces with us"; the relational relay of carrying "a little suitcase of LGBT with you, but it doesn't go very far when you're out in the world [because] you need other people" (FG1 Mississauga, 20 October 2019). The figure's solitariness also symbolizes the individual search for co-presence that is amplified in queerburbia: "The onus is on you to define those communities, to find your way to them, to connect with them to figure out what they call themselves, and

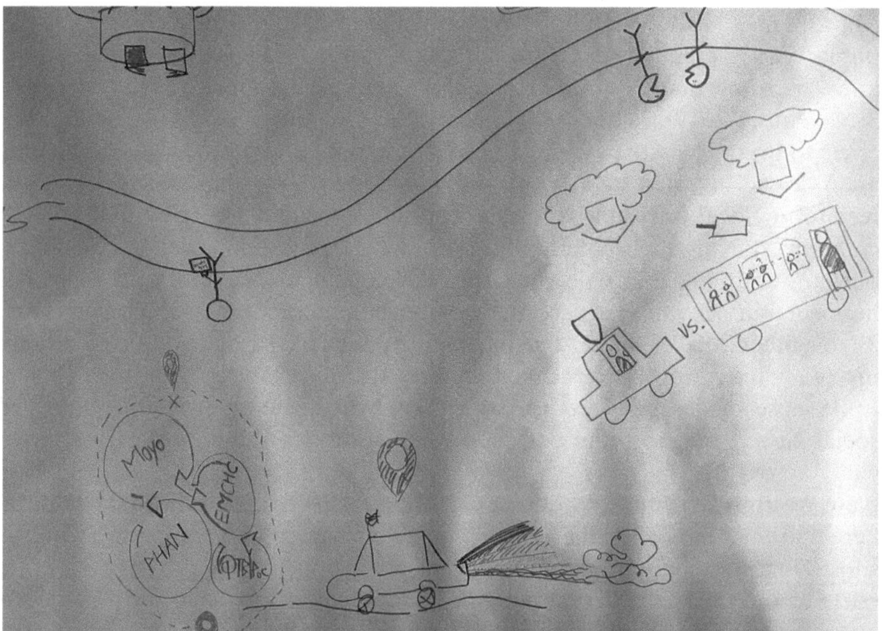

Figure 8.4 Nodular islands of safe space and suburban modes of transportation, 2019 (*Source:* Alison Bain).

where they meet" (FG1 Mississauga, 20 October 2019). While there is pressure "to look and find what you need to do" (FG4 Mississauga, 20 October 2019), queerburbia also requires resiliency, to "survive as an individual" in the absence of gathering places (FG6 Mississauga, 20 October 2019). Queerburbia necessitates the creation of "our own organic spaces as opposed … to there being places built for us … it becomes a space only because of the people that are in it" (FG1 Mississauga, 20 October 2019).

For many queerburbanites, such nodes simply represent "a space where I exist" – places attended frequently to buy groceries or borrow a book or have a coffee, places where another "queer person is there a lot of the time" (FG6 Brampton, 19 October 2019). These are places "where I've managed to be … I've made little pockets of places that I'm comfortable" (FG1 Mississauga, 20 October 2019). One such surprising pocket in Burnaby is the Metrotown shopping mall. Here there are transgender store managers and cross-dressing staff, gay men "looking at guys going by" and lesbians holding hands as part of their morning coffee routine making it "a nice comfortable ordinary-feeling kind of place"; it is "fun in there" (FG1 Burnaby, 8 June 2019). An Iranian-Canadian lesbian also identifies Metrotown as "a place where I felt you could be gay", describing herself as "one of those lesbians walking around holding hands" because "everyone is … minding their own business" (FG2 Burnaby, 8 June 2019). Across the country in Montreal's West Island, Fairview Pointe-Claire shopping mall is another queer-friendly space because the men's washrooms are known for cruising, and the new gender-neutral washroom signals inclusion in the face of civic reluctance to provide additional gender-neutral facilities.

Other chain and independent commercial spaces are identified as queer-friendly due to the co-presence of strangers. For a middle-aged gay man, a suburban Tim Hortons coffee shop offers a break from family life because of the "queer people working there … who try to pick you up" (FG2 Beaconsfield, 12 October 2023). A cluster of suburban bars near Pointe-Claire Village are also considered queer-friendly because of the crowds, morphological density, and relative anonymity. Suburban pubs and nightclubs provide "conditional acceptance" for queer young adults, especially those hosting drag shows, even if it is mostly straight audiences who "go and laugh at the queers" (FG1, Beaconsfield, 12 March 2020). This "conditional acceptance" can be undermined when suburban heterocisnormativity rises to the surface as when a lesbian couple was thrown out for kissing at a Pointe-Claire nightclub in 2013 or when bars that embrace "gays as the fashionable thing" have staff grappling with "actual diversity – people who are non-binary, pansexual, gender fluid, and gender queer – then it's out the window" (FG1, Beaconsfield, 12 March 2020).

Navigating queer-friendly suburban spaces is especially challenging for trans people due to varying levels of visibility and the potential for public misrecognition and transphobia. Yet even small acts of recognition and care are welcomed. For an unhoused white non-binary young adult, a youth shelter in Ajax was an important node. An LGBTQ2S-identified staff member helped to make it feel like "a place of comfort because everyone just accepted who I was" and "allowed me to go to the guy's closet, if I wanted to", for clothing that suited them (FG3 Ajax, 2 November 2019). A trans-friendly West Island pharmacy is another micro-node of care. A trans white senior makes monthly visits to pick up her hormone treatments, appreciating that the "people saw me transitioning, and the people behind the counter were really nice", even remarking, "'Oh, it's been a long time since I've seen you'" (FG1, Beaconsfield, 12 March 2020). Apart from service encounters, however, queerburbanites co-create care by gravitating towards one another, reading clues ranging from the more overt rainbow symbols on clothing and cars to the more subtle registers of eye contact, movements, and embodied similitudes. As a queerburbanite remarked, "it's a self-fulfilling prophecy of if we don't display our queerness, then other people won't see it, and they won't display it" (FG2 Mississauga, 20 October 2019). Small acts of recognition and care may extend into friendship, stretching relational sustenance from the material world to the virtual by staying connected on social media platforms and sharing queer resources online.

The transposition of virtual nodes onto collective IRL maps underlines the importance of online quests – through Google searches, Facebook feeds, and Instagram posts – for queerness in suburbia. These nodes are represented with pin drops, Wi-Fi signals, desktop computer terminals and keyboards, and smartphones, the listing of the apps with words like "Discord" and "Tumbler" or the virtual meeting site "Queer Minds" hosted by the Canadian Mental Health Association on the Microsoft Teams platform (Figure 8.5).[2] For queerburbanites, these resources are "hidden" and under-promoted when "there isn't even really much to begin with" (FG1 Ajax, 2 November 2019). Such fragments of queerness that "come across my feed" grab attention and remarks like "'look at this, this is happening, let's go and do that!'" (FG1 Markham, 3 November 2019). Material fragments of queer culture, however rare in suburbia, serve as landmarks that orient queerburbanites towards social infrastructure.

Landmarks: LGBTQ2S social infrastructure

There is a sparseness to queerburban collective maps. They cover large distances and lack social infrastructural density. As one participant remarks, "I struggled to think of a single LGBTQ-specific thing in Mississauga" (FG2 Mississauga, 20 October 2019). But sharing resources and knowledge by mapping queerburbia together enabled participants to multiply their sense of LGBTQ2S community and sociability. One Mississauga resident offered Studio 89 to the group as an understated, queer-friendly social enterprise space tucked away in a strip mall.[3] It was the first place in queerburbia "I got to meet other folks from the community and just talk and chat". Others shared references to "queer positive" and "non-judgmental" local programming as signalled by the presence of resources and pamphlets for LGBTQ2S-identified people and gender-neutral washrooms (FG2 Brampton, 19 October 2019) hosted in and by STI testing locations (e.g., Peel Healthy Sexuality and the

Figure 8.5 Virtual nodes of queerness emanating from Brampton, 2019
(*Source:* Alison Bain).

East Mississauga Community Health Centre), public libraries, and community centres, which have become queerburbia's "landmarks".

Parks and libraries, ordinary spaces intended for societal reproduction, are often interpreted as landmarks for LGBTQ2S people because they are semi-public, non-consumptive spaces of anomie and escape intended for all citizens. A permanent resident of Canada, originally from Columbia, identifies the Surrey Central Library as her most significant LGBTQ2S social infrastructure because "you can get whatever type of information you want without being judged" (FG3 Burnaby, 8 June 2019). It was here that she freely read her first lesbian book and found her first LGBTQ2S meetup group. A white New Westminster lower-income lesbian senior concurs, describing her local library branch as a place to affordably print and to read the local newspaper. Suburban LGBTQ2S community organizations (e.g., PFLAG and the youth Camp Rainbow Phoenix that it sponsors in the Toronto city-region) are added to collective maps along with youth groups in community centres, and the mental health and medicalized supports provided by health clinics. Other key landmarks are the public parks that host Pride events but are also enfolded in personal exercise routines and socializing rituals. They represent places of possibility through collective reclamation or "because it's no one's space, there's a little more freedom to be ... relaxed" (FG4 Mississauga, 20 October 2019).

Across the peripheries of Vancouver and Toronto, the most consistent landmarks are rainbow crosswalks, as in Brampton's Vivian Lane pedestrian passage or New Westminster's crosswalk in front of the former Heritage Grill (Figure 8.6). Some appreciate these markers of civic identity and pride for generating feelings of inclusion, safety, and well-being. Others have a "love-hate relationship" with them because although they are "a really cool idea", when rainbow crosswalks are vandalized, it reveals "the hatred that still exists ... and how far we still have to go with respect to human rights" (FG1 New West, 6 June 2019). They are also "beacon[s] for news stories and debates about where public funds should go", and provocateurs of homotransnegative comments such as "'where is the straight-people's crosswalk'?" (FG1 New West, 6 June 2019). Surrey's crosswalk is a landmark precisely "because it was so hated" yet endured. "Every time I go shopping, I walk across it because it is there, and I will not be ignored" or disrespected (FG2 Surrey, 8 June 2019). Another Surrey resident feels "vindicated" by the persistent presence of the rainbow crosswalk because "the large community of awfully negative people who are against freedom of this kind ... have to look at it every day, whether they like it or not" (FG1 Surrey, 8 June 2019).

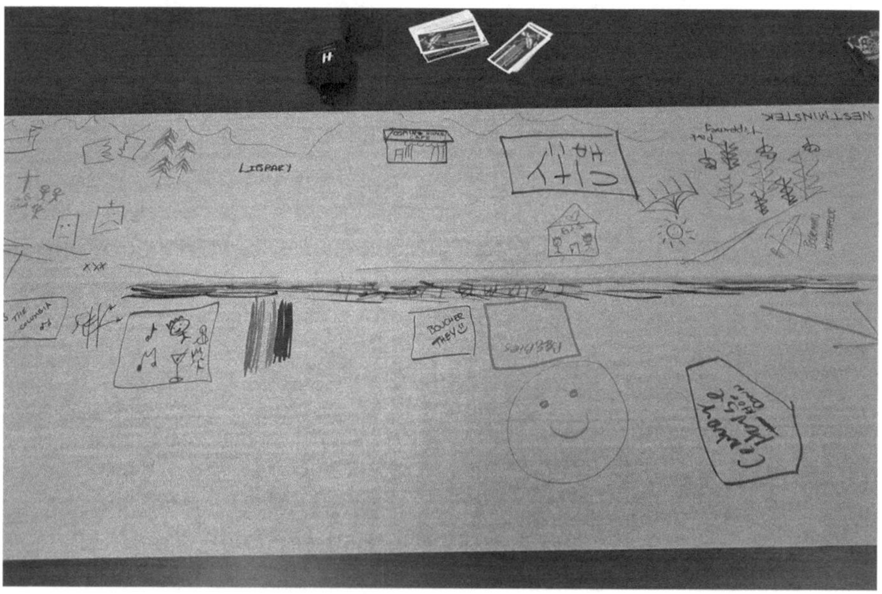

Figure 8.6 Landmarks of queerness along Columbia Street in New Westminster, 2019 (*Source:* Alison Bain).

Given the absence of rainbow crosswalks in Montreal's suburbs, queerburban landmarks are different. The West Island LGBTQ+ centre is one, a lone-star of queer support and safety in an ocean of absence repeatedly drawn as a rainbow church or using its "rainbow jeans" logo. For an out-of-province trans woman migrant, this location is "the biggest and thus far only sort of queer-centric space I've been to in Montreal", an important "place for the people who come here to feel safe" (FG3 Beaconsfield, 13 October 2023). For youth, many of whom come after school, it "is a place where they can be who they are as they're figuring themselves out" while for adults and seniors it is "a place for people to form community" that continues across stages of the life cycle (FG3 Beaconsfield, 13 October 2023). Locally, it is "a little bastion of 'if you're here, you're queer, you're safe'" (FG3 Beaconsfield, 13 October 2023). Montreal is the only city-region to have such suburban LGBTQ2S community service infrastructure.

New Westminster is the only peripheral municipality where a suburban artery, Columbia Street, is identified for its year-round queerburban social infrastructure that includes "at least four gay establishments … three bars and a steam bath" (FG1 Burnaby, 8 June 2019). There are weekly and monthly Rainbow Event Network (REN) activities that make "gay life in the suburbs pretty busy" such as Dining Under the Rainbow ("which I see on Facebook every day" (FG1 Burnaby, 8 June 2019)). REN events are often hosted at the local LGBTQ2S-friendly restaurants, pubs, and bars of Columbia Street: "Most of the fun things that I do in New West happen on [this] street" (FG2 New West, 5 June 2019). Described as "our gay walkway" (FG1 New West, 6 June 2019), Columbia Street is where "queer stuff happens" (FG1 New West, 6 June 2019). Represented by a martini glass with an olive and musical notes, the former Heritage Grill is not just a place to "sling drinks – we were there to build the foundation of the community" (FG1 New West, 5 June 2019). While the Coming Home Café in New Westminster's Uptown takes up a lot of cartographic real estate as "a super gay space" (FG2 New West, 6 June 2019), Columbia Street's gay-owned B-Bombshell Salon is depicted with Pride flags in the windows and described as inviting people to experiment with "look[ing] more masculine … and more feminine" and to "talk loud and be yourself" (FG4 Burnaby, 8 June 2019). New Westminster's city hall is the only civic site of political power with landmark designation on any queerburban map. Whether young and recently married or retired and volunteering as a resident on municipal committees, it is consistently identified as "a place of true openness, help, acceptance, and encouragement. They're all behind us" (FG2 New West, 5 June 2019). A clear proponent of Pride, city hall is positioned as "an extended part of my queer community" (FG1 New West, 6 June 2019).

With multiple landmarks that include the street infrastructure and city hall, queerburbia is highly legible in New Westminster. Across the three city-regions, however, queerburban landmarks, like the rare nodes of queerburban sociability, remain sparse and exceptional, limited to the diffuse presence of rainbow crosswalks or to the rare institutional infrastructure developed in particular suburbs. Columbia Street and the West Island LGBTQ+ Centre had enough singularity and contrast with the rest of the suburban landscape to be imageable and communicable. However, in most cases, the insufficiency and obscurity of LGBTQ2S queerburban landmarks limit collective identification and ultimately wayfinding.

Conclusion: Remaking places together

The queerburban counter-topographies represented in the yourLGBTQ2Scity collective mapping signal desires to remake place through social dreaming. A queer art teacher who grew up in Mississauga (FG4, 20 October 2019) in a lesbian-headed household asks, if there remains only one queer-friendly gathering place in a peripheral municipality approaching one million inhabitants, "are we willing to work to establish those places, or are we all just going to move away?" She could move to Toronto or Hamilton where sexual and gender minority communities are more visible, but she is politically committed to establishing such places locally *with* other people, rather than alone: "That's a huge reason why I wanted to come here today, I just wanted to see the faces of people" (FG4, 20 October 2019). In the absence of queer social infrastructure, queerburbanites often remain invisible to, and separate from, each other. The temporality of queerburban social dreaming is often stunted by a lot of "sitting around waiting" to find like-minded others (FG2 New West, 6 June 2019).

The collective social dreams of queerburbanites also provide exemplars of possibility. New Westminster's queer social life, for example, is a suburban beacon. A trans masculine college student in neighbouring Surrey (24 June 2019) sees it as a place that "I could have a future". Older queerburbanites in Surrey (4 July 2019) see New Westminster as a model, "my vision for here, but I don't know how to get there". Without an equivalent to New Westminster, Toronto's queerburbanites see possibility in events that bring queers together: "'we're going to this [event], we're going to find the queers!'" (FG1 Markham, 3 November 2019). The dream is to find a sense of intergenerational and intercultural queerburban community in "places that we could create" (e.g., in coffee shops, bookstores, theatres, and restaurants) (FG1 Ajax, 2 November 2019). A non-binary young adult (FG2 Ajax, 2 November 2019) "would love to see spaces that are not medicalized for queer people because a lot of these spaces are centered around health". The aspirational phrase "I wish" began many sentences as queerburbanites envisioned local improvements: "I wish we had people of different ages sharing their experiences of being queer" (FG1 New West, 6 June 2019); and "I wish there were more spaces … where it felt fun or it felt vibrant" (FG1 Mississauga, 20 October 2019). While social dreaming is hopeful, there remain many paths not yet taken that queerburbanites could navigate together, if only they could find each other and stay connected.

This chapter's topological approach has revealed the paths, edges, districts, nodes, and landmarks of queerburbia as represented through collective maps and in some cases the persistent gravitational pull of central-city gay villages as sites of agency, intimacy, and concentration. Fluid rather than static, these maps capture symbolic navigational elements in queerburban everyday lives, revealing places that are liked or disliked, feared, or called home. This collective dreaming exercise is the forward-looking, practiced-based desire that centres creative imagination and transformation from an LGBTQ2S subject position. Crafted over lifetimes, social dreaming involves more than merely wanting to fill the emptiness of blank paper and make it feel plentiful with its queer "lines of being" and desire (Probyn, 1996). Social dreaming "is closely tied to, or may even be, our wisdom" (Tuck, 2009, 418).

Notes

1 Colloquially known as "Dyke Valley" (FG1 Burnaby, 8 June 2019), Commercial Drive in East Vancouver is the only lesbian-oriented neighbourhood represented in the three city-region collective maps. It has a relative concentration of stores and restaurants with rainbow flags in the windows, queer programming in institutional facilities, public parks that support queer parent and child-based socializing, and is the site of a Dyke March and East Vancouver Pride. For one lesbian parent of colour, "the draw is friends and other families and because it feels so incredibly safe" (FG4 Surrey, 9 June 2019).

2 At Queer Minds, LGBTQ2S young people 16+ are invited to virtually connect for support and knowledge exchange (Canadian Mental Health Association – Peel Dufferin, 2024).

3 This fair-trade café and non-profit community hub originated from a grassroots youth-led parent organization to empower youth activism by hosting public events, programmes, and campaigns. It offers free event and workshop space in support of community dialogue, intercultural communication, and artistic expression (Studio.89, 2024).

9 Re-placing

Introduction

Through homologous comparison of different suburban places on the peripheries of Canada's three largest city-regions, this book has analyzed the relational practices of representing, including, serving, resisting, organizing, living, and dreaming to theorize queerburban place-making/unmaking/remaking. It has investigated the representations, institutional dynamics, individual and collective activism, and everyday practices of living and dreaming that make queerburbia. In so doing, it has revealed the neglected "place" of LGBTQ2S subjects – both as subordinated subjectivities and disadvantaged citizens – in urban governance and planning but also the construction of queerburban places as ongoing sites of intervention and political struggle. Queerburbanites are rarely the targets of inclusive policies, programmes, and services planned, structured, and financed by peripheral municipal governments, nor do they passively receive them. They emerge as "end-user innovators" (Andres et al., 2021, 3), hackers of suburban social infrastructure who intervene with activism, socialization, and embodied inhabitation to drive queer place-making/unmaking/remaking.

While globally renowned for its LGBTQ2S equality gains at the federal level (Smith, 2005), Canada remains a highly regional and ethno-culturally diverse country, especially in its suburbs. Beyond heterocisnormativities, queerburbia is shaped by ethno-social disjunctures between the order afforded by legislative policy frameworks and lives lived locally on the peripheries of metropolitan centres. Canadian suburbs are inhabited by multiple Indigenous, diasporic, and linguistic communities, as well as social class groups and household structures, all of which vary geographically and draw upon particular emplaced networks (Massey, 2005). Across the Toronto, Montreal, and Vancouver city-regions, there is significant regional intra-metropolitan variation in print media and same-sex household census representation, social inclusion policy, social service provision, LGBTQ2S activist resistance, and community organization of queer social infrastructure. This metropolitan variability signals important disconnections that are further disaggregated within individual peripheral municipalities wherein place-based social relations condition capacities to re-place queerness in suburbs.

DOI: 10.4324/9781032632544-9

Posing fundamental questions about the geographies of LGBTQ2S lives, this book has demonstrated the interconnections between queerburban liveability and place-making. The kinds of place-making associated with queerburbanism parallel that of other peripheral forms – such as rural and small cities – including the use of strategic surplus visibility, the repurposing of social infrastructure, and the practice of conviviality and care through resourceful community resilience. However, queerburban place-making has its own specificities informed by the role of peripheral municipalities in reproducing social inclusion and a predominance of issue-specific LGBTQ2S activism. These nuances are undergirded by suburbia's domestic and familial privacy, rapid residential land use redevelopment, more youthful populations, the predominance of institutions in public spaces (places of religious worship, schools, community and recreation centres, libraries, and parks), and social infrastructural voids, especially public gathering and commercial entertainment spaces. The primary agents of queerburban place-making also differ from their central-city counterparts. Less based on historic grassroots urban social movements and the gayborhoods they formed, queerburban activism involves smaller groups of community leaders and allies drawing on and locally adapting central city referents, and responding to suburban political opportunity structures and resource landscapes. Consequently, queerburban activism is more individualized, often institutionally idiosyncratic, and reliant upon the presence of insider activists, especially allies, who act as LGBTQ2S "champions". The transformation of suburban micro-public spaces for everyday interactions is also limited, often interiorized in domestic and para-public institutional spaces. Although queerburban inclusion and activism rely heavily on surplus visibility – extroverted moments of progressive performance in municipal Pride proclamations, celebrations, and Two-Spirit ceremonial events or the infrastructural animation of rainbow flags, crosswalks, and lighting – its results are more limited, short-term, and often less transformative. As LGBTQ2S subjects, however, queerburbanites are not necessarily "behind", "invisible", nor any "less queer" than their urban counterparts; they embrace where they live and lead liveable lives all the while dreaming of queerer suburban futures.

Contributions to suburban studies, geographies of sexualities, and urban place-making

Re-placing queerness from the suburban periphery has much to contribute to suburban studies, geographies of sexualities, and urban theories of place-making. For suburban studies, queerburbanism multiplies suburbia's "diversity" by implicating the complexities of sexuality and disrupting suburbia's heretofore binary gender relations. Queerburbia disrupts suburbia's heterocisnormative privatism and familism, reconfiguring it with new family structures and household forms, that, in turn, provoke infrastructural and institutional adaptations of its built form, and the programming of its schools, seniors, youth, cultural and community centres, and the public realm. By engaging with sexuality's registers, the book challenges the urban/queer–suburban/straight binary largely unquestioned in suburban studies scholarship by demonstrating that the materiality of suburbia is not, nor was it ever, in ontological terms, much like its discursively reductionist "straight" representation.

For the geographies of sexualities, the book has contributed to the rupturing of the metronormative fixation that limits scholarly understandings of queer lives and their socio-spatial configurations to gayborhoods and homonormative subjects (Nash & Gorman-Murray, 2014). Inserting itself between the queer urban and rural, this middling project has populated the metropolitan periphery with queer bodies and interrogated the worldings and politics of such inhabitations. In contrast with the existing literature that spatially confines the possibility of LGBTQ2S dispersal into queer-friendly neighbourhoods and inner-suburbs as a key metropolitan dynamic of queer urban studies (Gorman-Murray & Nash, 2018; Gorman-Murray & Waitt, 2009), this book has excavated the layers of the Canadian queerburban palimpsest and its regionally varied and unique place-making processes. Arguing for an analysis of place-making from the outside in, it provides conceptual tools and language for the ongoing investigation of the significant diversity of suburban places, within and beyond Canada, as but one of many iterations of queer peripheries. More than a locational adjustment, such an intellectual centrifugal shift to peripheral places, demands cultural and paradigmatic reorientations to interrogate queerness as embedded in (and not divorced from) the historical and contemporary suburban in all of its domesticated heterocisnormativities and familial morphologies.

To contribute to geographical scholarship on place, this book built upon Seamon's (2018) phenomenological inquiries into place-making, empirically demonstrating that it is indeed a multi-layered process. Crucially, it has offered an investigation into place-making that reveals the intimate ties between unmaking and remaking in a trifecta of place-making/unmaking/remaking. Place-making certainly involves "dedicated individuals who, understanding their place" and in "recognizing its needs, envision improvements in that place's environmental ensemble that, in turn, strengthen common presence" (Seamon, 2018, 151). Place-making is, however, undermined when it misunderstands or ignores local needs, resulting in "inappropriate, or useless policies, designs, and actions that weaken place by misinterpreting what it is and thereby negating its core workings and possibilities" (Seamon, 2018, 151). Nowhere is this dynamic of erasure and neglect more apparent than with respect to suburban sexual and gender minorities.

Queerburban place-making is conducted by supposedly mis-placed subjects who are situated in stigmatized "non-places" that are enveloped in heterocisnormativity (Augé, 2009). This peripheral process provides an illustration of the persistent longevity of queer place-making and its iterative enfolding in sexual and gender minority practices. There are also manifold registers of embodied civic and LGBTQ2S labour that make queerburban place-making/unmaking/remaking possible. These layers extend formally from the top down via national census and print media representations, municipal inclusion policies, and civic service. They also extend informally from the bottom up via vernacular resistance and convivial organizing that fosters individual and community capacities to live and dream otherwise. In the face of misrepresentation, exclusion, impoverished service regimes, and limited political opportunity structures, queerburban life and places persist in significant part because of sexual and gender non-normative co-presence, mutual recognition, and activism.

Learning from queerburbia

The book proposed and elaborated upon the term queerburbia. A counterargument to metronormativity, this neologism captures the necessary reconceptualization of the queer suburban and brings into dialogue its synergies. It conceptually captures queer suburban place-making for urban scholars while practically offering civic leaders, urban planners, and policymakers a comparative and overarching perspective on the societal misrecognition of queerness and brings into focus how cursory and idiosyncratic attempts to rectify it often have been. Thus, queerburbia is more than a critique. It re-envisions the metropolitan periphery as a site of multiple queer futures that could more actively and directly support the fulsome living and dreaming of its queerburban citizens. In the abstract, queerburbia would therefore be a place of sexual and gender non-normative citizenship in which the "right" to the suburbs is claimed through an embedding in the governance and inhabitation subprocesses of redistribution, recognition, and encounter (Fincher & Iveson, 2008).

Across ten peripheral municipalities in Canada's largest cities, this book has demonstrated that claims to queerburbia's place-based rights are contingent. Although universally designated as "suburban", each case study site is simultaneously discreet and relational. They are all places with their own "stories so far" (Massey, 2005, 9), yet they are embedded in wider queer city-regional topographies that are spatially and temporally unbounded. Each peripheral municipality – Ajax, Brampton, Burnaby, Laval, Longueuil, Markham, Mississauga, New Westminster, Surrey, and the West Island – raised critical questions about who participates in queer "rights" to place-making, and whose imaginary of place shapes its unmaking and remaking. These locally specific place-making/unmaking/remaking dynamics depend on embedded political opportunity structures and their necessary provision of discursive and material resource landscapes (Lepofsky & Fraser, 2003). Unmaking and remaking queerburbia is an ongoing "world dismantling effort" that requires commitment and persistence (Ahmed, 2019, 229). In sum, the book has constructed a portrait of queerburbia, but not before deconstructing its place-making dimensions of representing, including, and serving, its unmaking through resisting and organizing, and remaking in living and dreaming.

Representing

Formal processes of queerburban place-making begin with popular and more official representational narratives of emplaced suburban queerness in the public sphere. Such limited and often distorted representations are found within census data and print media (and also in urban planning and policy documents) that seldom specify queerburban subjects. As Chapter 2 demonstrated, the national census can only ever provide a partial suburban geography of same-sex cisgender couples, erasing transgender people, LGBTQ+ and non-binary youth, adults in non-monogamous unsanctioned relationships, and the multiplicity of queer household formations that extend across metropolitan areas. Print media representations of LGBTQ+ suburbanites span extremes, with highly pathologizing imagery characterizing the pre-equalities era, followed by the individual rights activism led by white gay men demanding suburban equalities, and finally, reports of the rainbow-washed progressive performances of municipalities and allied para-public institutions. While print media debates about suburban politicians and same-sex marriage give voice to mainstream discomforts revealing a suburban topography of (in)tolerance, the print media also contains stories of activism, giving imagery and text to reports of queerburbanites fighting discrimination in schools, churches, and neighbourhoods.

Both the census and the media have been transformed in recent years, making it possible to tell other queerburban stories. The 2021 inclusion of trans and non-binary couples creates the potential to remap and develop an intersectional analysis of queer and transburbanites on the peripheries of Canada's rapidly diversifying and growing city-regions. No longer bound by hierarchical information dissemination through national print media and television reporting, social media circulates autonomously constructed and curated queerburban narratives directly, offering fruitful ground for further analysis. More established media also has the responsibility to engage queerburbanites in the stories they tell adding to the public record of LGBTQ2S stories taking place in locations outside of gay villages. Rather than leaving queerburban stories to fledgling online local suburban news sources, the national journalistic imagination could enhance societal dialogue across the queer and suburban rendering it more legible and enhancing its significance.

Including

One of the most challenging questions posed by queerburban place-making/unmaking/remaking is whether it forwards an "inclusive" LGBTQ2S agenda. Chapter 3 showed that the enactment of suburban planning and policy governance frameworks perpetuates misrecognition of LGBTQ2S constituencies because there are so few policies naming sexual and gender minorities or directly articulating their needs. Queerburbanites are largely tokenized, only referenced in civic lists of marginalized citizens due to national non-discrimination legislation. Inter-municipal competition for adherence to diversity paradigms internally creates hierarchies of social group "winners" and "losers" based on their perceived capacity to synergistically align with civic efforts to commodify difference in performatively progressive ways. Social inclusion is not independent of the diversity place-branding process. A more palatable discursive rallying point than its more critically inflected counterpart social exclusion, social inclusion homogenizes and subordinates its counterpublics to conform to urban neoliberal understandings of difference (Duplan, 2023). In settler Canada, LGBTQ2S inclusion is further marginalized in suburbs since their municipal governments and planning departments prioritize the rapid accommodation of outward expansion to the administrative neglect of the suburban social realm (Grant, 2009).

As Chapter 3 showed, the extent to which suburban municipal governments address LGBTQ2S social inclusion in policy depends upon a range of factors (i.e., size, density, centrality, political ideology, internal leadership, and local activism). LGBTQ2S inclusion "tool kits", like the *LGBTQ2+ Inclusiveness: Toolkit for Inclusive Municipalities in Canada and Beyond* (Coalition of Inclusive Municipalities, 2019), provide opportunities to enumerate social inclusion initiatives and to share them for replication purposes with diverse local administrations. They inform civic leaders of the utility of enhanced LGBTQ2S inclusion to stem population loss and attract creative-city investment. They give municipalities new vocabularies and "best practice" training while providing a "how to" guide for civic rebranding performances and the symbolic transformation of civic infrastructure and events calendars. The selective adoption of their recommendations in practice, however, often results in symbolic and performative outcomes. Other limitations of suburban LGBTQ2S municipal social inclusions include the targeting of youth and seniors, an occasional dovetailing with plans to support newcomers, and an emphasis on funding for events rather than more substantive programming and policy changes.

The "quick wins" of LGBTQ2S toolkits were rarely realized in Canadian queerburbia whether this be gender-neutral facilities, inclusive language audits, pronoun preferences, or civic statements of commitment (Conseil québécois LGBT, 2024). The most common adoptions were ceremonial recognition, provision of gender-neutral facilities, interfacing with local community groups, supporting employees, and the development of anti-harassment policies. The more substantive allocation of responsibilities for LGBTQ2S concerns to politicians and advisory committees, the execution of an LGBTQ2S policy analysis, and the integration of sexuality and gender diversity issues into strategic plans were unevenly adopted. Much remains to be done to forward municipal LGBTQ2S social inclusion in suburban Canada where adoption remains selective, piecemeal, and irregular.

Serving

As Chapter 4 demonstrated, peripheral municipalities offer limited access to local political opportunity structures for their sexual and gender minority residents. They rely upon a narrowly defined set of infrastructural references such as the now ubiquitous rainbow crosswalk or the multiple material and luminescent iterations of the rainbow on flag and street poles. Rainbowization, through increasingly automated temporary rituals, is neutralized as part of a set of annual inclusion performances that municipalities select to forward their own distinctive place-brands. These performances are primarily structured around annual Pride and other LGBTQ2S events (e.g., International Day of Pink or Journée internationale contre l'homophobie, transphobie et biphobie (IDAHOBIT)). This chapter concluded that the ephemeral temporality of civic LGBTQ2S first-order institutional allyship practices revolve primarily around the Pride summer calendar and other events of limited duration that, while providing a degree of visibility, curtail the criticality of allyship and ultimately unmake peripheral municipalities as LGBTQ2S-focused places for the remainder of the year.

Instead of these symbolic rituals, municipalities should strengthen and increase access to political opportunity structures for sexual and gender minorities. This could be enabled with the designation of an LGBTQ2S portfolio to an elected official working with a targeted, yet diverse, citizen advisory board to collaborate on the political championing of sexual and gender diverse inclusion. Politically empowered as decision-makers, such leadership would advance social change and equity outcomes by dismantling the structural impediments to social inclusion innovation. Political leadership can advance LGBTQ2S inclusion not only by simply reminding the municipality to recognize important dates, histories, and language, but also by setting municipal agendas for more substantive inclusions and filtering them through annual departmental and committee programming and investment initiatives, policies, and practice.

In these ways, the performative and superficial first-order changes of institutional allyship can be reconfigured into more substantive and critical second-order allyship. By taking shared responsibility for LGBTQ2S institutional allyship, the individuated labour and "burden" of the insider activist is lessened, and the onus to enact change shifts to the institution. Such critical allyship, as modelled by select suburban para-public institutions in Chapter 5, requires the (re)creation of institutional spaces through a careful process of organizational reflection and consensus building, the unbounding and selective rebounding of spaces to create safety and signal recognition, and intentional and sustained participation in the co-creation of autonomous empowerment through the provision of services.

Resisting

Chapter 5 focused on the public authority and legitimacy of para-public institutions. It demonstrated that they often compensate for peripheral municipal misrecognition, yet vary in their motivations and tactics of LGBTQ2S allyship. As places of enlightenment, learning, and questioning, educational institutions enfold LGBTQ2S allyship into knowledge production and mentorship of future citizens. In public schools, adult-led GSAs and Pride proms ground queerburbia by intentionally creating spaces for LGBTQ2S students and their allies. Such spatial specificity is also a valuable affordance of suburban colleges and universities where young people are autonomously empowered in the leadership of clubs and collectives. The institutional allyship of public libraries is an extension of their professional mandate to equitably circulate information as a public good. They offer queerburbanites free and relatively anonymous access to multiple queer worlds through collections, displays and finding-aids, and LGBTQ2S-specific programming. Police forces practise LGBTQ2S allyship as "good public relations" by ceremoniously ameliorating their image in the face of anti-police movements, diversifying their replenishing ranks of officers while maintaining their central role as purveyors of public order and safety through reconciliation with criminalized and neglected populations. Independent of government and decentred by secularism, mainstream Protestant churches perfect their institutional allyship often in the absence of LGBTQ2S populations, focusing on congregation renewal through affirmation, representation, and accommodation. Beyond designated "safer" spaces, LGBTQ2S checklists, and rainbow rituals, the United Church is noteworthy for its congregational social accountability to sexual and gender minorities. The affirmation process requires an intentional, sustained, and participatory organizational reflection on LGBTQ2S allyship that re-makes place in ways that tangibly improve queerburban lives. The deep critical reflection of the affirmation process is a model for other civic institutions to better cohere their coalitional moments into something more substantively reliable for queerburbanites.

Suburban para-public institutions are, moreover, important LGBTQ2S place-unmakers because they internally resist suburbia's heterocisnormativities by mobilizing already existing institutionalized resources. Para-public institutional allyship, like its municipal counterpart, remains, however, as infinitely illusive as it is perfectible. On the one hand, it relies on the insights, energies, and experience of insider activists who work in a dynamic interplay between the insides and outsides of political institutions to build consensus for political change. These embedded LGBTQ2S insiders initiate change by taking on a social movement cause and proactively working through their access to the resources, power, and policymaking/implementation process. On the other hand, a heavy reliance on insider activists results in idiosyncratic applications of policy that are destabilized by events as small as individual career shifts, interpersonal conflict, and the selectivity of personal networks to community groups.

The para-public over-reliance on local and available insider activists as leaders of institutional allyship results in uneven and unreliable solidarities that are rarely scaled up. In queerburbia, there are many other para-public institutions that have yet to play a more active role in LGBTQ2S allyship, most especially at the nexus of religion, education, recreation, and health. Suburban temples, mosques, and other church denominations along with their private and religious schools, colleges, and universities could elaborate their LGBTQ2S allyship. Community, recreation, and neighbourhood centres might adopt a more intersectional approach to their resource provision that ensures informal, intergenerational, and cross-cultural social connections for LGBTQ2S populations in service of queerburban vitality and well-being. Limited adoption of such para-public institutional allyship means that queerburban community life remains primarily the purview of LGBTQ2S organizations. At a time of society-wide "silent withdrawal from social intercourse" that "has affected our propensity to pitch in on common tasks" (Putnam, 2000, 124), LGBTQ2S community organizations use improvisational scavenging, vernacular creativity, and strategic collaboration to create their own socio-cultural infrastructure in the fissures of suburbia.

Organizing

Queerburban organizing, as Chapter 6 revealed, occurs in relational dynamism between the underground, middleground, and upperground of peripheral municipalities to unmake and remake suburbs as queer. Such organizing appropriates and adapts existing spaces to convene in condominium recreation rooms, bars, youth service hubs, aging hotels, empty church halls and basements, community and women's centres, university boardrooms and classrooms, bowling alleys, coffee shops, parking lots, parks, and shopping malls. It runs events such as Pride, IDAHOBIT, PFLAG meetings, rainbow rooms, and summer camps. It convenes people through cultural organizations such as the Sisters of Perpetual Indulgence, the International Imperial Court System, Radical Faeries, Bears in the Burbs, and drag houses. From art workshops and discussion groups to condom packing and drag bingo to coffee-shop meet-ups and bare-chested harnessed men in dining rooms, this chapter revealed the vernacular creativity of queerburban organizing.

While such organizing is performative and embodied and may gather local traction because of the persistent and charismatic energy and social networks of particular individuals, it is reliant upon "participation and coordination that does not produce careers or notoriety for most participants" (Spade, 2020, 136). The devalued and uncompensated reproductive labour of community organizing – maintaining friendships, providing emotional support, coordinating rides to events, and ensuring that people are included and noticed – is the backbone of queerburbia. Thus queerburbia manifests daily in individual and collective practices of negotiating sexual and gender differences, forging and sustaining interpersonal connections, and fostering community through care-full and fun-full practices.

Although queerness and gender transitivity are portable in and through the assembly of bodies and the performative and decorative work of vernacular material culture, they deserve a greater degree of permanence in suburban built environments. As Gilbert et al. (2018, 30) note, "[i]n urban and suburban landscapes, the agency of ordinary citizens to creatively make places and even individual buildings has declined significantly" due to "a weakened local state and a planning system under extreme strain" giving more power to "commercial developers and corporations in place-making" who seldom prioritize sexual and gender minority resource needs. In both the Toronto and Vancouver city-regions, queerburban Prides are important organizing nodes for LGBTQ2S activists but peripheral municipalities need to commit to being more than upperground partners and annual event co-hosts. The suburbs of the Vancouver city-region are exemplary for their rich queerburban social infrastructure developed largely from the vernacular creativity of the underground. While organizing Pride and programming involves strategic collaboration with the intermediary middleground and occasional policy and funding support from the upperground, the invitation to civic leaders is to financially and spatially commit to providing permanent LGBTQ2S social infrastructure to help sustain queerburban lives.

Living

Public queerburban life is hacked together, as Chapter 7 showed, from social infrastructure which affords different capacities for social reproduction and the unmaking and remaking of place. Where available and affordable, the home remains a central nexus of queerburban social reproduction. This "giving space" (Darling, 2011) offers opportunities to care for oneself and others, thereby enabling outward extensions into the "integrative multiplicity" of suburban public spaces to remake them as sites of the LGBTQ2S community (Basu & Fiedler, 2017). Public parks and social venues undergird and augment these reproductive capacities of the domestic realm, while distinctions between private cars and public transportation networks create varied (im)mobilities for queerburbanites.

These grassroots practices of unmaking and remaking queerburban social infrastructure out of the discards of heterocisnormative worlds are rarely simple hacks. They necessitate confronting and working around micro-aggressions and discrimination – those negative place interactions that undermine place by "disrupting[ing] the co-presence" of its sexual and gender non-normative "users and generat[ing] distress, fragmentation, unseemliness, and decline" (Seamon, 2018, 94). Beyond homes and parks, the sociability of adapted storefronts and chain restaurants, condominium party rooms, shopping mall washrooms, and food courts, the parking lots of big box stores in conjunction with libraries, schools, colleges, universities, police stations, and churches (all essentially created to socially reproduce the long-established strictures of suburban heterocisnormativity) become the fleeting social infrastructure of queerburban co-presence and friendship.

Such hacks do, however, produce their own "bugs" and community exclusions that, in turn, demand further repair and revision. It is in the "first-hand engagement" and "emotional bonds" (Seamon, 2018, 106) of queer social reproduction within and between homes through the juxtaposition and blurring of private and public acts and spaces that such repair occurs. It unmakes and remakes hetercisonormative suburban social infrastructure, however temporarily, into spaces of anti-normative being and connection, propelling further actions by queerburbanites to appropriate suburbs as their own. In these quiet and seemingly mundane everyday ways, queer and trans public life exists and persists in often disconnected fragments beyond central cities.

Dreaming

Chapter 8 concluded that the persistence of queerburban life is predicated upon social dreaming as an articulation of a resistant imagination. Social dreaming involves overcoming, if only temporarily and in the representational abstract, the spatial discontinuities and disorientations that are the everyday realities of queerburban lives, simultaneously hyper-localized and city-regionally extensive. In so doing, the (in)congruities of individual lived experiences bring new ways of relating and being together in the world that invite affirmation by queerburban systems of responsibility and reciprocity. With its paths, edges, districts, nodes, and landmarks as represented in collective maps, queerburbia exists in the minds and hearts of sexual and gender minority residents. It is realized every day through their imaginations, embodied presence, and tireless activism to realize different futures. It demands ongoing challenges to "normative spaces and ways of being" and necessitates a collective reflection on the "paths not yet taken" to remake places (Kinkaid, 2020, 16). Queerburbanites recognize that "[b]y investing in these visions together" they can "stabilize these new spaces, these other worlds and within them, find ways to mutually recognize and affirm new modes of being beyond the confines of the present social and political order" of places (Kinkaid, 2020, 17).

Social dreaming is an underutilized vernacular creative resource for place-remaking in peripheral municipalities. As a participatory, collective, and inductive activity, it taps into local queerburban knowledge, visually communicating and potentially opening up resources that can turn these dreams into possibilities. The visuality, geometry, and language of counter-mapping materialize LGBTQ2S experiences and render them doable and sayable within the technocratic frameworks of municipal governance and urban planning. Collective maps illustrate unexpected LGBTQ2S uses of social infrastructure, making it possible to reconfigure suburbia's component parts to meet the unanticipated needs of an already marginalized population. Social dreaming, therefore, has the potential not only to synergize with urban planning's visionary capacity but also to inspire living social inclusion innovation. The collective identification of LGBTQ2S landmarks can be seen as an invitation to civic leaders to create consistent queerburban spaces to gather, be co-present, and realize dreams together. Like the components of the cognitive image, such spaces must be reliable, sustained, and convivial sites of queerburban inclusion to which LGBTQ2S people can return and re-network and from which they can satellite to other spaces.

Conclusion

More than a mere compendium of institutional allyship practices and organizational activism, this book has considered the layered ways in which queerburbanites give meaning to suburban places through their presence, subjective experiences, and political organizing. Beyond reductive representations of queerburbia as mostly absence, it has shown the nuanced and complex ways in which governance and policy frameworks and the resource landscapes of places inform queerburban possibility, and, in turn, how the ongoing embodied labour of allies and sexual and gender minorities remakes the peripheries of city-regions. It invites ongoing scholarly and public conversations about what queerburbia looks like elsewhere and how queerburban place-making/unmaking/remaking can be more explicitly fostered in the here-and-now.

A major goal of queerburban place-making/unmaking/remaking is "thinking through practical ways whereby vicious circles of place interaction might become virtuous circles" for suburban sexual and gender minorities (Seamon, 2018, 94). Through this place-making/unmaking/remaking process, the "vicious circles" of heterocisnormativity that condition the social infrastructure of suburbia's homes and neighbourhoods can become queerer "virtuous circles" of LGBTQ2S affirmation and queerburban life. While first-tier allyship and the upperground appear virtuous, there is a different geometry to the critical second-tier allyship of the middleground and the underground. One cyclically performs progressiveness, repeating proclamations and celebrations in community event calendars, while the other is constituted year-round in the quiet everyday actions of inhabitation and activism. It is living and dreaming that draw attention to this realm and the agency it affords queerburbanites.

Located within one settler nation and embedded in its national legislation, Canadian queerburbia is more different than it is similar. While there are queerburban commonalities across this vast nation, the case study peripheral municipalities differ in terms of language, demographics, political ideology, social class, and socio-cultural concentration. There are surely important differences between this Canadian story and the urban peripheries of European or Anglosphere nations that are yet to be examined. The call for further research extends ever outwards beyond the urban West to comparison and inter-referencing with more global enactments of peripheralized informalities embedded in "massive suburbanization" (Güney et al., 2019) as well as eco- and smart new towns and gated communities where queer life surely also exists in manifold complexity. A world of queerburbia awaits investigation.

References

Abrutyn, S. (2016). Institutional spheres: The macro-structure and culture of social life. In S. Abrutyn (Ed.), *Handbook of contemporary sociological theory* (pp. 207–228). Springer.

Access Alliance Multicultural Health & Community Services (2009). *Needs assessment of LGBTTTQQI youth in Peel Region: Final report*. Associated Youth Services of Peel.

Acosta, K. L. (2020). Religion in the lives of LGBTQ-parent families. In A. E. Goldberg & K. R. Allen (Eds.), *LGBTQ-parent families: Innovations in research and implications for practice* (pp. 215–227). Springer.

Addie, J. P., & Fraser, J. C. (2019). After gentrification: Social mix, settler colonialism, and cruel optimism in the transformation of neighbourhood space. *Antipode, 51*(5), 1369–1394.

Affirm United. (2019). *Steps to becoming an affirming ministry*. https://affirmunited.ause.ca/wp-content/uploads/sites/2/2020/02/StepsToBecomingAffirming-twoPages-July2019.pdf

Affirm United. (2021a). *Affirming ministries program*. https://affirmunited.ause.ca/affirming-ministries-program/

Affirm United. (2021b). *Our history*. https://affirmunited.ause.ca/history/

Agnew, J. (1987). *Place and politics*. Allen and Unwin.

Agnew, J. (2002). *Place and politics in modern Italy*. University of Chicago Press.

Ahmed, S. (2004). *The cultural politics of emotion*. Routledge.

Ahmed, S. (2006). *Queer phenomenology*. Duke University Press.

Ahmed, S. (2019). *What's the use? On the uses of use*. Duke University Press.

Ahmed-Ullah, N. (2016). Brampton, a.k.a Browntown. In J. Pitter & J. Lorinc (Eds.), *Subdivided: City-Building in an age of hyper-diversity* (pp. 242–253). Coach House Books.

Alarie, M. (2024, May 17). Journée internationale contre l'homophobie et la transphobie. Ceclia Macedo, conseillère municipale de Margot. Facebook. https://www.facebook.com/share/p/jN5hgkgV9D94ds3o/?mibextid=oFDknk

Alexander, L. B., & Miselis, S. D. (2007). Barriers to GLBTQ collection development and strategies for overcoming them. *Counterpoise, 11*(3/4), 73–79.

Alsayel, A., de Jong, M., & Fransen, J. (2022). Can creative cities be inclusive too? How do Dubai, Amsterdam and Toronto navigate the tensions between creativity and inclusiveness in their adoption of city brands and policy initiatives?. *Cities, 128*, 103786.

Alvarez Leon, L. F. (2022). From glitch epistemologies to glitch politics. *Dialogues in Human Geography, 12*(3), 384–388.

Aly, H. (2005, December 23). Tories say stand on gay marriage tightens race in Brampton. *The Globe and Mail*, A15.

Anacker, K., & Morrow-Jones, H. A. (2005). Neighborhood factors associated with same-sex households in US cities. *Urban Geography*, *26*(5), 385–409.
Anderson, B. (1983). *Imagined communities: Reflections on the origin and spread of nationalism*. Verso.
Andres, L. (2013). Differential spaces, power hierarchy and collaborative planning: A critique of the role of temporary uses in shaping and making places. *Urban Studies*, *50*(4), 759–775.
Andres, L., Bakare, H., Bryson, J. R., Khaemba, W., Melgaço, L., & Mwaniki, G. R. (2021). Planning, temporary urbanism and citizen-led alternative-substitute place-making in the Global South. *Regional Studies*, *55*(1), 29–39.
Andrucki, M. J. (2021). Queering social reproduction: Sex, care and activism in San Francisco. *Urban Studies*, *58*(7), 1364–1379.
Andrucki, M. J., & Kaplan, D. J. (2018). Trans objects: materializing queer time in US transmasculine homes. *Gender, Place & Culture*, *25*(6), 781–798.
Architizer.com. (2021). *Surrey city hall + civic centre*. https://architizer.com/projects/surrey-city-hall/
Arefi, M. (2014). *Deconstructing placemaking: Needs, opportunities, and assets*. Routledge.
Arrests, heated exchanges mark rallies over LGBTQ school policies. (2023, September 20). *CBC News*. https://www.cbc.ca/news/canada/rallies-gender-schools-1.6972606
Askins, K. (2015). Being together: Everyday geographies and the quiet politics of belonging. *ACME: An International Journal for Critical Geographies*, *14*(2), 470–478.
Audunson, R. (2005). How do politicians and central decision-makers view public libraries? The case of Norway. *IFLA Journal*, *31*(2), 174–182.
Augé, M. (2009). *Non-places: An introduction to supermodernity*. Verso Books.
Austin, I., & Spencer, K. (1997, March 18). Surrey official anti-gay. *The Province*, A4.
Bailey, M. M. (2014). Engendering space: Ballroom culture and the spatial practice of possibility in Detroit. *Gender, Place & Culture*, *21*(4), 489–507.
Bailey, R. C. (1999). *Gay politics, urban politics: Identity and economics in the urban setting*. Columbia University Press.
Bain, A. L. (2013). *Creative margins: Cultural production in Canadian suburbs*. University of Toronto Press.
Bain, A. L. (2022). Queer affordances of care in suburban public libraries. *Emotion, Space and Society*, *45*, 100923. https://doi.org/10.1016/j.emospa.2022.100923
Bain, A. L., & Nash, C. J. (2007). The Toronto Women's Bathhouse Raid: Querying queer identities in the courtroom. *Antipode*, *39*(1), 17–34.
Bain, A. L., Payne, W., & Isen, J. (2015). Rendering a neighbourhood queer. *Social & Cultural Geography*, *16*(4), 424–443.
Bain, A. L., & Podmore, J. A. (2020a). Challenging heteronormativity in suburban high schools through "surplus visibility": Gay-straight alliances in the Vancouver city-region. *Gender, Place & Culture*, *27*(9), 1223–1246.
Bain, A. L., & Podmore, J. A. (2020b). Scavenging for LGBTQ2S public library visibility on Vancouver's periphery. *Tijdschrift voor economische en sociale geografie*, *111*(4), 601–615.
Bain, A. L., & Podmore, J. A. (2021a). Linguistic ambivalence amidst suburban diversity: LGBTQ2S municipal "social inclusions" on Vancouver's periphery. *Environment and Planning C: Politics and Space*, *39*(7), 1644–1672.
Bain, A. L., & Podmore, J. A. (2021b). More-than-safety: Co-creating resourcefulness and conviviality in suburban LGBTQ2S youth out-of-school spaces. *Children's Geographies*, *19*(2), 131–144.
Bain A.L., & Podmore J. A. (2021c). Placing LGBTQ+ urban activisms. *Urban Studies*, *58*(7), 1305–1326.

Bain A.L., & Podmore J. A. (2021d). Relocating queer: Comparing suburban LGBTQ2S activisms on Vancouver's periphery. *Urban Studies*, *58*(7), 1500–1519.
Bain, A. L., & Podmore, J. A. (2022). The scalar arrhythmia of LGBTQ2S social inclusion policies: An analysis of the peripheral municipalities of a "progressive" city-region. *International Journal of Urban and Regional Research*, *46*(5), 784–806.
Bain, A. L., & Podmore, J. A. (2025). Rainbow infrastruggles: The infrapolitics of LGBTQ2S surplus visibility in suburban infrastructure. *Journal of Urban Affairs*, *47*(2), 428–450.
Bain, A. L., Podmore, J. A., & Rosenberg, R. (2020). "Straightening" space and time? Peripheral moral panics in print media representations of Canadian LGBTQ2S suburbanites, 1985–2005. *Social & Cultural Geography*, *21*(6), 839–861.
Baluja, T., & Hammer, K. (2011, March 18). Mississauga high school bans gay-straight alliance: Teens barred from launching group at St. Joseph's as church doctrine conflicts with public education policy. *The Globe and Mail*, A13.
Banerjea, N., & Browne, K. (2023). *Liveable lives: Living and surviving LGBTQ equalities in India and the UK*. Bloomsbury Academic.
Bassi, C. (2006). Riding the dialectical waves of gay political economy: A story from Birmingham's commercial gay scene. *Antipode*, *38*(2), 213–235.
Basu, R., & Fiedler, R.S. (2017). Integrative multiplicity through suburban realities: Exploring diversity through public spaces in Scarborough. *Urban Geography*, *38*(1), 25–46.
BCTF. (2023). *History of the BCTF*. https://www.bctf.ca/topics/services-information/about-the-bctf/bctf-history
Beam, M. (2018). *Gay, Inc.: The nonprofitization of queer politics*. University of Minnesota Press.
Beasley, L. (2019). *Vancouverism*. On Point Press.
Beaudet, G. (2021). *Banlieue, dites-vous? La suburbanisation dans la région métropolitaine de Montréal*. Presses de l'Université Laval.
Bell, D. (1995). Pleasure and danger: The paradoxical spaces of sexual citizenship. *Political Geography*, *14*(2), 139–153.
Bell, D. (2000). Eroticizing the rural. In R. Phillips, D. Shuttleton, & D. Watt (Eds.), *De-centering sexualities: Politics and representations beyond the metropolis* (pp. 91–110). Routledge.
Bell, M. M. (1997). The ghosts of place. *Theory and Society*, *26*(6), 813–836.
Berlant, L. (2016). The commons: Infrastructures for troubling times. *Environment and Planning D: Society and Space*, *34*(3), 393–419.
Berlant, L. (2022). *On the inconvenience of other people*. Duke University Press.
Berlant, L., & Warner, M. (1998). Sex in public. *Critical Inquiry*, *24*(2), 547–566.
Biddulph, M. (2014). Drawing and thinking: Representing place in the practice of place-making. *Journal of Urban Design*, *19*(3), 278–297.
Billig, M. (1995). *Banal nationalism*. Sage.
Bitterman, A. (2021). The rainbow connection: A time-series study of rainbow flag display across nine Toronto neighborhoods. In A. Bitterman & D. B. Hess (Eds.), *The life and afterlife of gay neighborhoods: Renaissance and resurgence* (pp. 117–140). Springer.
Block, I. (1999, November 25). Gay couple challenges tax. *The Gazette*, A6.
Blunt, A., & Dowling, R. (2022). *Home* (2nd Edition). Routledge.
Boisvert, Y. (2001, May 28). Plus jamais seul. *La Presse*, E1.
Booth, M. (2014, July 10). Pride flag flap reveals flaws in city policies. *The Tri-Cities Now*, A11. https://www.surreynowleader.com/community/booth-pride-flag-flap-reveals-flaws-in-surrey-policies/
Boudreau, J-A. (2003). The politics of territorialization: Regionalism, localism and other isms… The case of Montreal. *Journal of Urban Affairs*, *25*(2), 179–199.
Boudreau, J.-A., Keil, R., & Young, D. (2009). *Changing Toronto: Governing urban neoliberalism*. University of Toronto Press.

Bourne, L. S. (1996). Reinventing the suburbs: Old myths and new realities. *Progress in Planning*, *46*(3), 163–184.
Bourne, L. S., Britton, J. N. H., & Leslie, D. (2011). The Greater Toronto Region: The challenges of economic restructuring, social diversity, and globalization. In L. S. Bourne, T. Hutton, R. Shearmur & J. Simmons (Eds.), *Canadian urban regions: Trajectories of growth and change* (pp. 236–268). Oxford University Press.
Bourne, L. S., & Hulchanski, D. (2020). Inequality and neighbourhood change: Context, concept, and process. In J. Grant, A. Walks, & H. Ramos (Eds.), *Changing neighbourhoods: Social and spatial polarization in Canadian cities* (pp. 5–30). UBC Press.
Boussalem, A. (2021). In, out, or somewhere else entirely: Going beyond binary constructions of the closet in the lives of LGBTQ people from a Muslim background living in Brussels. *Transactions of the Institute of British Geographers*, *46*(2), 435–448.
Bowlby, S. (2011). Friendship, co-presence, and care: Neglected spaces. *Social & Cultural Geography*, *12*(6), 605–622.
Braun, E., Eshuis, J., Klijn, E. H., & Zenker, S. (2018). Improving place reputation: Do an open place brand process and an identity-image match pay off?. *Cities*, *80*, 22–28.
Brennan-Horley, C. (2010). Multiple work sites and city-wide networks: A topological approach to understanding creative work. *Australian Geographer*, *41*(1), 39–56.
Brickell, C., & Collard, J. (Eds.) (2019). *Queer objects*. Rutgers University Press.
Brim, M., & Ghaziani, A. (2016). *Imagining queer as method.* New York University Press.
Broad, K. L. (2002). Social movement selves. *Sociological Perspectives*, *45*(3), 317–336.
Broad, K. L., Alden, H., Berkowitz, D., & Ryan, M. (2008). Activist parenting and GLBTQ families. *Journal of GLBT Family Studies*, *4*(4), 499–520.
Broad, K. L., Crawley, S. L., & Foley, L. (2004). Doing "real family values": The interpretive practice of families in the GLBT movement. *Sociological Quarterly*, *45*(3), 509–527.
Brodyn, A., & Ghaziani, A. (2018). Accounting for new forms of inequity in the gayborhood. *City & Community*, *17*(2), 307–329.
Brown, G. (2007). Mutinous eruptions: Autonomous spaces of radical queer activism. *Environment and Planning A*, *39*(11), 2685–2698.
Brown, G. (2012). Homonormativity: A metropolitan concept that denigrates "ordinary" gay lives. *Journal of Homosexuality*, *59*(7), 1065–1072.
Brown, M. (1997). *Replacing citizenship: AIDS activism and radical democracy*. Guilford Press.
Brown, M. (2000). *Closet space: Geographies of metaphor from the body to the globe.* Routledge.
Brown, M. (2003). Sex, scale and the "new urban politics": HIV-prevention strategies from Yaletown, Vancouver. In D. Bell & G. Valentine (Eds.), *Mapping desire: Geographies of sexualities* (pp. 222–240). Routledge.
Brown, M., & Knopp, L. (2006). Places or polygons? Governmentality, scale, and the census in the gay and lesbian atlas. *Population, Space, and Place*, *12*, 223–242.
Brown, M., & Knopp, L. (2008). Queering the map: The productive tensions of colliding epistemologies. *Annals of the association of American Geographers*, *98*(1), 40–58.
Brown, M., & Knopp, L. (2016). Sex, drink, and state anxieties: Governance through the gay bar. *Social & Cultural Geography*, 17(3), 335–358.
Brown, M., Knopp, L., & Morrill, R. (2005). The culture wars and urban electoral politics: Sexuality, race, and class in Tacoma, Washington. *Political Geography*, *24*(3), 267–291.
Browne, K., & Bakshi, L. (2013). Insider activists: The fraught possibilities of LGBT activisms from within. *Geoforum*, *49*, 253–262.

Browne, K., Banerjea, N., McGlynn, N., Bakshi, L., Beethi, S., & Biswas, R. (2021). The limits of legislative change: Moving beyond inclusion/exclusion to create "a life worth living". *Environment and Planning C: Politics and Space, 39*(1), 30–52.
Brown-Saracino, J. (2018). *How places make us: Novel LBQ identities in four small cities.* University of Chicago Press.
Brown-Saracino, J., & Rumpf, C. (2011). Diverse imageries of gentrification: Evidence from newspaper coverage in seven US cities, 1986–2006. *Journal of Urban Affairs, 33*(3), 289–315.
Bruce, K. M. (2016). *Pride parades: How a parade changed the world.* New York University Press.
Bunnell, T., Yea, S., Peake, L., Skelton, T., & Smith, M. (2012). Geographies of friendship. *Progress in Human Geography, 36*(4), 490–507.
Burns, R., & Berbary, L. A. (2021). Placemaking as unmaking: Settler colonialism, gentrification, and the myth of "revitalized" urban spaces. *Leisure Sciences, 43*(6), 644–660.
Burton, P. (2015). The Australian good life: The fraying of a suburban template. *Built Environment, 41*(4), 504–518.
Buser, M., Bonura, C., Fannin, M., & Boyer, K. (2013). Cultural activism and the politics of place-making. *City: Analysis of urban trends, culture, theory, policy, action, 17*(5), 606–627.
Cadge, W., & Wildeman, C. (2008). Facilitators and advocates: How mainline Protestant clergy respond to homosexuality. *Sociological Perspectives, 51*(3), 587–603.
Cameron, A. (2007). Geographies of welfare and exclusion: Reconstituting the "public". *Progress in Human Geography, 31*(4), 519–526.
Campkin, B. (2023). *Queer premises: LGBTQ+ venues in London since the 1980s.* Bloomsbury Publishing.
Canaday, M. (2009). *The straight state: Sexuality and citizenship in twentieth-century America.* Princeton University Press.
Canadian Heritage. (2022). *National flag of Canada etiquette.* https://www.canada.ca/en/canadian-heritage/services/flag-canada-etiquette.html
Canadian Mental Health Association – Peel Dufferin. (2024). *Queer minds.* https://cmhapeeldufferin.ca/events/queer-minds/
Carter, D. (2004). *Stonewall: The riots that sparked the gay revolution.* St. Martin's Press.
Casey, E. S. (2001). Between geography and philosophy: What does it mean to be in the place-world? *Annals of the Association of American Geographers, 91*(4), 683–693.
Casey, E. S. (2002). *Representing place: Landscape painting and maps.* University of Minnesota Press.
Casey, E. S. (2009). *Getting back into place: Toward a renewed understanding of the place-world.* (2nd Edition). Indiana University Press.
Catungal, J. P. (2013). Ethno-specific safe houses in the liberal contact zone: Race politics, place-making and the genealogies of the AIDS sector in global multicultural Toronto. *ACME: An International E-Journal for Critical Geographies, 12*(2), 250–278.
Catungal, J. P., Klassen, B., Ablenas, R., Lambert, S., Chown, S., & Lachowsky, N. (2021). Organising care and community in the era of the "gay disease": Gay community responses to HIV/AIDS and the production of differentiated care geographies in Vancouver. *Urban Studies, 58*(7), 1346–1363.
Catungal, J. P., & McCann, E. J. (2010). Governing sexuality and park space: Acts of regulation in Vancouver, BC. *Social & Cultural Geography, 11*(1), 75–94.
Cavicchia, R. (2021). Are Green, dense cities more inclusive? Densification and housing accessibility in Oslo. *Local Environment, 26*(10), 1250–1266.
Centre des femmes de Longueuil. (2023) *Embarquer-vous? Rapport d'activités, 2022–2023.* https://www.centredefemmeslongueuil.org/_files/ugd/5fde2c_1b1d7a92a7e44a49820881d97b6cfe99.pdf

Chamberlain v. Surrey School District No. 36 (2002). 4 S.C.R. 710, 2002 SCC 86. https://www.canlii.org/en/ca/scc/doc/2002/2002scc86/2002scc86.html

Chapman, T. (2011). Queering the political-economy: Anti-discrimination law and the urban regime in Orlando, Florida. In P. L. Doan (Ed.), *Queerying planning: Challenging heteronormative assumptions and reframing planning practice* (pp. 145–166). Routledge.

Charrieras, D., Darchen, S., & Sigler, T. (2018). The shifting spaces of creativity in Hong Kong. *Cities*, *74*, 134–141.

Chattopadhyay, S., & White, J. (2014). City halls: Civic representation and public space. In S. Chattopadhyay & J. White (Eds.), *City Halls and civic materialism: Towards a global history of urban public space* (pp. 3–13). Routledge.

Chávez, K. R. (2013). *Queer migration politics: Activist rhetoric and coalitional possibilities*. University of Illinois Press.

Cilliers, E., & Timmermans, W. (2015). The importance of creative participatory planning in the public place-making process. *Environment and Planning B: Planning and Design*, 41(3), 413–429.

Cisneros, J., & Bracho, C. (2019). Coming out of the shadows and the closet: Visibility schemas among undocuqueer immigrants. *Journal of Homosexuality*, *66*(6), 715–734.

Cisneros, J., & Bracho, C. (2020). Undocuqueer stress: How safe are "safe" spaces, and for whom? *Journal of Homosexuality*, *67*(11), 1491–1511.

City of Brampton. (2018a). *Culture Master Plan*. Retrieved from: https://www.brampton.ca/EN/Arts-Culture-Tourism/CulturalSrvs/Documents/CulturalMaster-Plan/City%20of%20Brampton%20-Final%20Culture%20Master%20Plan.pdf.

City of Brampton. (2018b). *Living the mosaic: Brampton 2040 vision*. https://geohub.brampton.ca/pages/brampton-2040-vision

City of Brampton. (2023). *Corporate identity*. https://www.brampton.ca/EN/City-Hall/Pages/Corporate-identity.aspx

City of Brampton. (2024a). *Committees*. https://www.brampton.ca/EN/City-Hall/Council-Committees/Pages/Inclusion-and-Equity-Committee.aspx

City of Brampton. (2024b). Protocol Office: Flag raising, half masting. https://www.brampton.ca/EN/City-Hall/Protocol-Office/Pages/Flag-Raising-Half-Mastings.aspx#:~:text=Providing%20the%20flag%20one%20week,exceeds%20the%20allotted%201%20hour

City of Burnaby. (2011). Burnaby social sustainability strategy. www.burnaby.ca/Assets/city±services/planning/Social±Sustainability±Strategy.pdf

City of Burnaby. (2016). *Proclamation, International Day of the Pink*. Office of the Mayor.

City of Burnaby. (2018). *Proclamation, International Day against Homophobia and Transphobia*. Office of the Mayor.

City of Burnaby. (2019a, May 13). *Minutes, city council*. https://pub-burnaby.escribemeetings.com/filestream.ashx?DocumentId=43291

City of Burnaby. (2019b, May 27). *Minutes, city council*. https://pub-burnaby.escribemeetings.com/filestream.ashx?DocumentId=43476

City of Burnaby. (2019c). *Proclamation, Pride Week in the City of Burnaby*. Office of the Mayor.

City of Burnaby. (2019d, June 17). *Report from the Director of Engineering to the City Manager on rainbow crosswalks*. https://pub-burnaby.escribemeetings.com/filestream.ashx?DocumentId=43606

City of Burnaby. (2020, May 11). *Finance department, financial plan, 2020–2024*. https://www.burnaby.ca/sites/default/files/acquiadam/2021-07/2020-2024-Financial-Plan.pdf

City of Dorval. (2023). *Modifications to the sustainable master plan and to municipal by-laws*. https://www.ville.dorval.qc.ca/medias/files/en/miscellaneous/Presentation_assemblee_20231011_eng.pdf

City of Markham. (2008). *Everyone welcome: Markham diversity action plan* https://www.markham.ca/wps/wcm/connect/markham/d9def2b1-32f9-4ed2-984e-a39d5482455e/Markham-DAP-FINAL.pdf?MOD=AJPERES&CONVERT_TO=url&CACHEID=ROOTWORKSPACE.Z18_2QD4H901OGV160QC8BLCRJ1001-d9def2b1-32f9-4ed2-984e-a39d5482455e-msYKPSY

City of Markham. (2018). *Markham's inclusion charter: Presentation to general committee*. https://www.markham.ca/wps/portal/home/about/deia/COM-Inclusion-Charter/02-COM-Inclusion-Charter

City of Mississauga. (2009a). *Mississauga youth plan!* City of Mississauga.

City of Mississauga. (2009b). *Mississauga culture master plan*. City of Mississauga.

City of Mississauga. (2009c). *City of Mississauga strategic plan*. https://www.mississauga.ca/wp-content/uploads/2018/10/03085738/StrategicPlan_Web_04_22_2009.pdf

City of Mississauga. (2014). *Master plan for recreation: Final report*. https://www7.mississauga.ca/Departments/Rec/future-directions/pdf/2014-master-plans/recreation-master-plan.pdf

City of Mississauga. (2015). *Terms of reference for advisory committees: Diversity and Inclusion Advisory Committee*. https://www7.mississauga.ca/documents/Communications/2014/Terms_Of_Reference_Diversity_Inclusion_Committee_Dec_2015.pdf

City of Mississauga. (2017). *Workforce diversity and inclusion strategy*. https://www.mississauga.ca/wp-content/uploads/2021/04/27095710/Workforce_Diversity_and_Inclusion_Strategy.pdf

City of Mississauga. (2019). *Celebrate Pride month*. https://campaign.mississauga.ca/pride-month/

City of Mississauga. (2022a). *Diversity and Inclusion Advisory Committee*. https://www.mississauga.ca/council/committees/diversity-and-inclusion-advisory-committee/

City of Mississauga. (2022b). *Terms of reference for the Combating Racism, Discrimination and Hatred Advisory Committee*. https://www.mississauga.ca/wp-content/uploads/2024/02/28074249/combating-racism-discrimination-and-hatred-committee-terms-of-reference.pdf

City of Mississauga. (2024). *Clock tower lighting: Pride month–rainbow colours*. https://www.mississauga.ca/events-and-attractions/events-calendar/clock-tower-lighting-pride-month---pride-colo-71?eventdate=2023-06-01T20%253A00%253A00

City of New Westminster. (2010, January 25). *Minutes*. Regular Meeting of City Council.

City of New Westminster. (2017). *New Westminster age-friendly community strategy*. https://www.newwestcity.ca/database/files/library/Age_Friendly_Community_Strategy___reduced_size.pdf

City of New Westminster. (2019, November 5). *Minutes*. Community and Social Issues Committee. https://www.newwestcity.ca/database/files/library/CSI_2019_Nov_5_Agenda_Package.pdf

City of Pointe-Claire. (2020). *Pointe-Claire: A diverse and inclusive community*. https://www.pointe-claire.ca/en/news/pointe-claire-a-diverse-and-inclusive-community

City of Pointe-Claire. (2023). *Pointe-Claire Strategic Plan, 2023–2030*. https://www.pointe-claire.ca/content/uploads/2023/08/pc_planstrategique-an_web.pdf

City of Surrey. (2011, June 27). *Public hearing minutes*. Regular Council. https://www.surrey.ca/sites/default/files/minutes/MIN_RCPH_2011-06-27.pdf

City of Surrey. (2014a, June 23). *Corporate report, Flag policy, City Clerk*. https://www.surrey.ca/sites/default/files/corporate-reports/RPT_2014-R104.pdf

City of Surrey. (2014b, June 9). *Minutes, council-in-committee*. https://www.surrey.ca/sites/default/files/minutes/MIN_CIC_2014-06-09.pdf

City of Surrey. (2014c). *Refugee newcomers in Surrey: Approaches to building a more welcoming and inclusive community. Surrey welcoming communities project.* https://www.surreylip.ca/sites/default/files/Approaches_to_Building_a_More_Welcoming_and_Inclusive_Community_-_Surrey_-_March_2014.pdf

City of Surrey. (2016a, November 2). *Corporate report, General manager, Parks, recreation and culture and General Manager, human resources.* https://www.surrey.ca/sites/default/files/corporate-reports/RPT_2016-R246.pdf

City of Surrey. (2016b, June 13). *Public hearing minutes, Regular Council.* https://www.surrey.ca/sites/default/files/minutes/MIN_RCPH_2016-06-13.pdf

City of Surrey. (2016c). *Sustainability charter 2.0.* https://www.surrey.ca/files/SustainabilityCharter.pdf

City of Surrey. (2017, January 16). *Corporate report. True Colours of Surrey–We Are Surrey anti-racism public awareness campaign.* https://www.surrey.ca/sites/default/files/corporate-reports/RPT_2017-R013.pdf

Clark, G. (1997, April 30). What do we tell the kids?: Lesbian Surrey mom hints at rights action. *The Province*, A5.

Clay, S. (2019). Queer smartphones. In C. Brickell & J. Collard (Eds.), *Queer objects* (pp. 387–390). Rutgers University Press.

Coalition of Inclusive Municipalities. (2019). *LGBTQ2+ Inclusiveness: Toolkit for Inclusive Municipalities in Canada and Beyond.* https://en.ccunesco.ca/search#q=lgbtq2%2B

Coblence, A. (2021). What do place-makers actually do to sustain knowledge dynamics? Place-making practices in a Czech suburban knowledge location. *European Planning Studies*, *29*(10), 1886–1905.

Cohendet, P., Grandadam, D., & Laurent, S. (2010). The anatomy of the creative city. *Industry and Innovation*, *17*(1), 91–111.

Coleman, E. J. (2013). *Tongzhi politics of same-sex eroticism in Chinese societies.* Taylor & Francis.

Colley, T. (2005, June 15). Gay flag won't fly at city hall. *Surrey Now*, 1.

Collier, S. (2023). Statement from Mayor Shaun Collier regarding the support for the 2SLGBTQIA communities. https://www.ajax.ca/en/statement-from-mayor-shaun-collier-regarding-the-support-for-the-2slgbtqia-communities.aspx

Collin, J.-P.., & Poitras, C. (2002). La fabrication d'un espace suburbain: La Rive-Sud de Montréal. *Recherches sociographiques*, *43*(2), 275–310.

Collins, A. (2004a) Sexual dissidence, enterprise, and assimilation: Bedfellows in urban regeneration. *Urban Studies*, *41*(9), 1789–1806.

Collins, D. (2006). Culture, religion and curriculum: Lessons from the "three books" controversy in Surrey, BC. *The Canadian Geographer*, *50*(3), 342–357.

Collins, R. (2004b). *Interaction ritual chains.* Princeton University Press.

Collister, E., & Noel, A. (1993, February 12). Gays fear serial killer stalks them; Ready to carry arms after police warning. *The Gazette*, A1.

Community Development Council Durham. (2011). *Service competency for the LGBTQ Community: A community asset mapping exercise in Durham region.* Ajax: Community Development Council Durham.

Compton, D. R., & Baumle, A. K. (2012). Beyond the Castro: The role of demographics in the selection of gay and lesbian enclaves. *Journal of Homosexuality*, *59*(10), 1327–1356.

Concordia University. (2017). *Territorial acknowledgement.* https://www.concordia.ca/indigenous/resources/territorial-acknowledgement.html

Conseil québécois LGBT. (2024). *Guide pratique pour des municipalités québécoises inclusives des personnes LGBTQIA2+.* https://conseil-lgbt.ca/wp-content/uploads/2024/02/MUNI_Gude-complet_fevrier-2024.pdf

Cooper, D. (2006). Active citizenship and the governmentality of local lesbian and gay politics. *Political Geography*, *25*(8), 921–943.

Cooper, D. (2018). Possessive attachments: Identity beliefs, equality law and the politics of state play. *Theory, Culture & Society*, *35*(2), 115–135.
Cooper, D., & Monro, S. (2003). Governing from the margins: Queering the state of local government. *Contemporary Politics*, *9*(3), 229–255.
Corner, J. (2011). The agency of mapping: Speculation, critique and invention. In M. Dodge, R. Kitchin, & C. Perkins (Eds.), *The map reader: Theories of mapping, practice, and cartographic representation* (pp. 89–101). Wiley-Blackwell.
Cornwell, T. (2022). Mississauga permanently recognizes June as Pride month, to raise Progress flags at every city facility. Mississauga.com. https://www.mississauga.com/news/council/mississauga-permanently-recognizes-june-as-pride-month-to-raise-progress-flags-at-every-city-facility/article_0bf3ac6b-38bb-5b1d-8c92-6f5dcf37b02e.html
Corporation de développement communautaire de l'agglomération de Longueuil. (2021). *Bienvenu*. https://www.cdcal.org/outils/approches-inclusives-et-transversales-coffre/
Costello, C. (2020). Towards a politics of municipal homonationalism in Brampton, Ontario. Unpublished Major Research Paper. Faculty of Environmental and Urban Change. York University.
Council of the Regional Municipality of York. (2018, June 28). *Minute Nos. 98, 102 and 109: Inclusion charter for York Region*. Corporate Services, Regional Clerk's Office.
Cowen, D. (2005). Suburban citizenship? The rise of targeting and the eclipse of social rights in Toronto. *Social & Cultural Geography*, *6*(3), 335–356.
Cravens, Royal G. III (2015). Morality politics and municipal LGBT policy adoption: A rare-event analysis. *State and Local Government Review*, *47*(1), 15–25.
Crawford, J. B., & Herland, K. (2014). Sex garage: Unspooling narratives, rethinking collectivities. *Journal of Canadian Studies*, *48*(1), 106–131.
Cresswell, T. (1996). *In place/Out of place: Geography, ideology, and transgression*. University of Minnesota Press.
Cresswell, T. (2014). *Place: An introduction*. John Wiley & Sons.
D'Angelo, P., & Kuypers, J. A. (Eds.) (2010). *Doing news framing analysis: Empirical and theoretical perspectives*. Routledge.
Dagenais, D. (2020). *Grossières indécences: Pratiques et identités homosexuelles à Montréal, 1880-1929*. McGill-Queen's Press.
Daniels, S. (1992). Place and the geographical imagination. *Geography: Journal of the Geographical Association*, *77*(4), 310–322.
Darling, J. (2011). Giving space: Care, generosity and belonging in a UK asylum drop-in centre. *Geoforum*, *42*(4), 408–417.
Davis, J. S. (2005). Representing place: "Deserted isles" and the reproduction of Bikini Atoll. *Annals of the Association of American Geographers*, *95*(3), 607–625.
Davis, K. D. (2021). Transnational blackness at Toronto Pride: Queer disruption as theory and method. *Gender, Place & Culture*, *28*(11), 1541–1560.
De Filippi, F., Coscia, C., & Guido, R. (2019). From smart-cities to smart-communities: How can we evaluate the impacts of innovation and inclusive processes in urban context?. *International Journal of E-Planning Research*, *8*(2), 24–44.
Depictions divide educators, clerics. (1997, August 1). *The Province*, A3.
Di Feliciantonio, C. (2020). Migration as an active strategy to escape the 'second closet' for HIV-positive gay men in Barcelona and Rome. *Social & Cultural Geography*, *21*(9), 1177–1196.
Dickey Young, P. (2008). Two by two: Religion, sexuality and diversity in Canada. In L. Beaman & P. Beyer (Eds.), *Religion and diversity in Canada* (pp. 85–103). Brill Academic Publishers.
DiManno, R. (1994, July 27). Burlington tavern lets its red neck shine through. *Toronto Star*, A7.

Dines, M. (2005). Sacrilege in the sitting room: Contesting suburban domesticity in contemporary gay literature. *Home Cultures*, *2*(2), 175–194.
Dines, M. (2009). *Gay suburban narratives in American and British culture: Homecoming queens*. Palgrave Macmillan.
Distress Centre Durham. (2021). *LGBT prideline*. https://distresscentredurham.com/gethelp/lgbtprideline/
DIVERSEcity. (2018). *Mapping lesbian, gay, bisexual, trans & queer migrant needs in Surrey, BC: A report commissioned by DIVERSCity*. https://pflagsurrey.ca/wp-content/uploads/2021/05/DiverseCity-last-final.pdf
Doan, P. L. (2010). The tyranny of gendered spaces: Reflections from beyond the gender dichotomy. *Gender, Place & Culture*, *17*(5), 635–654.
Doan, P. L. (Ed.) (2015). *Planning and LGBTQ communities: The need for inclusive queer spaces*. Routledge.
Doan, P. L. (2016). To count or not to count: Queering measurement and the transgender community. *WSQ: Women's Studies Quarterly*, *44*(3), 89–110.
Dobele, A. R., Ferguson, S., Hartman, A. E., & Schuster, L. (2022). Gender equity in the marketing academy: From performative to institutional allyship. *Australasian Marketing Journal*, *30*(3), 185–194.
Dostie, A. (2022, October 6). Animaux domestiques et communauté LGBTQ+ parmi les nouveaux enjeux du caucus Fournier. *Le courrier du sud*. https://www.lecourrierdusud.ca/animaux-domestiques-et-communaute-lgbtq-parmi-les-nouveaux-enjeux-du-caucus-fournier/
Downs, R. M., & Stea, D. (2011) Cognitive maps and spatial behaviour: Process and products. In M. Dodge, R. Kitchin & C. Perkins (Eds.), *The map reader* (pp. 312–317). John Wiley & Sons.
Drescher, J., Shidlo, A., & Schroeder, M. (2002). *Sexual conversion therapy: Ethical, clinical and research perspectives*. The Haworth Medical Press.
Druschel, B. E. (2016). In service of camp or the campiness of service: "The Court" as queer civic and fraternal organization. *The Journal of American Culture*, *39*(1), 33–40.
Duff, C. (2010). On the role of affect and practice in the production of place. *Environment and Planning D: Society and Space*, *28*(5), 881–895
Dufour, V. (2001, May 28). Pointe-Claire: Un defile gay, version banlieue. *Le Devoir*, A1.
Dumont, M.-E. (2014, July 29). Longueuil: Une famille reçoit une lettre de menace homophobe. *Le Journal de Montréal*, 12.
Duncan, J., & Ley, D. (Eds.) (1993). *Place/culture/representation*. Routledge.
Duplan, K. (2023). What would an inclusive city for gender and sexual minorities be like? You need to ask queer folx!. *Social Inclusion*, *11*(3), 138–149.
Dwyer, C., Gilbert, D., & Ahmed, N. (2015). Building the sacred in suburbia: Improvisation, reinvention, and innovation. *Built Environment*, *41*(4), 477–490.
Dwyer, C., Gilbert, D., & Shah, B. (2013). Faith and suburbia: Secularisation, modernity, and the changing geographies of religion in London's suburbs. *Transactions of the Institute of British Geographers*, *38*(3), 403–419.
Ebbensgaard, C. L. (2020). Standardised difference: Challenging uniform lighting through standards and regulations. *Urban Studies*, *57*(9), 1957–1976.
Edelman, L. (2004). *No future: queer theory and the death drive*. Duke University Press.
Edensor, T. (2015). The gloomy city: Rethinking the relationship between light and dark. *Urban Studies*, *52*(3), 422–438.
Elwood, S. (2021). Digital geographies, feminist relationality, Black and queer code studies: Thriving otherwise. *Progress in Human Geography*, *45*(2), 209–228.
Entman, R. (1993). Framing: Toward clarification of a fractured paradigm. *Journal of Communication*, *43*(4), 51–58.

Entrikin, N. (1991). *The betweenness of place: Towards a geography of modernity*. Johns Hopkins University Press.
Escorcia Hernández, J. R., Torabi Moghadam, S., & Lombardi, P. (2023). Sustainability assessment in social housing environments: An inclusive indicators selection in Colombian post-pandemic cities. *Sustainability*, *15*(3), 2830.
Ettlinger, N. (2010). Bringing the everyday into the culture/creativity discourse. *Human Geography*, *3*(1), 49–59.
Fava, S. F. (1956). Suburbanism as a way of life. *American Sociological Review*, *21*(1), 34–37.
Fejes, F. (2008). *Gay rights and moral panic: The origins of America's debate on homosexuality*. St. Martin's Press.
Ferbrache, F., & Knowles, R. D. (2017). City boosterism and place-making with light rail transit: A critical review of light rail impacts on city image and quality. *Geoforum*, *80*, 103–113.
Fetner, T. (2024). *Sex in Canada: The who, why, when, and how of getting down up north*. UBC Press.
Fetner, T., & Kush, K. (2008). Gay-straight alliances in high schools: Social predictors of early adoption. *Youth & Society*, *40*(1), 114–130.
Field, T. L., & Mattson, G. (2016). Parenting transgender children in PFLAG. *Journal of GLBT Family Studies*, *12*(5), 413–429.
Filion, P. (2015). Suburban inertia: The entrenchment of dispersed suburbanism. *International Journal of Urban and Regional Research*, *39*(3), 633–640.
Fincher, R., & Iveson, K. (2008). *Planning and diversity in the city: Redistribution, recognition and encounter*. Palgrave Macmillan.
Fincher, R., Pardy, M., & Shaw, K. (2016). Place-making or place-masking? The everyday political economy of "making place". *Planning Theory and Practice*, *17*(4), 516–536.
Fine, A. G. (2012). *Tiny publics: A theory of group action and culture*. Russell Sage Foundation.
Fine, G. A., & Corte, U. (2017). Group pleasures: Collaborative commitments, shared narrative, and the sociology of fun. *Sociological Theory*, *35*(1), 64–86.
Fischer, F. (2003). *Reframing public policy: Discursive politics and deliberative practices*. Oxford University Press.
Flicker, S., Wilson, M., Travers, R., Bereket, T., McKay, C., van der Meulen, A., Guta, A., Cleverley, S., & Rourke, S. B. (2009). Community-based research in AIDS-service organizations: What helps and what doesn't? *AIDS Care*, *21*(1), 94–102.
Flint, J., & Robinson, D. (Eds.). (2008). *Community cohesion in crisis?: New dimensions of diversity and difference*. Policy Press.
Florida, R. (2002). *The rise of the creative class: And how it's transforming work, leisure, community and everyday life*. Basic Books.
Florida, R. (2005). *Cities and the creative class*. Routledge.
Fondation Émergence. (2020). *May 17: International Day Against Homophobia and Transphobia*. https://en.may17mai.com/
Fondation Émergence. (2022). *Rapport d'activités 2021–2022*. https://en.fondationemergence.org/_files/ugd/cdd9d7_9e550bb7bb394f1db0cde2889a298135.pdf
Forest, B. (1995). West Hollywood as symbol: The significance of place in the construction of a gay identity. *Environment and Planning D: Society and Space*, *13*(2), 133–157.
Forrest, J., Gorman-Murray, A., & Siciliano, F. (2019). The geography of same-sex couples and families in Australia: An empirical review. *Australian Geographer*, *50*(4), 493–509.
Fraser, N. (2000). Rethinking recognition. *New Left Review*, *3*, 107–120.

Freeman, E. (2010). *Time binds: Queer temporalities, queer histories*. Duke University Press.
Freeman, L. M., & Blomley, N. (2019). Enacting property: Making space for the public in the municipal library. *Environment and Planning C: Politics and Space*, *37*(2), 199–218.
Frisch, M. (2021). A queer reading of the United States census. In A. Bitterman & D. Baldwin Hess (Eds.), *The life and afterlife of gay neighborhoods: Renaissance and resurgence* (pp. 61–85). Springer.
Frisque, G. (2018, August 14). Brampton Inclusion and Equity Committee plagued by attendance issues. *Brampton Guardian*. https://www.bramptonguardian.com/news-story/8811406-brampton-inclusion-and-equity-committee-plagued-by-attendance-issues/
Frum, D. (2004, July 20). What went wrong for the Tories. *The National Post*, A16.
Futch, V. A., & Fine, M. (2014). Mapping as a method: History and theoretical commitments. *Qualitative Research in Psychology*, *11*(1), 42–59.
Gamarel, K., Walker, J., Rivera, L., & Golub, S. (2014). Identity safety and relational health in youth spaces: A needs assessment with LGBTQ youth of color. *Journal of LGBTQ Youth*, *11*(3), 289–315.
Gans, H. J. (1967). *Levittown and America*. Pantheon.
Gates, G. J. (2007). *Geographic trends among same-sex couples in the US Census and the American Community Survey*. The Williams Institute. https://escholarship.org/uc/item/51w3b4z9
Gates, G. J., & Cooke, A. M. (2011). *United States Census snapshot: 2010*. The Williams Institute. https://escholarship.org/uc/item/4j23r1rx.
Gates, G. J., & Ost, J. (2004). *The gay and lesbian atlas*. Urban Institute Press.
Gay book ban backed. (1998, February 19). *The Globe and Mail*, A20.
Gay couple's neighbours join march to protest harassment. (2001, May 28). *The Record*, D10.
Germain, A., & Rose, D. (2000). *Montreal: Quest for a metropolis*. John Wiley and Sons.
Gerometta, J., Haussermann, H., & Longo, G. (2005). Social innovation and civil society in urban governance: Strategies for an inclusive city. *Urban Studies*, *42*(11), 2007–2021.
Ghaziani, A. (2014a). Measuring urban sexual cultures. *Theory and Society*, *43*(3), 371–393.
Ghaziani, A. (2014b). *There goes the gayborhood?*. Princeton University Press.
Ghaziani, A. (2019). Cultural archipelagos: New directions in the study of sexuality and space. *City & Community*, *18*(1), 4–22.
Gieseking, J. J. (2013). Where we go from here: The mental sketch mapping method and its analytic components. *Qualitative Inquiry*, *19*(9), 712–724.
Gieseking, J. J. (2018a). Operating anew: Queering GIS with good enough software. *The Canadian Geographer*, *62*(1), 55–66.
Gieseking, J. J. (2018b). Size matters to lesbians, too: Queer feminist interventions into the scale of big data. *The Professional Geographer*, *70*(1), 150–156.
Gieseking, J. J. (2020a). *A queer New York: Geographies of lesbians, dykes, and queers*. New York University Press.
Gieseking, J. J. (2020b). Mapping lesbian and queer lines of desire: Constellations of queer urban space. *Environment and Planning D: Society and Space*, *38*(5), 941–960.
Gilbert, D., Dwyer, C., Ahmed, N., Cuch, L., & Hyacinth, N. (2018). The hidden geographies of religious creativity: Place-making and material culture in West London faith communities. *Cultural Geographies*, *26*(1), 23–41.
Gilligan, K. (2015). Ajax has a new coordinator to reach out to diverse population. https://www.durhamregion.com/news/ajax-has-new-coordinator-to-reach-out-to-diverse-population/article_853a57db-4372-5b16-9b15-9b8f08d5c588.html
Gin, J., & Taylor, D. E. (2010). Movements, neighborhood change, and the media: Newspaper coverage of anti-gentrification activity in the San Francisco Bay Area,

1995–2005. In D. E. Taylor (Ed.), *Environment and social justice: An international perspective* (pp. 75–114). Emerald Group Publishing.
Goffman, E. (1974). *Frame analysis: An essay on the organization of experience*. Harper and Row.
Gohier-Drolet, A. (2018, September 21). Sida-Vie Laval devient Sphère. *Courrier Laval*. https://courrierlaval.com/sida-vie-laval-devient-sphere/
Gonzalvez, K., Foth, M., & Caldwell, G. A. (2021). Radical placemaking: Utilizing low-tech AR/VR to engage in communal placemaking during a pandemic. *Interaction, Design, and Architecture(s) Journal, 48*, 143–164.
Goodwin, C. (1994). Professional vision. *American Anthropologist, 96*(3), 606–633.
Gopinath, G. (2005). *Impossible desires: Queer diasporas and South Asian public cultures*. Duke University Press.
Gorman, M. R. (2014). *The empress is a man: Stories from the life of Jose Sarria*. Routledge.
Gorman-Murray, A. (2006). Homeboys: Uses of home by gay Australian men. *Social & Cultural Geographies, 7*(1): 53–69.
Gorman-Murray, A. (2011). Queerying planning in Australia: The problems and possibilities of multiscalar governance for LGBT sexual minorities. In P. L. Doan (Ed.), *Queerying planning: Challenging heteronormative assumptions and reframing planning practice* (pp. 129–143). Routledge.
Gorman-Murray, A. (2013). Straight–gay friendships: Relational masculinities and equalities landscapes in Sydney, Australia. *Geoforum, 49*: 214–223.
Gorman-Murray, A., Brennan-Horley, C., McLean, K., Waitt, G., & Gibson, C. (2010). Mapping same-sex couple family households in Australia. *Journal of Maps, 6*(1), 382–392.
Gorman-Murray, A., & Nash, C. J. (2017). Transformations in LGBT consumer landscapes and leisure spaces in the neoliberal city. *Urban Studies 54*(3): 786–805.
Gorman-Murray, A., & Nash, C. J. (2018). Queer suburbs: (Sub)urban spaces and sexualities in the global north. In B. Hanlon & T. Vicino (Eds.), *The Routledge companion to the suburbs* (pp. 135–146). Routledge.
Gorman-Murray, A., & Waitt, G. (2009). Queer-friendly neighbourhoods: Interrogating social cohesion across sexual difference in two Australian neighbourhoods. *Environment and Planning A, 41*(12), 2855–2873.
Grant, J. L. (2009). Theory and practice in planning the suburbs: Challenges to implementing new urbanism, smart growth, and sustainability principles. *Planning Theory & Practice, 10*(1), 11–33.
Grant, R., Amos, N., Lyons, A., McNair, R., Power, J., Carman, M., Hill, A., & Bourne, A. (2024). Out in suburbia: Associations between residential location, mental health, and community connectedness among LGBTQ Australians. *Social & Cultural Geography, 25*(8), 1272–1290. https://doi.org/10.1080/14649365.2023.2296472
Gray, M. L., Johnson, C. R., & Gilley, B. J. (Eds.) (2016). *Queering the countryside: New frontiers in rural queer studies*. New York University Press.
Grundy, J., & Smith, M. (2005). The politics of multiscalar citizenship: The case of lesbian and gay organizing in Canada. *Citizenship Studies, 9*(4), 389–404.
Güney, K. M., Keil, R., & Üçoğlu, M. (Eds.). (2019). *Massive suburbanization: (Re) building the global periphery*. University of Toronto Press.
Halberstam, J. (2005). *In a queer time and place: Transgender bodies, subcultural lives*. New York University Press.
Halberstam, J. (2011). *The queer art of failure*. Duke University Press.
Halberstam, J. (2018). Unbuilding gender: Trans* anarchitectures in and beyond the work of Gordon Matta-Clark. *Places Journal*. https://placesjournal.org/article/unbuilding-gender/
Hall, M., & Lee, B. (2010). How diverse are US suburbs?. *Urban Studies, 47*(1), 3–28.

Hall, T. (1997). Images of industry in the post-industrial city: Raymond Mason and Birmingham. *Ecumene*, *4*(1), 46–68.

Hamilton, G. (2001, May 23). Mayor requests a dress code for gay protest: Harassed by neighbours, couple accuse city hall of prejudice. *National Post*, A5.

Hanhardt, C. B. (2013). *Safe space: Gay neighborhood history and the politics of violence*. Duke University Press.

Hanrahan, S. (2020, April 4). *These 11 spots around Quebec are lighting up in rainbow support*. Mtl Blog.com. https://www.mtlblog.com/montreal/11-quebec-spots-doing-the-rainbow-trend-on-a-large-scale

Haritaworn, J. (2019). On these bones: The queer regenerations of the Toronto gay village serial killings. *Topia: Canadian Journal of Cultural Studies*, *40*, 15–41.

Harney, N. D. (2006). The politics of urban space: Modes of place-making by Italians in Toronto's neighbourhoods. *Modern Italy*, *11*(1), 25–42.

Harrington, A. (2006). Lifeworlds. *Theory, Culture & Society*, *23*(2–3), 341–342.

Harris, R. (2015). Using Toronto to explore three suburban stereotypes, and vice versa. *Environment and Planning A*, *47*(1), 30–49.

Harris, R., & Hendershott, K. (2018). How newspapers portray suburbs: A paradox. *Journal of Urban Affairs*, *40*(7), 974–991.

Harris, R., & Vorms, C. (Eds.) (2017). *What's in a name?: Talking about urban peripheries*. University of Toronto Press.

Hartt, M., DeVerteuil, G., & Potts, R. (2023). Age-unfriendly by design: Built environment and social infrastructure deficits in Greater Melbourne. *Journal of the American Planning Association*, *89*(1), 31–44.

Harvey, D. (1989). *The condition of postmodernity: An enquiry into the origins of cultural change*. Blackwell.

Hayslett, K. L., & Kane, M. D. (2011). "Out" in Columbus: A geospatial analysis of the neighborhood-level distribution of gay and lesbian households. *City & Community*, *10*, 131–156.

Hayter, R., Patchell, J., & Rees, K.G. (2024). *The thin edge of innovation: Metro Vancouver's evolving economy*. UBC Press.

Hermes, K. A. (2019). Political buttons. In C. Brickell & J. Collard (Eds.), *Queer objects* (pp. 315–319). Rutgers University Press.

Herring, S. (2010). *Another country: Queer anti-urbanism*. NYU Press.

Hess, D. B., & Bitterman, A. (2023). Community support organizations in gay neighborhoods: Assessing engagement during the COVID-19 pandemic. *Urban Planning*, *8*(2), 235–248.

Hewidy, H., & Lilius, J. (2022). In the blind spot: Ethnic retailing in Helsinki and the spontaneous placemaking of abandoned places. *European Planning Studies*, *30*(8), 1493–1513.

Hillier, B., & Abbruzzese, T. (2023). Smart cities: Big data and surveillance. In A. L. Bain & L. Peake (Eds.), *Urbanization in a global context* (2nd Edition) (pp. 197–215). Oxford University Press.

Hines, S. (2013). *Gender diversity, recognition and citizenship: Towards a politics of difference*. Palgrave Macmillan.

Hodkinson, P. (2017). Bedrooms and beyond: Youth, identity and privacy on social network sites. *New Media & Society*, *19*(2), 272–288.

Holloway, L., & Hubbard, P. (2014). *People and place: The extraordinary geographies of everyday life*. Routledge.

Houlbrook, M. (2005). *Queer London: Perils and pleasures in the sexual metropolis, 1918–1957*. University of Chicago Press.

Hubbard, P. (2013). Kissing is not a universal right: Sexuality, law and the scales of citizenship. *Geoforum*, *49*, 224–232.

Hulchanski, J. D. (2010). *The three cities within Toronto: Income polarization among Toronto's neighbourhoods, 1970–2005*. The Cities Centre, University of Toronto.
Human Rights Campaign. (2023). *LGBTQ+ Pride flags*. https://www.hrc.org/resources/lgbtq-pride-flags
Hutchby, I. (2001). Technologies, texts and affordances. *Sociology*, *35*(2), 441–456.
Ingemarsdotter, J. (2017). Normal cars and queer driving: Or, why Charlie loved to speed. *Lamda Nordica*, *22*(1), 38–70.
Insauga.com. (2019, February 11). Brampton has this new service for the LGBT community. https://www.insauga.com/brampton-has-this-new-service-for-the-lgbt-community/)
Insauga.com. (2023, June 15). City scrambles to bring Pride crosswalks to Brampton before end of June. https://www.insauga.com/city-scrambles-to-bring-pride-crosswalks-to-brampton-before-end-of-june/
Ivison, J. (2005, January 21). Harper lands his "value" issue: Conservatives set to launch ad campaign against gay marriage. *National Post*, A6.
Jackson, P. (1989). *Maps of meaning: An introduction to cultural geography*. Routledge.
Jennings, R. (2006). The Gateways club and the emergence of a post-Second World War lesbian subculture. *Social History*, *31*(2), 206–225.
Jessop, B., Brenner, N., & Jones, M. (2008). Theorizing sociospatial relations. *Environment and Planning D: Society and Space*, *26*(3), 389–401.
Johnson, D. (2007). "This is political!": Negotiating the legacies of the first school-based gay youth group. *Children, Youth and Environments*, *17*(2), 380–387.
Johnson, L. C. (2006). Style wars: revolution in the suburbs?. *Australian Geographer*, *37*(2), 259–277.
Johnson, O. (1997, July 7). For minorities, timing is everything. *The Globe and Mail*, A14.
Johnston, L., & Valentine, G. (1995). "Wherever I lay my girlfriend, that's my home": The performance and surveillance of lesbian identities in domestic environments. In D. Bell & G. Valentine (Eds.), *Mapping desire* (pp. 99–113). Routledge.
Jones, T. (2011). A sexuality education discourses framework: Conservative, liberal, critical, and postmodern. *American Journal of Sexuality Education*, *6*(2), 133–175.
Junos, K., & Bowman, G. (2023, January 13). Coquitlam drag queen story time event braces for protests. *CityNews*. https://vancouver.citynews.ca/2023/01/13/bc-drag-queen-storytime-protests/
Jukes, J. (forthcoming). LGBTQ+ friendship and spatial commitments: Queer socialities in the making of a new rural. In A. L. Bain, J. A. Podmore, & C. Arun-Pina (Eds.), *Queer geographies: Key debates and contending perspectives*. Edward Elgar.
Kalbfuss, E. (2003, August 7). Priest strikes a nerve. *The Gazette*, A1.
Kalogerakis, G. (1999, August 3). Judge rules transsexual convict will serve time in women's prison: Jail officials say former guard will be in constant danger. *National Post*, A4.
Kanai, J. M., & Kenttamaa-Squires, K. (2015). Remaking South Beach: Metropolitan gayborhood trajectories under homonormative entrepreneurialism. *Urban Geography*, *36*(3), 385–402.
Kapica, J. (1993, February 18). Church plays down nude-photo lawsuit: Minister sues for reinstatement. *The Globe and Mail*, A10.
Kataoka, S. (2009). Vancouverism: Actualizing the livable city paradox. *Berkeley Planning Journal* *22*(1), 42–57.
Katz, C. (2004). *Growing up global: Economic restructuring and children's everyday lives*. University of Minnesota Press.
Kaufman, M. (2001). *The Laramie project*. Vintage Books.
Kay, E. S., & Musgrove, K. (2020). From HIV to Coronavirus: AIDS service organizations adaptive responses to COVID-19, Birmingham, Alabama. *AIDS and Behavior*, *24*(9), 2461–2462.

Kearns, A., Kearns, O., & Lawson, L. (2013). Notorious places: Image, reputation, stigma. The role of newspapers in area reputations for social housing estates. *Housing Studies*, *28*(4), 579–598.

Kearns, A., & Parkinson, M. (2001). The significance of neighbourhood. *Urban Studies 38*(12), 2103–2110.

Keil, R. (2017). *Suburban planet: Making the world from the outside in*. Polity.

Keil, R. (2020). After Suburbia: Research and action in the suburban century. *Urban Geography*, *41*(1), 1–20.

Keil, R., & Üçoğlu, M. (2021). Beyond sprawl? Regulating growth in Southern Ontario: Spotlight on Brampton. *disP-The Planning Review*, *57*(3), 100–118.

Keith, M., & Pile, S. (Eds.) (1993). *Place and the politics of identity*. Routledge.

Kelaita, P. (2024). Suburban vogue and other queer survival strategies. *Cultural Studies*, *38*(2), 273–298.

Kelly, J., Ruther, M., Ehresman, S,. & Nickerson, B. (2017). Placemaking as an economic development strategy for small and midsized cities. *Urban Affairs Review*, *53*(3), 435–462.

Ketchum, A. D. (2020). Lost spaces, lost technologies, and lost people: Online history projects seek to recover LGBTQ+ spatial histories. *Digital Humanities Quarterly*, *14*(3), n.p. https://www.digitalhumanities.org/dhq/vol/14/3/000483/000483.pdf.

Kinkaid, E. (2018). (en)Vision(ing) otherwise: Queering visuality and space in Lefebvre's production. *GeoHumanities*, *4*(2), 438–461.

Kinkaid, E. (2020). Re-encountering Lefebvre: Toward a critical phenomenology of social space. *Environment and Planning D: Society and Space*, *38*(1), 167–186.

Kirkey, K., & Forsyth, A. (2001). Men in the valley: Gay male life on the suburban-rural fringe. *Journal of Rural Studies*, *17*(4), 421–441.

Klapeer, C. M. & Laskar, P. (2018). Transnational ways of belonging and queer ways of being: Exploring transnationalism through the trajectories of the rainbow flag. *Identities*, *25*(5), 524–541.

Klinenberg, E. (2018). *Palaces for the people: How social infrastructure can help fight inequality, polarization, and the decline of civic life*. Crown.

Knopp, L., & Brown, M. (2021). Travel guides, urban spatial imaginaries and LGBTQ+ activism: The case of Damron guides. *Urban Studies*, *58*(7), 1380–1396.

Knox, P., & Pinch, S. (2014). *Urban social geography: An introduction*. Routledge.

Kojima, D., Catungal, J. P., & Diaz, R. (2017). Introduction: Feeling queer, feeling Asian, feeling Canadian. *Topia*, *38*, 69–80.

Korinek, V. J. (2004). The history of public discourses of homosexuality in Saskatchewan, 1971–93. In J. McNinch & M. Cronin (Eds.), *"I could not speak my heart": Education and social justice for gay and lesbian youth* (pp. 109–137). University of Regina Press.

Korinek, V. J. (2018). *Prairie fairies: A history of queer communities and people in Western Canada, 1930–1985*. University of Toronto Press.

Kramberger, A. (2013, August 28). Gay-rights activist makes Pointe-Claire council bid. *The Gazette*, D2.

Kunstler, J. H. (1993). *Geography of nowhere: The rise and decline of America's man-made landscape*. Simon and Schuster.

L'Ecuyer, J. & MacLennan, M. (2004). *Prom queen: The Marc Hall story*. Screendoor.

L'R des Centres des femmes du Québec. (2016). *Base d'unité politique*. https://rcentres.qc.ca/base-dunite-politique/

Lamey, M. (2006, August 5). Homophobic jokes mar Outgames: Laval athletes asked to leave. *National Post*, A2.

Langford, B. (2000). Margins of the city: Towards a dialectic of suburban desire. In R. Phillips, D. Shuttleton & D. Watt (Eds.), *De-centering sexualities: Politics and representations beyond the metropolis* (pp. 64–80). Routledge.

Lapointe, P.-L. (2022). Laval. *The Canadian encyclopedia*. https://www.thecanadianencyclopedia.ca/en/article/laval
Laramée, F. (2022, October 16). Longueuil nomme une élue responsable des relations avec la communauté LGBTQ+. *La relève*. https://www.lareleve.qc.ca/2022/10/16/longueuil-nomme-une-elue-responsable-des-relations-avec-la-communaute-lgbtq/
Laskar, P., Johansson, A., & Mulinari, D. (2017). Decolonising the rainbow flag. *Culture unbound: Journal of current cultural research*, *8*(3), 192–217.
Laszczkowski, M. (2015). Scraps, neighbors, and committees: Material things, place-making, and the state in an Astana apartment block. *City & Society*, *27*(2), 136–159.
Latham, A., & Layton, J. (2019). Social infrastructure and the public life of cities: studying urban sociality and public spaces. *Geography Compass*, *13*(7), e12444.
Laws needed to protect homosexuals from violence, MP says after incident. (1988, February 29). *The Vancouver Sun*, A1.
Leclerc, J.-C. (2008, September 8). Le Vatican et le prêtre-député: L'abbé Raymond Gravel a-t-il fait le meilleur choix? *Le Devoir*, B6.
Lefebvre, H. (1991). *The production of space*. Blackwell.
Lenc, X. (2020). Queer placemaking, settler colonial time, and the desert imaginary in Palm Springs. In C. Courage, T. Borrup, M. R. Jackson, K. Legge, A. Mckeown, L. Platt, & J. Schupbach (Eds.), *The Routledge handbook of placemaking* (pp. 81–91). Routledge.
Leorke, D., Wyatt, D., & McQuire, S. (2018). 'More than just a library': Public libraries in the 'smart city'. *City, Culture and Society*, *15*, 37–44.
Lepofsky, J., & Fraser, J. C. (2003). Building community citizens: Claiming the right to place-making in the city. *Urban Studies* *40*(1), 127–142.
Lessard, R., & Raynault, M-F. (2009). Public health and poverty. *Canadian Journal of Public Health*, *100*(4), 247–248.
Levitsky, S. (2007). Niche activism: Constructing a unified movement identity in a heterogenous organizational field. *Mobilization*, *12*(3), 271–286.
Lewis, N. M. (2013). Ottawa's Le/The Village: Creating a gaybourhood amidst the "death of the village". *Geoforum*, *49*, 233–242.
Lewis, N. M. (2017). Canaries in the mine? Gay community, consumption, and aspiration in neoliberal Washington, DC. *Urban Studies* *54*(3), 695–712.
Ley, D. (2010). *Millionaire migrants: Trans-Pacific lifelines*. Blackwell-Wiley.
Ley, D. (2021). A regional growth ecology, a great wall of capital and a metropolitan housing market. *Urban Studies*, *58*(2), 297–315.
Ley, D., & Lynch, N. (2012). *Divisions and disparities in Lotus-Land*. Cities Centre, Univ. Toronto. Discussion Paper, 223.
Li, W. (2009). *Ethnoburb: The new ethnic community in urban America*. University of Hawai'i Press.
Liang, D., De Jong, M., Schraven, D., & Wang, L. (2022). Mapping key features and dimensions of the inclusive city: A systematic bibliometric analysis and literature study. International *Journal of Sustainable Development & World Ecology*, *29*(1), 60–79.
Lichterman, P. (1996). *The search for political community: American activists reinventing commitment*. Cambridge University Press.
Linteau, P-A. (2007). *Brève histoire de Montréal*. Éditions Boréal.
Liu, C. Y., Hu, F. Z., & Jeong, J. (2020). Towards inclusive urban development? New knowledge/creative economy and wage inequality in major Chinese cities. *Cities*, *105*, 102385.
Liu, S., & Blomley, N. (2013). Making news and making space: Framing Vancouver's Downtown Eastside. *The Canadian Geographer*, *57*(2), 119–132.

Lo, L., Preston, V., Anisef, P., Basu, R., & Wang, S. (2015). *Social infrastructure and vulnerability in the suburbs*. University of Toronto Press.

Lomash, E. F., Brown, T. D., & Galupo, M. P. (2019). "A whole bunch of love the sinner hate the sin": LGBTQ microaggressions experienced in religious and spiritual context. *Journal of Homosexuality*, *66*(10), 1495–1511.

Longueuil police are stepping up foot, motorcycle and car patrols at a park they say is being used by gay men having sex on the sly. (1991, June 6). *The Gazette*, G1.

Longueuil police keeping close eye on gays. (1988, August 25). *The Gazette*, G4.

Loukaitou-Sideris, A., Liggett, R., & Sung, H. G. (2007). Death on the crosswalk: A study of pedestrian-automobile collisions in Los Angeles. *Journal of Planning Education and Research*, 26(3), 338–351.

Lubitow, A., Carathers, J., Kelly, M., & Abelson, M. (2017). Transmobilities: mobility, harassment, and violence experienced by transgender and gender nonconforming public transit riders in Portland, Oregon. *Gender, Place & Culture*, *24*(10), 1398–1418.

Lynch, F. R. (1987). Non-ghetto gays: A sociological study of suburban homosexuals. *Journal of Homosexuality*, *13*(4), 13–42.

Lynch, K. (1960). *The image of the city*. MIT Press.

MacDonnell, J. A., & Daley, A. (2015). Examining the development of positive space in health and social service organizations: A Canadian exploratory study. *Journal of Gay & Lesbian Social Services*, *27*(3), 263–301.

Maddux, S. (2011). *Gen silent*. Interrobang Productions.

Manalansan, M. F. IV (2015). Queer worldings: The messy art of being global in Manila and New York. *Antipode*, *47*(3), 566–579.

Marchbank, J. & Muller-Myrdahl, T. (2019). Queering Timmies: Theorising LGBTQ youth claiming and making space in Surrey, BC, Canada. In S. Habib & M. Ward (Eds.), *Youth, place and theories of belonging* (pp. 39–50). Routledge.

Marrades, R., Collin, P., Catanzaro, M., & Mussi, E. (2021). Planning from failure: Transforming a waterfront through experimentation in a placemaking living lab. *Urban Planning*, *6*(1), 221–234.

Martin, D. (2003). "Place-framing" as place-making: Constituting a neighborhood for organizing and activism. *Annals of the Association of American Geographers*, *93*(3), 730–750.

Mascoll, P. (1993, June 10). Police fight to reclaim "'perverts' playground". *Toronto Star*, SD2.

Mason, G. (2005, October 4). Sudbury stages the play that Surrey shunned. *The Globe and Mail*, A10.

Mason, K., & Whitehead, M. (2012). Transition urbanism and the contested politics of ethical place making. *Antipode*, *44*(2), 493–516.

Massey, D. (1991). The political place of locality studies. *Environment and Planning A*, *23*(2), 267–281.

Massey, D. (2005). *For space*. Sage.

Massey, D. (2007). *World city*. Polity Press.

Mattson, G. (2016). Digital project: Mapping lost lesbian bars 2006-2016. https://greggormattson.com/2016/09/17/mapping-lost-lesbian-bars-2006-2016/

McCann, E. (2002). The cultural politics of local economic development: Meaning-making, place-making, and the urban policy process. *Geoforum*, *33*, 385–398.

McConnell, L. (2002, May 16). Oshawa teen Marc Hall won the right to take his boyfriend to prom 20 years ago. *Insauga.com*. https://durham.insauga.com/oshawa-teen-marc-hall-won-the-right-to-take-his-boyfriend-to-prom-20-years-ago/

McGlynn, N. (2021). Bears in space: Geographies of a global community of big and hairy gay/bi/queer men. *Geography Compass*, *15*(2), e12553.

McManus, T. (2022, July 12). New Westminster to begin phasing out Royal City in city's branding. *New Westminster Record.* https://www.newwestrecord.ca/local-news/new-westminster-to-begin-phasing-out-royal-city-in-citys-branding-5572873

McNaughton, P. (2004, December 14). Letter: New face of the 'burbs: "We just live life like you". *Toronto Star*, A23.

McRobbie, A. & Garber, J. (1976). Girls and subcultures. In S. Hall and T. Jefferson (Eds.), *Resistance through rituals: Youth subcultures in post-war Britain* (pp. 209–222). Harper Collins.

Melili, A. (2023, January 30). The 2023 rainbow book list. https://glbtrt.ala.org/rainbowbooks/archives/1444

MetroVancouver. (2023). *Services*. https://metrovancouver.org/services

Miller, V. (2005). Intertextuality, the referential illusion, and the production of a gay ghetto. *Social & Cultural Geography*, *6*(1), 61–79.

Mills, C. (2011, October 21). Students want schools to act on cyber-bullying: Survey shows they and parents strongly support gay-straight alliances to promote acceptance. *The Globe and Mail*, A13.

Misgav, C. (2015). With the current, against the wind: constructing spatial activism and radical politics in the Tel-Aviv Gay Center. *ACME: An International Journal for Critical Geographies*, *14*(4), 1208–1234.

Montgomery, A. (2016). Reappearance of the public: Placemaking, minoritization, and resistance in Detroit. *International Journal of Urban and Regional Research*, *40*(4), 776–799.

Moon, M., Sedgwick, E.K., Gianni, B., & Weir, S. (1994). Queers in (single-family) space. *Assemblage*, *24*, 30–37.

Moreira, L. (2020). Doing intimate citizenship: Resisting the heterosexual matrix across and beyond intimacy. *Sexuality & Culture*, *24*(6), 1875–1892.

Morin, E. (1987, April 4). Killer's mental state key in parole eligibility: Judge. *The Gazette*, A8.

Morrish, L. & O'Mara, K. (2011). Queering the discourse of diversity. *Journal of Homosexuality*, *58*(6–7), 974–991.

Moten, F., & Harney, S. (2013). *The undercommons: Fugitive planning & Black study*. Wivenhoe.

MP pleads for acceptance of gays. (1988, March 1). *The Globe and Mail*, A5.

Muller Myrdahl, T. M. (2011). Queerying creative cities. In P. L. Doan (Ed.), *Queerying planning: Challenging heteronormative assumptions and reframing planning practice* (pp. 157–166). Routledge.

Muller Myrdahl, T. M. (2013). Ordinary (small) cities and LGBQ lives. *ACME: An International Journal for Critical Geographies*, *12*(2), 279–304.

Muller Myrdahl, T. M. (2016). Visibility on their own terms? LGBTQ lives in small Canadian cities. In G. Brown & K. Browne (Eds.), *The Routledge research companion to geographies of sex and sexualities* (pp. 37–44). Routledge.

Muñoz, J. E. (1999). *Disidentifications: Queers of color and the performance of politics*. University of Minnesota Press.

Muñoz, J. E. (2009). *Cruising utopia: The then and there of queer futurity*. New York University Press.

Murphy, D. (1997, April 20). Homophobia 101. *The Province*, A40.

Murray, C. (2015). Queering Vancouver: Where LGBTQism meets Vancouverism in the creative city. *BC Studies: The British Columbian Quarterly*, *188*, 55–80.

Namaste, K. (1996). Genderbashing: Sexuality, gender, and the regulation of public space. *Environment and Planning D: Society and Space*, *14*(2), 221–240.

Namaste, V. (2005). *C'était du spectacle!: L'histoire des artistes transsexuelles à Montréal, 1955–1985*. McGill-Queen's Press.

Nash, C. J. (2006). Toronto's gay village (1969–1982): Plotting the politics of gay identity. *Canadian Geographer*, *50*(1), 1–16.

Nash, C. J. (2012). The age of the "post-mo": Toronto's gay village and a new generation. *Geoforum, 49*, 243–252.
Nash, C. J., & Browne, K. (2020). *Heteroactivism: Resisting lesbian, gay, bisexual and trans rights and equalities*. Bloomsbury Publishing.
Nash, C. J., & Gorman-Murray, A. (2015). Recovering the gay village: A comparative historical geography of urban change and planning in Toronto and Sydney. *Historical Geography 43*, 84–105.
Nash, C. J., & Gorman-Murray, A. (2014). LGBT neighbourhoods and "new mobilities": Towards understanding transformations in sexual and gendered urban landscapes. *International Journal of Urban and Regional Research, 38*(3), 756–772.
Nelson, V. (2015). Place reputation: Representing Houston, Texas as a creative destination through culinary culture. *Tourism Geographies, 17*(2), 192–207.
Newman, L. (1991). *Belinda's bouquet*. Alyson Books.
Nicholls, W. (2009). Place, networks, space: Theorising the geographies of social movements. *Transactions of the Institute of British Geographers, 34*(1), 78–93.
Niedt, C. (Ed.) (2013). *Social justice in diverse suburbs: History, politics, and prospects*. Temple University Press.
Nixon, S. A. (2019). The coin model of privilege and critical allyship: implications for health. *BMC Public Health, 19*(1), 1637. https://doi.org/10.1186/s12889-019-7884-9
O'Neil, P. (2005, May 18). MP wants rights act amended: Transgendered need special protection, NDP's Siksay says. *The Vancouver Sun*, A17.
Oakley, S., & Johnson, L. (2013). Place-taking and place-making in waterfront renewal, Australia. *Urban Studies, 50*(2), 341–355.
O'Connor, E. (2004, November 15). Moms take trustees to tribunal: They say homophobic remarks were way out of line. *The Province*, A14.
Oldenburg, R. (1999). *The great good place: Cafés, coffee shops, bookstores, bars, hair salons, and other hangouts at the heart of the community*. Marlowe.
Orne, J. (2011). "You will always have to 'out' yourself": Reconsidering coming out through strategic outness. *Sexualities, 14*(6), 681–703.
Osborne, T., & Rose, N. (2004). Spatial phenomenotechnics: Making space with Charles Booth and Patrick Geddes. *Environment and Planning D: Society and Space, 22*(2), 209–228.
Oswin, N. (2012). The queer time of creative urbanism: Family, futurity, and global city Singapore. *Environment and Planning A, 44*(7), 1624–1640.
Özogul, S., & Tasan-Kok, T. (2018). Exploring transformative place-making within the comprehensive spatial governance of Toronto. *DisP - The Planning Review, 54*(4), 59–73.
Paasi, A. (2002). Bounded spaces in the mobile world: Deconstructing regional identity. *Tijdschrift voor Economische en Sociale Geografie, 93*(2), 137–148.
Paganoni, M. C. (2012). City branding and social inclusion in the glocal city. *Mobilities, 7*(1), 13–31.
Palermo, P. C., & Ponzini, D. (2015). *Place-making and urban development: New challenges for contemporary planning and design*. Routledge.
Papillon, M. (2002). *Immigration, diversity and social inclusion in Canada's cities*. Discussion Paper F|27 Family Network. Canadian Policy Research Networks.
Parisi, P., & Holcomb, B. (1994). Symbolizing place: Journalistic narratives of the city. *Urban Geography, 15*(4), 376–394.
Parker, B. (2017). *Masculinities and markets: Raced and gendered urban politics in Milwaukee*. The University of Georgia Press.
Parker, D., & Karner, C. (2010). Reputational geographies and urban social cohesion. *Ethnic and Racial Studies, 33*(8), 1451–1470.
Parkes, D., & Shepherd, H. (2005, July 16). Baptisms in doubt for gays' children. *The Gazette*, A8.

Pateman, J., & Vincent, J. (2016). *Public libraries and social justice*. Routledge.
Paulmann, R. (2023). New city brand for Mississauga. https://www.ci-portal.de/new-city-brand-for-mississauga/
Paulse, M., & Elwin, R. (1990). *Asha's mums*. Women's Press.
Peake, L. (2016). The twenty-first-century quest for feminism and the global urban. *International Journal of Urban and Regional Research*, *40*(1), 219–227.
Peck, J. (2005). Struggling with the creative class. *International Journal of Urban and Regional Research*, *29*(4), 740–770.
Peel District School Board. (2024). *About us*. https://peelschools.org/aboutus/equity/Pages/default.aspx
Peel HIV/AIDS Network. (2017). *PHAN strategic plan*. https://uploads-ssl.webflow.com/628a60939076f36ac291d442/62af3fb2cd95df6b2d24f58c_PHAN_StrategicPlan_2017_2022.pdf
Peel Region. (2023). *Diversity, Equity, and Inclusion survey asks for input from Peel community*. https://www.peelregion.ca/news/archiveitem.asp?year=2023&month=0&day=3&file=202303a.xml
Peloso, L. (2002). Prom fight: The Marc Hall story.
Peritz, I. (2001, May 28). Residents rally for gay couple. *The Globe and Mail*, A6.
Pettinicchio, D. (2012). Institutional activism: Reconsidering the insider/outsider dichotomy. *Sociology Compass*, *6*(6), 499–510.
PFLAG Canada. (2024). *Family for all*. https://pflagcanada.ca/
PFLAG Canada–Durham Region. (n.d.). *About*. https://www.pflagdurhamregion.com/about
PFLAG Canada–York Region. (2016). *Request for proclamation and flag raising for York Region's Pride Week: June 11-19, 2016*. https://pub-markham.escribemeetings.com/filestream.ashx?documentid=9902
Phillips, R., Shuttleton, D., & Watt, D. (Eds.) (2000). *De-centering sexualities: Politics and representations beyond the metropolis*. Routledge.
Pierce, J., Martin, D., & Murphy, J. T. (2011). Relational place-making: The networked politics of place. *Transactions of the Institute of British Geographers*, *36*(1), 54–70.
Pierce, J., Williams, O., & Martin, D. (2016). Rights in place: An analytical extension of the right to the city. *Geoforum*, *70*, 79–88.
Pincetl, S., & Gearin, E. (2005). The reinvention of public green space. *Urban Geography*, *26*(5), 365–384.
Podmore, J. (2016). Contested dyke rights to the city: Montreal's 2012 dyke marches in time and space. In K. Browne & E. Ferreira (Eds.), *Lesbian geographies: Gender, place and power* (pp. 71–90). Routledge.
Podmore, J. A. (2001). Lesbians in the crowd: Gender, sexuality and visibility along Montréal's Boul. St-Laurent. *Gender, Place & Culture*, *8*(4), 333–355.
Podmore, J. A. (2006). Gone "underground"? Lesbian visibility and the consolidation of queer space in Montréal. *Social & Cultural Geography*, *7*(4), 595–625.
Podmore, J. A. (2015). From contestation to incorporation: LGBT activism and urban politics in Montréal. In M. Tremblay (Ed.), *Queer mobilizations: Social movement activism and Canadian public policy* (pp. 187–207). UBC Press.
Podmore, J. A., & Bain, A. L. (2019). On the edge of urban "equalities": Framing millennial suburban LGBTQ+ activisms in Canada. *Geography Research Forum*, *39*, 43–66.
Podmore, J. A., & Bain, A. L. (2020). "No queers out there"? Metronormativity and the queer suburban. *Geography Compass*, *14*(9), e12505.
Podmore, J. A., & Bain, A. L. (2021). Whither queer suburbanisms? Beyond heterosuburbia and queer metronormativities. *Progress in Human Geography*, *45*(5), 1254–1277.
Podmore, J. A., & Bain, A. L. (2023). Redistributing more than the LGBTQ2S Acronym? Planning beyond recognition and rainbows on Vancouver's periphery. *Urban Planning*, *8*(2), 208–222.

Poindexter, C. C. (2007). Management stresses and successes for AIDS service organizations. *Administration in Social Work, 31*(3), 5–28.
Poitras, C., & Linteau, P-A. (2020). Longueuil. *The Canadian encyclopedia*. https://www.thecanadianencyclopedia.ca/en/article/longueuil
Porteous, D., & Smith, S. E. (2001). *Domicide: The global destruction of home*. McGill-Queen's Press.
Pratt, M. L. (1992). *Imperial eyes: Travel writing and transculturation*. Routledge.
Pred, A. (1984). Structuration, biography formation, and knowledge: observations on port growth during the late mercantile period. *Environment and Planning D: Society and Space*, *2*(3), 251–275.
Preston, V., & Ray, B. (2020). Placing the second generation: A case study of Toronto. *The Canadian Geographer*, *64*(2), 215–231.
Prince, R. (2020). The geography of statistics: Social statistics from moral science to big data. *Progress in Human Geography*, *44*(6), 1047–1065.
Probyn E. (1996). *Outside belongings*. Routledge.
Puar, J. K. (2007). *Terrorist assemblages: Homonationalism in queer times*. Duke University Press.
Puleo, T. (2014). Art-making as place-making following disaster. *Progress in Human Geography*, *38*(4), 568–580.
Putnam, R. D. (2000). *Bowling alone: The collapse and revival of American community*. Simon and Schuster.
Pynn, L. (1988, November 22). Backers cheer Robinson: It's a win for gays, he tells rally. *The Vancouver Sun*, B1.
Qian, J. (2017). Beyond heteronormativity? Gay cruising, closeted experiences and self-disciplining subject in People's Park, Guangzhou. *Urban Geography*, *38*(5): 771–794.
QTBIPOC Sauga. (2024). *About*. https://qtbipocsauga.ca/about
Queer it Up! (2011). *Queer It Up–Pride March Mississauga*. https://www.youtube.com/watch?app=desktop&v=W01bMCN9h5U
Radis, B., Wenocur, K., Jin, J., & Keeler, C. (2022). A rainbow for reading: A mixed-methods exploratory study on drag queen reading programs. *Journal of Creativity in Mental Health*, *17*(3), 332–349.
Radke, H. R., Kutlaca, M., Siem, B., Wright, S. C., & Becker, J. C. (2020). Beyond allyship: Motivations for advantaged group members to engage in action for disadvantaged groups. *Personality and Social Psychology Review*, *24*(4), 291–315.
Rainbow Salad. (2023). *Queering the system*. https://www.rainbowsalad.ca/about
Ramírez, M. M. (2020). City as borderland: Gentrification and the policing of Black and Latinx geographies in Oakland. *Environment and Planning D: Society and Space*, *38*(1), 147–166.
Ramsey, M. (2005, September 22). School district axes play on town where gay man killed. *The Province*, A4.
Raphael, M. (2001, May 30). Census causes identity crisis for transsexual: "My gender was not listed". *National Post*, A7.
Rault, W. (2023). Same-sex unions in high-income countries: More widely recognized and more frequent. *Population & Societies*, *607*(1), 1–4.
Ray, B., & Preston, V. (2015). Working with diversity: A geographical analysis of ethno-racial discrimination in Toronto. *Urban Studies*, *52*(8), 1505–1522.
Raynor, K., Matthews, T., & Mayere, S. (2017). Shaping urban consolidation debates: Social representations in Brisbane newspaper media. *Urban Studies*, *54*(6), 1519–1536.
Rayside, D., & Wilcox, C. (Eds.) (2011). *Faith, politics and sexual diversity in Canada and the United States*. UBC Press.
Redshaw, S. (2007). Articulations of the car: The dominant articulations of racing and rally driving. *Mobilities*, *2*(1), 121–141.
Region of Peel. (2020, May 3). *Peel Pride resolution: Support for Peel Pride*. https://pub-peelregion.escribemeetings.com/filestream.ashx?DocumentId=1777

Reid, A. (2018, June 7). Surrey mayor shrugs off criticism about rainbow crosswalk. *Surrey Now-Leader*. https://www.surreynowleader.com/news/surrey-mayor-shrugs-off-criticism-about-rainbow-crosswalk-2935751
Relph, E. (1976). *Place and placelessness*. Pion.
Relph, E. (2014). *Toronto: Transformations in a city and its region*. University of Pennsylvania Press.
Renegade church ordains gay minister. (1999, February 16). *Standard*, A8.
Retzloff, T. (2015). Gay organizing in the "desert of suburbia" of metropolitan Detroit. In J. Archer, P. J. P. Sandul, & K. Solomonson (Eds.), *Making suburbia: New histories of everyday America* (pp. 51–62). University of Minnesota Press.
Rich, A. (1980). Compulsory heterosexuality and lesbian existence. *Signs: Journal of Women in Culture and Society*, *5*(4), 631–660.
Richardson, C. (2014). Orientalism at home: The case of "Canada's toughest neighbourhood". *British Journal of Canadian Studies*, *27*(1), 75–95.
Rio, M. (2020). Architecture is burning: An urbanism of queer kinship in ballroom culture. *Thresholds*, *48*, 122–132.
Rios, M., & Watkins, J. (2015). Beyond "place": Translocal placemaking of the Hmong diaspora. *Journal of Planning Education and Research*, *35*(2), 209–219.
Robinson's sex life called his own affair by most MPs. (1988, February 26). *The Vancouver Sun*, A6.
Rose, D. (1996). Economic restructuring and the diversification of gentrification in the 1980s: A view from a marginal metropolis. In J. Caulfield & L. Peake (Eds.), *City lives and city forms: Critical research and Canadian urbanism* (pp. 131–172). University of Toronto Press.
Rose, D., & Twigge-Molecey, A. (2013). *A city-region growing apart? Taking stock of income disparity in Greater Montréal, 1970-2005*. Research Paper 222, Cities Centre, University of Toronto.
Rose, G. (1993). *Feminism and geography: The limits of geographical knowledge*. University of Minnesota Press.
Rose, G. (2001). *Visual methodologies: An introduction to the interpretation of visual materials*. Sage.
Rosenberg, R. (2017). The whiteness of gay urban belonging: Criminalizing LGBTQ youth of color in queer spaces of care. *Urban Geography*, *38*(1), 137–148.
Rosenberg, R. (2021b). Psychic geographies of queer multiculturalism: Reading Fanon, settler colonialism and race in queer space. *Environment and Planning D: Society and Space*, *39*(6), 1129–1146.
Rosenberg, R. D. (2021a). Negotiating racialised (un)belonging: Black LGBTQ resistance in Toronto's gay village. *Urban Studies*, *58*(7), 1397–1413.
Ross, I. (1996, June 11). Presbyterians reject hiring of gay minister. *Toronto Star*, A8.
Royson, J. (1993, June 30). Cultural cash cut for gay groups. *Toronto Star*, A6.
Ruberg, B., & Ruelos, S. (2020). Data for queer lives: How LGBTQ gender and sexuality identities challenge norms of demographics. *Big Data & Society*, *7*(1). https://doi.org/10.1177/2053951720933286
Ruble, B. A., Tulchin, J. S., & Hanley, L. M. (2005). Moving towards inclusive cities. *The Brown Journal of World Affairs*, *11*(2), 69–77.
Rushbrook, D. (2002). Cities, queer space, and the cosmopolitan tourist. *GLQ: A Journal of Lesbian and Gay Studies*, *8*(1), 183–206.
Russell, E.K. (2018). Carceral Pride: The fusion of police imagery with LGBTI rights. *Feminist Legal Studies*, *26*(3), 331–350.
Russell, G. M., & Bohan, J. S. (2016). Institutional allyship for LGBT equality: Underlying processes and potentials for change. *Journal of Social Issues*, *72*(2), 335–354.
Russell, L. (2012). "Digital dualism and the glitch feminism manifesto." *The Society Pages*. https://thesocietypages.org/cyborgology/2012/12/10/digital-dualism-and-the-glitch-feminism-manifesto/

Russell, L. (2020). *Glitch feminism: A manifesto*. Verso Books.
Sainte-Anne-de-Bellevue. (2023). *Seniors policy of the City of Sainte-Anne-de-Bellevue: 2023 – 2025 action plan*. https://ville.sainte-anne-de-bellevue.qc.ca/medias/files/MADA/1-seniors-policy-sadb-2023-2025.pdf
Sandoval, E. (2018). More than violence: Undocuqueers' narratives of disidentification and world-making in Seattle, Washington. *Gender, Place & Culture*, *25*(12), 1759–1780.
Schneekloth, L. H., & Shibley, R. G. (1995). *Placemaking: The art and practice of building communities*. Wiley Publications.
Schrader, A. M. (2007). "I thought I'd find myself at the library": LGBTQ services and collections in public and school libraries. *PNLA Quarterly*, *72*, 4–9.
Schraven, D., Joss, S., & De Jong, M. (2021). Past, present, future: Engagement with sustainable urban development through 35 city labels in the scientific literature 1990–2019. *Journal of Cleaner Production*, *292*, 125924.
Schroeder, C. G. (2012). Making space for queer youth: Adolescent and adult interactions in Toledo, Ohio. *Gender, Place & Culture*, *19*(5), 635–651.
Schroeder, C. G. (2015). Sanctuary or prison: Queer youth and the family, household and home. *Social & Cultural Geography*, *16*(7), 783–797.
Seamon, D. (1979). *A geography of the lifeworld: Movement, rest and encounter*. Croom Helm.
Seamon, D. (2018). *Life takes place: Phenomenology, lifeworlds and place making*. Routledge.
Seitz, D. (2017). *A house of prayer for all people: Contesting citizenship in a queer church*. University of Minnesota Press.
Selley, C. (2011, June 16). Column: Full pundit. *The National Post*, A19.
Sexuality cited in office attack. (1988, February 29). *The Globe and Mail*, A5.
Sezer, C., & Maldonado, A. M. F. (2017). Cultural visibility and urban justice in immigrant neighbourhoods in Amsterdam. *Built Environment*, *43*(2), 193–213.
Shakibaei, S., & Vorobjovas-Pinta, O. (2024). Access to urban leisure: Investigating mobility justice for transgender and gender diverse people on public transport. *Leisure Sciences*, *46*(5), 639–657.
Sharp, M., & Nilan, P. (2017). Floorgasm: Queer(s), solidarity and resilience in punk. *Emotion, Space and Society*, *25*, 71–78.
Shearmur, R., & Rantisi, N. (2011). Montreal: Rising again from the same ashes. In L. S. Bourne, T. Hutton, R. Shearmur & J. Simmons (Eds.), *Canadian urban regions: Trajectories of growth and change* (pp. 173–201). Oxford University Press.
Sheller, M. (2018). *Mobility justice: The politics of movement in an age of extremes*. Verso.
Sheller, M., & Urry, J. (2006). The new mobilities paradigm. *Environment and Planning A*, *38*(2), 207–226.
Shepherd, H. (1988, May 4). Church leaders vote for delay of homosexual clergy debate. *The Gazette*, A3.
Shepherd, H. (1996, August 8). Gay group upset over park arrests: Crackdown on public sex is necessary: Police chief. *The Gazette*, E1.
Shepherd, H. (1999, February 13). Breakaway parish to ordain gay pastor. *Toronto Star*, L14.
Sher Vancouver. (2023). *For queer South Asians & friends, families & allies*. https://www.shervancouver.com
Shi, J. (2021). *Parenting in an uncertain future: The experiences of parents in the "PFLAG China" community*. [Doctoral dissertation, University of Essex]. University of Essex Research Repository. https://repository.essex.ac.uk/31059/
Shields, R. (1991). *Places on the margins: Alternative geographies of modernity*. Routledge.
Sibley, D. (1995). *Geographies of exclusion: Society and difference in the West*. Routledge.

Sizemore-Barber, A. (2017). Archival movements: South Africa's gay and lesbian memory in action. *Safundi: The Journal of South African and American Studies, 18*(2), 117–130.
Slade, L. (2005). Social dreaming for a queer culture. *Self & Society, 33*(3), 31–40.
Sliwinski, S. (2017). *Dreaming in dark times: Six exercises in political thought*. University of Minnesota Press.
Smart, M. J., & Whittemore, A. H. (2017). There goes the gayborhood? Dispersion and clustering in a gay and lesbian real estate market in Dallas, TX, 1986–2012. *Urban Studies, 54*, 600–615.
Smith, C. G. (2020). "Where u from, who u wit?!" Black pride festivals as itinerant hospitality. *Journal of Canadian Studies, 54*(2–3), 395–414.
Smith, M. (2004). Questioning heteronormativity: Lesbian and gay challenges to education practice in British Columbia, Canada. *Social Movement Studies, 3*(2), 131–145
Smith, M. (2005). Resisting and reinforcing neoliberalism: Lesbian and gay organizing and the federal and local levels in Canada. *Policy & Politics, 13*(1), 75–94.
Smith, W. R. (2015). *The 2011 National Household Survey: The complete statistical story*. https://www.statcan.gc.ca/en/blog-blogue/cs-sc/2011NHSstory
Soja, E. (1996). *Thirdspace: Journeys to Los Angeles and other real-and-imagined places*. Wiley Blackwell.
Spade, D. (2020). Solidarity not charity: Mutual aid for mobilization and survival. *Social Text, 38*(1), 131–151.
Spencer, K. (1997a, April 20). Homosexuality: Meet Rick: He's gay and living in fear. *The Province*, A14.
Spencer, K. (1997b, April 11). Surrey board vows gay books "not approved". *The Province*, A8.
Sphère. (2020). *Les groupes de soutien Pluriels*. https://www.spheressg.org/index.php/a-propos
Spring, A. (2013). Declining segregation of same-sex partners: Evidence from Census 2000 and 2010. *Population Research and Policy Review, 32*, 687–716.
Springgay, S., & Irwin, R. (2005). A/r/tography as living inquiry through art and text. *Qualitative Inquiry, 11*, 897–912.
Stasińska, A. (2022). Tender gestures in heteronormative spaces: Displaying affection in public by families of choice in Poland. *Gender, Place & Culture, 29*(2), 177–200.
Statistics Canada. (2009). *2006 Census: Family portrait: Continuity and change in Canadian families and households in 2006*. https://www12.statcan.gc.ca/census-recensement/2006/as-sa/97-553/p4-eng.cfm
Statistics Canada. (2012a). *2011 Census of Population: Families, households, marital status, structural type of dwelling, collectives*. https://www150.statcan.gc.ca/n1/daily-quotidien/120919/dq120919a-eng.htm#:~:text=The%20census%20counted%2064%2C575%20same,in%20Canada%20in%20July%202005.
Statistics Canada. (2012b). *Portrait of families and living arrangements in Canada: Families, households and marital status, 2011 Census of Population*. Catalogue no. 98-312-X2011001. https://www12.statcan.gc.ca/census-recensement/2011/as-sa/98-312-x/98-312-x2011001-eng.pdf
Statistics Canada. (2017). *Census in Brief: Same-sex couples in Canada in 2016*. https://www12.statcan.gc.ca/census-recensement/2016/as-sa/98-200-x/2016007/98-200-x2016007-eng.cfm
Statistics Canada. (2019). *Census Profile, 2016 Census*. https://www12.statcan.gc.ca/census-recensement/2016/dp-pd/prof/index.cfm?Lang=E
Statistics Canada. (2022). *Focus on geography series, 2021 Census of population*. https://www12.statcan.gc.ca/census-recensement/2021/as-sa/fogs-spg/page.cfm?lang=E&topic=4&dguid=2021A000011124

Statistics Canada. (2023a). Canada is the first country to provide census data on transgender and non-binary people. *The Daily*. https://www150.statcan.gc.ca/n1/daily-quotidien/220427/dq220427b-eng.htm

Statistics Canada. (2023b). Census Profile, 2021 Census of Population. https://www12.statcan.gc.ca/census-recensement/2021/dp-pd/prof/index.cfm?Lang=E

Statistics Canada. (2023c). Same-sex couples and sexual orientation… by the numbers. *The Daily*. https://www.statcan.gc.ca/en/dai/smr08/2015/smr08_203_2015

Stein, M. (2000). *City of sisterly and brotherly loves: Lesbian and gay Philadelphia, 1945–1972*. University of Chicago Press.

Stevens, Q. (2007). *The ludic city: Exploring the potential of public spaces*. Routledge.

Stevens, Q., & Dovey, K. (2023). *Temporary and tactical urbanism: (Re)assembling urban space*. Routledge.

Stevenson, D. (2003). *Cities and urban cultures*. McGraw-Hill Education.

Stone, A. (2018). The geography of research on LGBTQ life: Why sociologists should study the South, rural queers, and ordinary cities. *Sociology Compass*, *12*(11), e12638.

Strapagiel, L. (2015, April 1). Former Catholic school student's complaints of homophobia settled just before human rights hearing. *The National Post*. https://nationalpost.com/news/canada/human-rights-tribunal-of-ontario-to-hear-former-catholic-school-students-complaints-of-homophobia

Street cred: The rise of rainbow sidewalks. (2020). *The Homoculture*. https://www.thehomoculture.com/street-cred-the-rise-of-rainbow-sidewalks/

Streitmatter, R. (1995). *Unspeakable: The rise of the gay and lesbian press in America*. Faber and Faber.

Studio.89 (2024). *About studio.89*. https://studio89.org/about/

Sullivan, P. (1999, October 27). The trouble in Surrey City: Sex, sex, skinheads. *The Globe and Mail*, A19.

Sumerau, J. E., Forbes, T. D., Denise, E. J., & Mathers, L. A. (2021). Constructing allyship and the persistence of inequality. *Social Problems*, *68*(2), 358–373.

Summers, B. T. (2019). *Black in place: The spatial aesthetics of race in a post-chocolate city*. UNC Press Books.

Sundén, J. (2018). Queer disconnections: Affect, break, and delay in digital connectivity. *Transformations*, *31*, 63–78.

Table de quartier sud de l'Ouest-de-l'Île. (2023). *Your social development community plan*. https://tqsoi.org/communityplan-2/

Takeuchi, C. (2013, July 29). Vancouver unveils Canada's first permanent rainbow crosswalks. *The Georgia Straight*. https://www.straight.com/life/405041/vancouver-unveils-canadas-first-permanent-rainbow-crosswalks

Tarrow, S. (1998). *Power in movement: Social movements and contentious politics*. Cambridge University Press.

Task Force on Equity, Diversity, and Inclusion. (2021). *Putting the learner first: Executive summary and final report*. Lester B. Pearson School Board. https://www.lbpsb.qc.ca/wp-content/uploads/2021-Task-Force-Final-Report.pdf

Teixeira, C. (2007). Residential experiences and the culture of suburbanization: A case study of Portuguese homebuyers in Mississauga. *Housing Studies*, *22*(4), 495–521.

Thanh Ha, T. (2001, April 7). Fighting homophobia in their own backyard. *The Globe and Mail*, A3.

Theriault, D. (2014, September 27). Un parc devenu le lieu de rencontres gais. *Le Journal de Montréal*, 40.

Thomas, L. L. (2021). From childless tower to child-full density: Families and the evolution of Vancouverism. *Planning Perspectives*, *36*(3), 559–581.

Tillmann, S., Tobin, D., Avison, W., & Gilliland, J. (2018). Mental health benefits of interactions with nature in children and teenagers: A systematic review. *J Epidemiol Community Health*, *72*(10), 958–966.

Tin, L.-G. (2008). *The dictionary of homophobia: A global history of gay and lesbian experience*, (M. Redburn, Trans.). Arsenal Pulp Press.
Tongson, K. (2011). *Relocations: Queer suburban imaginaries*. New York University Press.
Toronto Board of Trade. (2020). *Shaping our future: A playbook for rebooting and reimaging the regional economy in Ontario's innovation corridor*. https://bot.com/Resources/Resource-Library/Shaping-Our-Future
Toronto Public Library. (2024, March 26). *Chasing rainbows: 2SLGBTQ+ in library collections*. Panel Discussion, Toronto Reference Library. https://www.youtube.com/watch?v=ju8fI9fdN2w
Town of Ajax. (2010). *Diversity and community engagement plan: Fostering a culture of inclusion*. https://www.ajax.ca/en/inside-townhall/resources/DACE/COUNCIL-APPROVED-AJAX---DACE-PLAN-2010.pdf
Town of Ajax. (2017). *Diversity and community engagement plan phase 2*. https://www.ajax.ca/en/inside-townhall/resources/DACE/DACE-Phase-2-Plan-Final.pdf
Town of Ajax. (2022). *TOA unveils new Pride crosswalk*. https://www.ajax.ca/en/toa-unveils-new-pride-crosswalk.aspx
Town of Baie-d'Urfé. (2022). *Strategic vision, 2022–2026: Highlights*. https://www.baie-durfe.qc.ca/medias/en/nouvelles/BaieDUrfe_Rapport_2022_EN_web_VF_compressed.pdf
Tracey, L. (2022). Harvey Milk's Gay Freedom Day speech. JSTOR Daily. https://daily.jstor.org/harvey-milks-gay-freedom-day-speech-annotated/
Transsexual sues alleging humiliation. (1987, August 11). *The Gazette*, A4.
Tuan, Y. F. (1974). *Topophilia: A study of environmental perception, attitudes, and values*. Columbia University Press.
Tuck, E. (2009). Suspending damage: A letter to communities. *Harvard Educational Review*, *79*(3), 409–428.
Tuohy, C. J. (1992). *Policy and Politics in Canada: Institutionalized Ambivalence*. Temple University Press.
Turesky, M., & Crisman, J. J. A. (2023). 50 Years of Pride: Queer spatial joy as radical planning praxis. *Urban Planning*, 8(2), 262–276.
Turner, J. H. (2010). *Theoretical principles of sociology* (Volume 1). Springer.
United Nations. (1996). *Habitat Agenda and Istanbul Declaration on Human Settlements: Summary road map to the future*. United Nations Centre for Human Settlements.
United Nations. (2015). *Transforming our world: The 2030 Agenda for Sustainable Development*. https://sdgs.un.org/publications/transforming-our-world-2030-agenda-sustainable-development-17981
University of Toronto. (2023). Positive space. https://positivespace.utoronto.ca/about/
Urry, J. (2002). Mobility and proximity. *Sociology*, *36*(2), 255–274.
Urry, J. (2004). The "system" of automobility. *Theory, Culture & Society*, *21*(4-5), 25–39.
Valentine, G. (2007). Theorizing and researching intersectionality: A challenge for feminist geography. *The Professional Geographer*, *59*(1), 10–21.
Valentine, J. (1994). *One dad, two dads, brown dad, blue dads*. Alyson Books.
Vallerand, O. (2013). Home is the place we all share: Building queer collective utopias. *Journal of Architectural Education*, *67*(1), 64–75.
Vallerand, O. (2020). *Unplanned visitors: Queering the ethics and aesthetics of domestic space*. McGill-Queen's University Press.
Valverde, M. (2012). *Everyday law on the street: City governance in an age of diversity*. University of Chicago Press.
Vanwynsberghe, R., Surborg, B., & Wyly, E. (2013). When the games come to town: Neoliberalism, mega-events and social inclusion in the Vancouver 2010 Winter Olympic Games. *International Journal of Urban and Regional Research*, *37*(6), 2074–2093.
Verenca, T. (2015, July 27). City shows its Pride with rainbow crosswalk. *New Westminster Record*. https://www.newwestrecord.ca/local-news/city-shows-its-pride-with-rainbow-crosswalk-3015541

Verma, A. (2017). Mississauga meetup: Queer organizing in the 905. In S. Chambers, J. Farrow, M. FitzGerald, E. Jackson, J. Lorinc, T. McCaskell, R. Sheffield, T. Taylor, & R. Thawer (Eds.), *Any other way: How Toronto got queer* (pp. 226–228). Coach House Press.

Victoria & Albert Museum. (2023). *The progress pride flag*. https://www.vam.ac.uk/articles/the-progress-pride-flag

Ville de Beaconsfield. (2024). *Analysis of citizens' needs and expectations in terms of culture, sports activities, community life, leisure activities and healthy living: Final report*. https://www.beaconsfield.ca/en/my-city/living-in-beaconsfield/surveys-and-studies

Ville de Laval. (2011a). *Introducing Laval*. https://www.laval.ca/Documents/Pages/En/About/introducing-laval.pdf

Ville de Laval. (2011b). *La Politique de l'urbanisme durable de Laval. ÉvoluCité. Vers l'équilibre urbain*. https://www.laval.ca/Documents/Pages/Fr/A-propos/politiquesmunicipales/Politique-urbanisme-durable-evolucite.pdf

Ville de Laval. (2015a). *Laval today: Reflection paper 2015: Baseline Study for rethinking Laval*. https://www.laval.ca/Documents/Pages/En/Citizens/strategic-vision-laval-2035-urban-in-nature.pdf

Ville de Laval. (2015b). *Urbaine de nature, Laval 2035: Une vision, une ville*. https://www.laval.ca/Documents/Pages/Fr/A-propos/politiques-municipales/urbaine-de-nature-VF.pdf

Ville de Laval. (2016a). *Mémoire de la ville de Laval: Consultation publique: Solidarité et inclusion sociale. Service de la culture, des loisirs, du sport et du développement social*. https://www.laval.ca/Documents/Pages/Fr/Nouvelles/memoire-consultation-solidarite-inclusion-sociale.pdf

Ville de Laval. (2016b, April 5). *Procès-verbal d'une séance ordinaire du Conseil municipal de la Ville de Laval*. https://www.laval.ca/GESDDocuments/CM/PV/CM_PV_5458_1.5.pdf

Ville de Laval. (2019). *Laval affiche ses nouvelles couleurs*. https://www.laval.ca/Pages/Fr/Nouvelles/nouvelle-image.aspx

Ville de Laval. (2020a). *Bilan des travaux: Fonds Place-du-Souvenir. Service de la culture, des loisirs, du sport et du développement social*. https://www.laval.ca/Documents/Pages/Fr/A-propos/fonds-place-du-souvenir/bilan-travaux-2020.pdf

Ville de Laval. (2020b, March 31). #ÇAVABIENALLER. Facebook. https://www.facebook.com/100064513487058/posts/3412164168797801/

Ville de Laval. (2021a). *Plan d'action intégré en matière de jeunesse, 2021-2025*. Service de la culture, des loisirs, du sport et du développement social. https://www.laval.ca/Documents/Pages/Fr/A-propos/fonds-place-du-souvenir/fonds-place-souvenir-plan-action-jeunesse-2021-2025.pdf

Ville de Laval. (2021b). *Politique d'accessibilité universelle*. https://www.laval.ca/Documents/Pages/Fr/A-propos/politiques-municipales/politique-accessibilitee-universelle.pdf

Ville de Laval. (2022). *Laval, ville inclusive et ouverte à la diversité*. https://www.laval.ca/Pages/Fr/Nouvelles/journee-internationale-contre-homophobie.aspx

Ville de Laval. (2024, May 7). Séance du conseil municipal. https://archivesvilledelaval.webtv.coop/conseil-de-ville-de-laval-du-7-mai-2024/

Ville de Longueuil. (2019). *Portrait social de l'arrondissement du Vieux-Longueuil*. https://developpementsocialvieuxlongueuil.files.wordpress.com/2019/06/portrait-social-vieux-longueuil.pdf

Ville de Longueuil. (2020). *Plan stratégique de la ville de Longueuil, 2025*. https://cms.longueuil.quebec/sites/default/files/medias/2021-07/plan_strategique_2025_vf-web_coll.pdf

Ville de Longueuil. (2023). *Longueuil centre-ville*. https://investir.longueuil.quebec/fr/centre-ville

Ville de Longueuil. (2024). *Plan d'action pour contrer le racisme et la discrimination et pour favoriser l'inclusion* 2024–2026. https://cms.longueuil.quebec/sites/default/files/medias/documents/2024-04/2024-04/Plan%20d%26%23039%3Baction%20%28PA-RDI%29_2024_1.pdf
Vyhnak, C. (2010, September 24). Candidate in flap over gay flag in Ajax. *Toronto Star*, GT2.
Waitt, G., & Head, L. (2002). Postcards and frontier mythologies: Sustaining views of the Kimberley's as timeless. *Environment and Planning D: Society and Space*, *20*(3), 319–344.
Walks, R. A. (2011). Economic restructuring and trajectories of socio-spatial polarization in the Twenty-First Century Canadian city. In L. S. Bourne, T. Hutton, R. Shearmur & J. Simmons (Eds.), *Canadian urban regions: Trajectories of growth and change* (pp. 125–159). Oxford University Press.
Walks, R. A. (2013). Suburbanism as a way of life, slight return. *Urban Studies*, *50*(8), 1471–1488.
Warf, B., & Winsberg, M. (2010). Geographies of mega churches in the United States. *Journal of Cultural Geography*, *27*(1), 33–51.
Warner, M. (1993). *Fear of a queer planet: Queer politics and social theory*. University of Minnesota Press.
Watson, S. (2009). Performing religion: Migrants, the church and belonging in Marrickville, Sydney. *Culture and Religion*, *10*(3), 317–338.
Weintrob, A., Hansell, L., Zebracki, M., Barnard, Y., & Lucas, K. (2021). Queer mobilities: Critical LGBTQ perspectives of public transport spaces. *Mobilities*, *16*(5), 775–791.
Weissinger, S. E., & Mack, D. A. (2018). *Law enforcement in the age of black lives matter: policing Brown and Black bodies*. Lexington Books.
Wells, V. S. (2018). LGBT in B.C.: All the rainbow crosswalks in British Columbia. *The Georgia Straight*. https://www.straight.com/life/1088456/lgbt-bc-all-rainbow-crosswalks-british-columbia
Weston, K. (1995). Get thee to a big city: Sexual imaginary and the great gay migration. *GLQ: A Journal of Lesbian and Gay Studies*, *2*(3), 253–277.
White, H. R. (2015). *Reforming Sodom: Protestants and the rise of gay rights*. University of North Carolina Press.
Whitehead, M. (2003). Love thy neighbourhood: Rethinking the politics of scale and Walsall's struggle for neighbourhood democracy. *Environment and Planning A*, *33*(2), 277–300.
Whittemore, A. H., & Smart, M. J. (2016). Mapping gay and lesbian neighborhoods using home advertisements: Change and continuity in the Dallas-Fort Worth Metropolitan Statistical Area over three decades. *Environment and Planning A: Economy and Space*, *48*(1), 192–210.
Whyte, W. H. (1956). *The organization man*. Simon and Schuster.
Williams, M. (2017). Care-full justice in the city. *Antipode*, *49*(3), 821–839.
Wolowic, J. M., Heston, L. V., Saewyc, E. M., Porta, C., & Eisenberg, M. E. (2017). Chasing the rainbow: Lesbian, gay, bisexual, transgender and queer youth and pride semiotics. *Culture, Health & Sexuality*, *19*(5), 557–571.
Wong, R., Madaio, M. A., & Merrill, N. (2023). Seeing like a toolkit: How toolkits envision the work of AI ethics. *Proceedings of the ACM on Human-Computer Interaction, 7(CSCW1)*, Article 145, 1–27.
Wyly, E. K. (2020). The new planetary suburban frontier. *Urban Geography*, *41*(1), 27–35.
York Pride. (2024). *24+ years: What makes York Pride, York Pride?* https://yorkpride.ca/new-page
Young, I. M. (1990). *Justice and the politics of difference*. Princeton University Press.
Young, I. M. (2000). *Inclusion and democracy*. Oxford University Press.
Youth 4 a Change. (2024). *Youth 4 a change*. https://youth4achange.ca/

Index

Pages in *italics* refer to figures and pages followed by "n" refer to notes.

FSC
www.fsc.org
MIX
Papier | Fördert gute Waldnutzung
FSC® C083411

Kolibri 360 GmbH
Ferdinand-Jühlke-Straße 7
99095 Erfurt, Deutschland
produktsicherheit@kolibri360.de